The Andean Wonder Drug

The Andean Wonder Drug

Cinchona Bark and Imperial Science in the Spanish Atlantic, 1630–1800

Matthew James Crawford

UNIVERSITY OF PITTSBURGH PRESS

Published by the University of Pittsburgh Press, Pittsburgh, Pa., 15260

Printed on acid-free paper
10 9 8 7 6 5 4 3 2 1

Cataloging-in-Publication data is available from the Library of Congress

ISBN 13: 978-0-8229-6814-6
ISBN 10: 0-8229-6814-2

Publisher: University of Pittsburgh Press, 7500 Thomas Blvd., 4th floor, Pittsburgh, PA 15260, United States, www.upittpress.org

EU Authorized Representative: Easy Access System Europe, Mustamäe tee 50, 10621 Tallinn, Estonia, gpsr.requests@easproject.com

IN MEMORIAM

James P. Crawford
1945–2010

John A. Marino
1946–2014

Dryden W. Hull
1977–2011

Contents

Acknowledgments

It was in the summer of 2004 that I first encountered the bundles of manuscripts labeled "Documents related to the discovery and development of quinine" in the Archivo General de Indias in Seville, Spain. Since that fortuitous encounter, I have incurred substantial debts to many individuals and institutions who helped bring this project to fruition. While historical research and writing can be rewarding on their own, I am most grateful for the supportive and engaging friends, colleagues, and scholars I met as a result of working on this project. I want to express my gratitude to all who made this book possible. First and foremost, I must acknowledge my mentors and colleagues at the University of California, San Diego. I am most indebted to my graduate advisers John A. Marino and Naomi Oreskes. During my time in graduate school and beyond, both John and Naomi have been generous with their time, advice, and, most important, their constructive criticism. More than that, they remain an inspiration to me as scholars, as teachers, and as colleagues. Special thanks must also go to Luce Giard, who first encouraged my interest in colonial Latin America and the Spanish Empire; Robert S. Westman, who taught me how to do the history of science; and Christine Hunefeldt, who always pushed me to consider the colonial Andean dimensions of this story. I am also grateful to Paula De Vos, Marcel Henaff, Eric Van Young, and Andrew Lakoff for their guidance and mentorship during my time in San Diego.

Various institutions have provided financial support for this project. A fellowship from the William J. Fulbright Program at the United States Department of State supported an extended research trip to Spain. Additional support for research trips and writing was provided by the National Science Foundation (NSF Grant SES-0349956: Proof, Persuasion, and Policy: A Research and Training Grant for the UCSD Science Studies Program), the All UC Group in Economic History, and the following institutions at the University of California, San Diego: the Department of History, the Science Studies Program, the Center for the Humanities, the Center for Iberian and Latin American Studies, and the Center for Interdisciplinary and Area Studies. More recently, I have received support from the University Research Council at Kent State University. Finally, a Theodore and Mary Herdegen Fellowship from the Chemical Heritage Foundation in Philadelphia, Pennsylvania, and a Barbara S. Mosbacher Fellowship from the John Carter Brown Library in Providence, Rhode Island, provided access not only to a wealth of historical materials but also to a stimulating intellectual community of fellows, researchers, and staff that encouraged me to take this project in new and fruitful directions.

The writing of history is not possible without primary sources. I owe a debt of gratitude to the archivists and librarians at the following institutions for their assistance and for their continuing efforts to conserve the traces of the past that are used by scholars and the public: the Archivo General de Indias in Seville, the Biblioteca Nacional de España in Madrid, the Archivo and Biblioteca del Palacio Real in Madrid, the Archivo and Biblioteca de la Real Academia Nacional de Medicina in Madrid, the Archivo and Biblioteca de la Real Jardín Botánico in Madrid, the Archivo General de Simancas in Valladolid, the Archivo Nacional de Ecuador in Quito, the Archivo General de la Nación in Bogotá, the Biblioteca Nacional in Colombia, the Rare Book Library at the Colegio de Nuestra Señora del Rosario in Bogotá, the John Carter Brown Library, the Beinecke Library at Yale University, and the Rare Book Library at the Chemical Heritage Foundation.

Beyond these institutions, I am appreciative of the wider network of colleagues and scholars who have provided feedback on my conference presentations, read parts of the manuscript, and provided advice and encouragement at various times and places in the United States, Spain, Ecuador, and Colombia. In particular, I thank Miruna Achim, James Amelang, Marcelo Aranda, Antonio Barrera-Osorio, Jeremy Baskes, Emily Berquist-Soule, Carin Berkowitz, Daniela Bleichmar, Christiana Borchart de Moreno, Benjamin Breen, Rainer Bussman, Hugh Cagle, Jorge Cañizares-Esguerra, Lina Maria del Castillo, Fiona Clark, Harold Cook, Susan Deans-Smith, Margaret Ewalt, Richard Fagoaga, Martha Few, Joseph Gabriel, Cathy Gere, Claire Gherini,

Pablo Gómez, Tamar Herzog, Ruth Hill, Ryan Kashanipour, Andrew Lipman, Bertie Mandelblatt, Kate Murphy, Mauricio Nieto Olarte, Christopher Parsons, Francisco Javier Puerto Sarmiento, Miguel Angel Puig Samper, Matthew Shindell, Gabriel Paquette, Maria Portuondo, José Carlos de la Puente, Evan Ragland, Paul Ramirez, David Ringrose, Julia Sarreal, Neil Safier, Suman Seth, Cristobal Silva, James Voekel, Timothy Walker, Ken Ward, Adam Warren, Susan Verdi Webster, and Bárbara Zepeda.

I also want to acknowledge the support and guidance that I have received from my colleagues at the Department of History at Kent State University. I especially want to recognize the enthusiasm and support of Ken Bindas, chair of the Department of History. Liz Smith-Pryor, Kevin Adams, Tim Scarnecchia, Ann Heiss, and Shane Strate were all kind enough to read portions of the manuscript and to provide advice and moral support as I navigated the publication process.

At the University of Pittsburgh Press, I have been fortunate to work with two excellent editors, Audra Wolfe and Abby Collier. Audra provided feedback on early drafts of the manuscript as well as useful insight into the world of publishing. I am especially grateful to Abby for her insightful comments on the completed manuscript and for her care and attention throughout the review and publication process. As a first time author, I could not have asked for a better experience. I also express my gratitude to Joel Coggins for his fine work on the design of the cover, to Huanyang (Patrick) Zhao for designing the map of the Andean world, and to the anonymous reviewers who provided rigorous and constructive feedback on the manuscript.

On a personal note, the completion of this book is bittersweet because three of the people with whom I most wanted to share the finished product have passed away. It is to them—my dad, James P. Crawford (1945–2010), my dissertation adviser, John A. Marino (1946–2014), and my best friend, Dryden Hull (1977–2011)—that I dedicate this book. Their loss has affected me profoundly, but it has also strengthened my appreciation for the love and support that I have received from my family and friends, especially my mom, Peggy Crawford, and my mother-in-law, Scotia MacKay. I am thankful as well to Matthew Shindell, Adam Beaver, Mark Waddell, Evan Ragland, Kelly Wisecup, and Ryan Kashanipour for their camaraderie and friendship over the years. Pete Finnerty was instrumental in helping me to get back on track so that I could complete the manuscript. My daughters, Leanor and Lucy, deserve acknowledgment for providing the right amounts of distraction and inspiration when needed. And, finally, the person whose encouragement and sacrifice has meant the most and who made it all possible is my wife, Greta. We share this achievement.

The Andean Wonder Drug

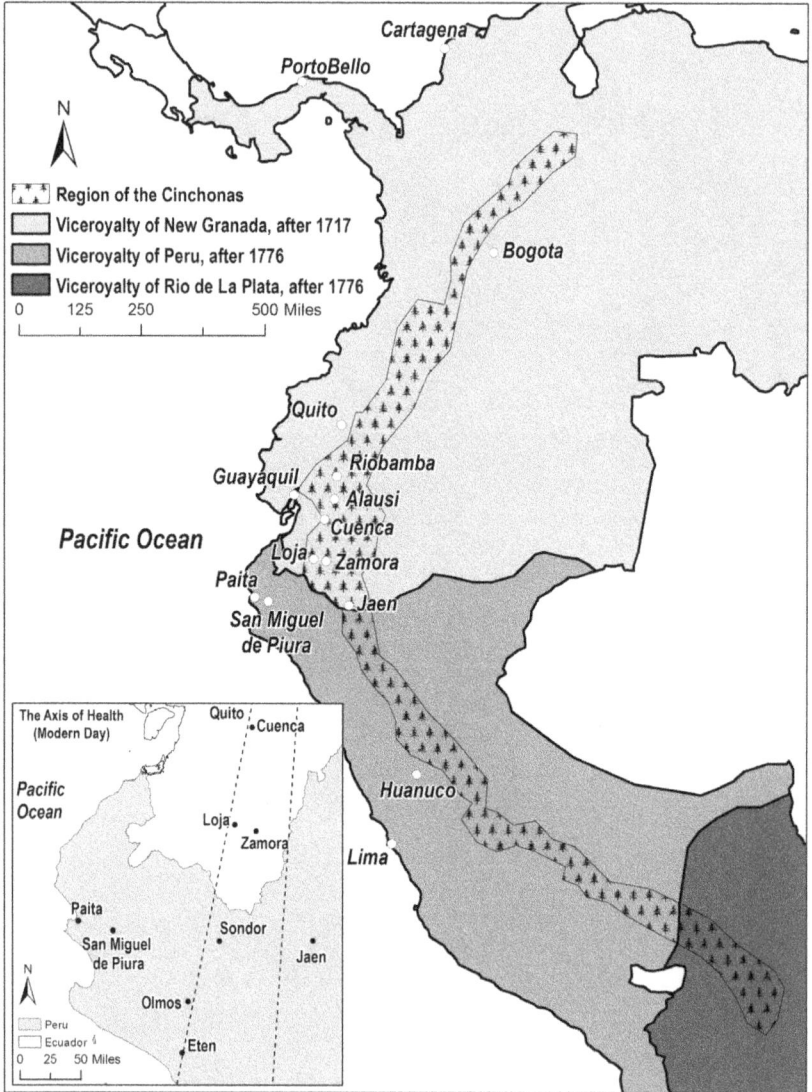

Figure 1. Map of the viceroyalties of Peru and New Granada in the eighteenth century, with inset of the Axis of Health in northern Peru and southern Ecuador. Prepared by Huanyang Zhao, Department of Geography, Kent State University.

The Power and Fragility of European Science in the Spanish Atlantic World

The *Esqueletos* of Empire

In November 1783, Pedro de Valdivieso received an unusual request. The archbishop-viceroy of New Granada needed some "skeletons."[1] As the *corregidor* (royal governor) of Loja, a province in the southern sierra of New Granada (figure 1), Valdivieso was obliged to fulfill this request from his superior in the Spanish colonial government. Fortunately, the skeletons in question were only plant remains. *Esqueletos,* the Spanish term, was a colloquialism for the dried specimens collected by eighteenth-century botanists in order to identify, compare, and classify plants. Although the transformation of plants into specimens was a routine activity for botanists, the production of esqueletos was not necessarily common practice for officials in the Spanish colonial government, who often worried more about transforming plants into useful commodities than into botanical specimens. In this instance, the archbishop-viceroy asked for specimens of the trees that produced Loja's most important export: a medicinal tree bark known as *quina* or *cascarilla*. He had made the request on behalf of the botanists of the Royal Botanical Expedition of New Granada in Bogotá, who needed the specimens in order to determine whether trees recently discovered near Bogotá were the same as Loja's quina trees.[2] Because the botanists were unable to make the nearly 700-mile journey south to Loja, they were counting on Valdivieso, who was in a unique position to help.

Much more was at stake than botanical classification though. In the preceding century, quina had become one of the most valuable and widely used medicaments in the Atlantic World on account of its efficacy in treating intermittent fevers, a prevalent and deadly cluster of ailments in the early modern world that we now recognize as symptoms of malaria.[3] In 1751, the Spanish Crown made this Andean wonder drug an object of empire by converting several of the hills near Loja into an *estanco* or royal reserve. The purpose of the royal reserve was to supply the Royal Pharmacy in Madrid with regular, annual shipments of the best quina available.[4] The trees that produced quina could be found throughout the Andean forests of Spain's viceroyalties of New Granada and Peru; however, quina from Loja was widely regarded as the best (figure 1). By the early 1780s, more than a century of intensive harvesting of the bark from wild trees had taken its toll. Loja's forests were exhausted and its quina trees increasingly scarce. Officials in Madrid and New Granada had come to believe that the empire desperately needed a new source of bark that was comparable to quina from Loja. It is not surprising then that the archbishop-viceroy described the study of quina trees as one of the "primary objectives" of the Royal Botanical Expedition in New Granada.[5]

Things looked different in Loja however. For landowners and laborers in the province, quina had served as major source of income, especially since the bark was one of the few exports connecting Loja to the broader trade networks of Spanish America and the Atlantic World.[6] Pedro de Valdivieso was well aware of this. Unlike other corregidores that were appointees from Spain, Valdivieso was born and raised in Loja as a member of a creole family that had been in South America for several generations by the late eighteenth century.[7] Moreover, many Valdiviesos were active in the quina trade and the wealth derived from bark trading had helped the family achieve a place of prominence among the small number of creole elites in a province of approximately 23,000 people with a majority indigenous population (53.9%) as well as significant populations of creoles and mestizos (23.6%) and free blacks (22.6%).[8] If botanists in Bogotá confirmed that "quina from Santa Fe," as bark from the forests near Bogotá was known, was equivalent to quina from Loja, it would have put the economic livelihood of Loja and its elites in jeopardy. In short, Pedro de Valdivieso must have had some serious reservations about fulfilling the archbishop-viceroy's request for the skeletons of Loja's quina trees.

Although he must have been aware of the stakes for his province, Valdivieso voiced a different concern in the letter that accompanied the skeletons he submitted to the archbishop-viceroy. He questioned whether botanists were the right choice for the job. Valdivieso made clear that he did not doubt "the intelligence of the Botanists" but he was not convinced that they had sufficient experience to perform the comparison properly.[9] Instead, Valdivieso urged the archbishop-viceroy to send him the "leaves and bark" of the quina trees from the forests near Bogotá.[10] With these materials, the corregidor would then compare them to samples from Loja's trees to determine whether quina from Santa Fé was equivalent to quina from Loja.

Undoubtedly, economic and personal interest motivated Valdivieso, but there is no reason to think that he was any more or less interested in the outcome of this comparison of quinas than the botanists in Bogotá, who had ties to the archbishop-viceroy and the Crown. More than that, Valdivieso *did* have a point. After a lifetime of ties to the local quina trade and more than a decade as the director of the royal reserve of quina in Loja, Valdivieso was one of the world's foremost experts on quina trees and their bark. Meanwhile, none of the members of the archbishop-viceroy's "Company of Botanists" had even visited Loja to observe its quina trees in situ.[11] In other words, the corregidor of Loja had good reason to propose that he was a better choice than botanists to perform a comparison of quinas for the archbishop-viceroy.

In the end, Valdivieso relented and submitted the skeletons of Loja's quina trees to the archbishop-viceroy. In an accompanying letter, he emphasized the importance of local knowledge to the production of these botanical speci-mens. Valdivieso noted that he relied on one of his sons to collect samples of the quina trees "in order to avoid any fraud." He knew that he could trust a family member to find the very best bark from Uritusinga, the name of one of the hills in Loja that produced "the most select cascarilla."[12] He also noted that the specimens were shipped to Bogotá only after they had been prepared and packaged according to his exacting standards. Such statements not only reinforced Valdivieso's reputation as a fastidious imperial servant but also signaled that these esqueletos were not simply products of nature. These sam-ples were artifacts of local knowledge and expertise. Valdivieso also included a set of instructions for the archbishop-viceroy's botanists explaining how to compare quinas.[13] If he could not perform the comparison himself, he could at least advise the botanists how to do it correctly. Ultimately, Valdivieso wanted to make sure that the archbishop-viceroy recognized that the local knowledge of bark collectors remained just as important to the empire as the learned knowledge of botanists.

A Microcosm of Science and Empire in the Atlantic World

Quina, a medicinal tree bark harvested from various species of cinchona trees in the Andean forests of South America, was a unique natural resource but its story reflects many of the important developments in the early mod-ern Atlantic World. In this opening vignette, the presence of botanists at the vice-regal court in New Granada illustrates one of these developments that would have significant consequences in modern world history: the entanglement of European sciences and empires. Yet, at the same time, Valdivieso's resistance to transforming Loja's quina trees into botanical specimens is emblematic of the challenges in transforming European botany into an imperial science. *The Andean Wonder Drug* elucidates these challenges and their significance by exploring an overlooked chapter in the entangled histories of science, empire, and nature in the early modern Atlantic World: the Spanish Empire's struggles in the second half of the eighteenth century to control the cinchona tree and its bark.

In 1751, when the Spanish Crown established the royal reserve of quina in Loja, the main purpose of this enterprise must have seemed quite modest and even admirable. In the royal order that established the royal reserve, the Crown emphasized the health of its people and their access to effective medicaments as central concerns.[14] The royal order further explained that by regulating the harvesting of bark in Loja, the royal reserve would ensure a regular supply of the highest quality quina so that the king's vassals—in both Spain and Spanish America—could have reliable access to this important medicament. Undoubtedly, this justification was saturated with prevailing European notions of a ruler's obligation to promote public health as well as ideas of the healing powers of royalty.[15]

How could anyone question, let alone reject, such a seemingly noble cause? Many did because, even with the best of intentions, the royal reserve was still an imperial enterprise. In that same royal order, the Crown drew on prevailing European notions of empire and asserted the Spanish Empire's right to exploit and derive direct benefit from the natural resources of its colonial territories in the Americas. Such rhetoric signaled that this project was part of the broader imperial reforms initiated under the auspices of the Bourbons, the new ruling family of Spain and its empire since the death of Charles II, the last Habsburg king, in 1700.[16] In this context, the Crown and its reformers imbued the royal reserve of quina with great political and symbolic meaning and it was understood as a pilot project in pursuit of a new vision of the Spanish Empire revitalized by science and commerce.[17] Like their contemporaries elsewhere in Enlightenment Europe, officials in Spain hoped that with the help of European science their imperial designs would have greater authority and greater efficacy.[18] Consequently, the royal reserve of quina became a microcosm for enacting a new synergy of science and empire as the Crown and its advisers in Madrid turned to Spanish pharmacists, physicians, and botanists to improve the royal reserve and strengthen imperial control over the cinchona trees and their bark. Ultimately this project failed. This book explains why and why understanding this failure offers new insight into the histories of science, empire, and the environment.

Why did science prove so ineffective in this Spanish imperial enterprise when other European empires successfully wielded science elsewhere in the eighteenth-century world? *The Andean Wonder Drug* makes two main arguments in response to this question. First, it argues that the Andean and Atlantic contexts mattered. In the mid-eighteenth century, when the Spanish Crown took an explicit interest in it, quina was not an unknown natural object waiting for science to make it intelligible and useful to empire. The bark was already a cultural artifact that was understood and evaluated in different ways at different locations as it made its way from Andean forests to European pharmacies. While *cascarilleros* (bark collectors) and *curanderos* (healers) in the Andes had their own methods to identify and evaluate different kinds of bark from various species of cinchona trees and various parts of trees, merchants in Atlantic ports as well as pharmacists and naturalists in Europe used other methods to identify and evaluate quina.[19] Certainly healers, bark collectors, merchants, government officials, pharmacists, and naturalists were well aware of the different ways of knowing quina that coexisted in the eighteenth-century Atlantic World. Yet these different groups did not necessarily agree on which was the *best* way to understand and evaluate the bark. As much as networks of trade and empire facilitated the circulation of the bark and knowledge of it, the different sites of production, exchange, and consumption—Andean forests, Atlantic markets, European pharmacies, and botanical gardens—tended to support and privilege distinctive ways of knowing the cinchona tree and its bark. In other words, quina's Atlantic itineraries gave rise to a distinctive geography of knowledge in which Andean forests and Atlantic ports were as much centers of knowledge production as European pharmacies and botanical gardens.[20]

In second half of the eighteenth century, when the Spanish Crown attempted to assert greater control over the cinchona tree and its bark, a central challenge was implementing a consistent set of practices and standards for identifying and evaluating the bark throughout the Spanish Empire. It is not surprising then that the Spanish Crown's main response to divergent understandings of the bark was to attempt to export the concepts and practices employed by European pharmacists and naturalists for evaluating botanical materia medica. By emphasizing quina's identity as both a natural object and a cultural artifact embedded in local ways of knowing that were connected by networks of trade and empire, this book argues that the apparent impotence of Spanish pharmacists and botanists as agents of empire was not the result of some inherent deficiency in science as practiced in the Iberian Atlantic but a result of the tenacity of bark collectors, merchants, and local officials in asserting that authority of their own knowledge and expertise. Moreover, by drawing on local and established bodies of knowledge and ways of knowing, healers, bark collectors, and even local government officials were able to resist and undermine the efforts of Spanish pharmacists, botanists, and imperialists to assert the superiority and universality of European science.

The mere existence of other ways of knowing nature, however, is not sufficient to explain why European science faltered as a tool of empire in the late eighteenth-century Spanish Atlantic World. In many cases, the practitioners of European science did not feel obliged to recognize or even engage with the claims made by indigenous peoples, merchants, or government officials, especially since early modern European sciences increasingly acquired their own institutions and networks of knowledge production such as the scientific societies established in European cities that became the centers of global correspondence and trade networks that supported the circulation of information and objects.[21] In the Spanish Atlantic World, the colonial government played an important role as a network of knowledge production and, as a result, provided an institutional context in which Spanish representatives of European pharmacy, medicine, and botany not only encountered but also had to engage with other ways of knowing the natural world that existed throughout the Spanish Atlantic. Consequently, *The Andean Wonder Drug* makes a second argument that the structure and style of Spanish colonial governance also undermined the authority and efficacy of European science, especially since science and empire were so profoundly intertwined in the Spanish Atlantic World.[22]

The production of knowledge about the Americas had been an important part of Spanish colonial rule since the early decades of exploration, conquest, and colonization in the late fifteenth and early sixteenth centuries. In order for the Crown as well as its officials in Spain and Spanish America to effectively govern the vast territories of the empire, they needed information and knowledge about the people, places, and potential resources of those regions. In the sixteenth century, the Crown and its Council of the Indies supported a number of initiatives including the first scientific expedition to the Americas under the auspices of Francisco Hernández, a Spanish physician; a project—known as the *relaciones geográficas*—designed to systematically collect geographic, natural historical, and anthropological information from all regions of the Americas, and the institutionalization of the collection and production of cartographic and cosmographic knowledge at the Casa de Contratación in Seville.[23] By the eighteenth century, the Spanish colonial government had developed an identifiable and prevalent "epistemic culture," a phrase that refers to the values and methods used by officials in the colonial government to collect, circulate, and certify knowledge.[24] It is important to recognize that this epistemic culture predated the integration of European science as a regular part of the Spanish imperial enterprise in the eighteenth century.[25] Although existing studies have highlighted the various ways in which the Crown and its Council of the Indies facilitated the production of knowledge about the Americas through new institutions, the case of quina is useful in that it provides a more holistic account of the ways in which this epistemic culture pervaded the institutions of Spanish colonial governance at all levels on both sides of the Atlantic.[26]

Historical records of the royal reserve of quina allow us to track generation, circulation, and evaluation of knowledge in the form of reports, letters, and bark samples that wended their way through the colonial government from local town councils in South America to the advisory juntas of viceroys as well as the Royal Pharmacy and the Ministry of the Indies in Spain. In particular, these records highlight three main characteristics of the epistemic culture of the Spanish colonial government that shaped and, at times, undermined the authority and efficacy of European science as part of the Spanish imperial enterprise. First, empiricism was a well-established characteristic of this epistemic culture. Since the sixteenth century, officials in the colonial government valued the reports of individuals or groups that had firsthand experience with the phenomenon in question. In the eighteenth century, this commitment to empiricism was reflected in the reports and testimonies that Spanish officials collected from those groups directly involved with quina and the quina trade. Second, this epistemic culture was often overtly political meaning that officials often solicited information and advice from anyone and everyone with relevant knowledge and experience as well as an interest in the matter at hand. As a result, the production of knowledge within the colonial government could be quite competitive and contentious as the representatives of indigenous, commercial, and scientific bodies of knowledge all claimed the authority to speak for nature and natural resources.

By soliciting information and advice about quina and other natural resources from different interested groups in colonial society, officials insured that the colonial government and the Crown played a central role in the production of knowledge, which is the third main feature of this epistemic culture. Like many other early modern European empires and states, a key component of colonial rule in the Spanish Empire was the consideration and adjudication of competing interests.[27] This feature became even more pronounced during the eighteenth-century Bourbon Reforms as the Spanish Crown and its advisers worked to reclaim royal power. Consequently, within the epistemic culture of the colonial government, the Crown and its representatives in the colonial government claimed the ultimate authority to decide between competing claims about natural phenomena like quina. Moreover, because the Spanish colonial government was comprised of many different institutions, officials in the government did not always agree—a feature of colonial governance that local officials in the quina-producing communities of South America recognized and tried to use to their advantage.

Taken together, the empirical, political, and royalist features of the epistemic culutre of the Spanish colonial government help to explain some of the difficulties and challenges that Spanish pharmacists and botanists faced as they came to play a larger role in the Spain's imperial enterprise in the Americas. Ultimately, the case of quina reveals that there was a significant epistemological dimension to Spanish colonial governance and shows how American nature served as a material substrate on which the tensions of empire were articulated and adjudicated through the politics of knowledge. Moreover, these features of the epistemic difficulties and obstacles that Spanish pharmacists and botanists encountered as they became a part of this epistemic culture in which the authority and efficacy of science was not taken for granted but still open for debate.

The contests over quina within the Spanish colonial government in the eighteenth century and their implications for our understanding of the history of science and empire in the Atlantic World have been largely overlooked in existing scholarship. With the exception of studies the story of cinchona bark as illustrative of developments in European botany or European pharmacology, the eighteenth-century history of cinchona has generally been ignored in favor of emphasizing the more obvious moments of high drama such as the early globalization of the bark in the seventeenth century and the global transplantation of cinchona trees in the nineteenth century.[28] In the history of medicine, most accounts of the history of cinchona have focused on the bark's introduction to Europe in the mid-seventeenth century and its impact on early modern European medicine. Beginning with the celebrations of 300th anniversary of the "discovery" of cinchona bark in the 1930s, the culmination of this strand of scholarship was the publication of Saul Jarcho's *Quinine Predecessor* in 1993, a work that meticulously traces the introduction and spread of the use of cinchona bark in Western Europe ending with the work of Italian physician Francesco Torti (1658–1741) in the early eighteenth century.[29]

In the history of empire, most scholarship has focused on the history of cinchona in the nineteenth century, when cinchona bark and the newly isolated antimalarial alkaloid quinine became tools of European imperialism as the British and the Dutch succeeded in transplanting cinchona trees from Latin America to their colonial territories in South and Southeast Asia.[30] Thus, quinine has received considerable attention because of its auspicious role in world history since the early nineteenth century. After all, this plant's alkaloid and its derivatives have alleviated the sufferings of countless individuals afflicted with malaria in the past two centuries, most recently in the developing world where malaria remains endemic, despite many efforts to eradicate the disease.[31] Yet this wonder drug of modern medicine has not always been put to the most benevolent uses. In the late nineteenth century, in one of the most notorious acts of botanical espionage in world history, agents of the British and Dutch governments succeeded in transplanting cinchona trees from South America to their colonies in India and Southeast Asia.[32] Once established, these cinchona plantations produced the cinchona bark and quinine that facilitated the advancement of European imperialism around the globe by safeguarding the health of soldiers and settlers.[33] As a result, the story of quina is often reduced to a seemingly inconsequential prelude in the history of quinine, an antimalarial alkaloid that two French pharmacists isolated from cinchona bark in 1820.[34]

The historical significance of quina as "quinine's predecessor" is undeniable, but there is much more to the bark's story, especially in the eighteenth century. Several studies of the trade in cinchona bark have demonstrated the significance of the bark to the economic and social history of specific regions of South America.[35] Meanwhile, other studies have demonstrated how the Royal Pharmacy in Madrid took on new importance in the eighteenth century, as it became the clearinghouse for all the cinchona bark imported under the auspices of the Crown.[36] Yet, with few exceptions, most of these studies have focused primarily on specific locations and provided little insight into how the story of quina might illuminate the ways in which developments at these different sites were impacted by long-distance connections fostered by the Spanish Empire and the Atlantic World.[37]

Drawing on a wide array of sources from archives and libraries in Spain and South America, *The Andean Wonder Drug* offers a more comprehensive account of quina as a product of local, imperial, and commercial networks in eighteenth-century Atlantic World. Instead of reducing quina to just *one* of its roles in the early modern world, this book embraces the bark's multiple identities as natural resource, commodity, and medicament in order to highlight the intimate interactions between the politics of knowledge and tensions of empire at various sites in the Spanish Atlantic World. Although the histories of quina and quinine are intertwined, this book demonstrates that their histories are not interchangeable and that greater emphasis on the story of quina can yield new insights into the history of science and empire. After all, it is important to recognize that these two substances—quina and quinine—were the products of distinctive regimes of knowledge and power. Both were powerful medicaments, but they took different therapeutic forms and had different meanings to people that used them. Both medicaments also became objects of empire, but with one important difference: while quinine became an effective tool for European imperialists in the nineteenth century, quina was an object that confounded European science and empire in the eighteenth century. The story of quina matters most because it highlights an overlooked, yet fundamental, aspect of European science in the early modern world: its fragility, especially in colonial contexts.

Putting Early Modern Science in Its Place

By setting European science in the context of the myriad ways of knowing nature in the early modern Atlantic World and in the context of the epistemic culture of the Spanish colonial government, *The Andean Wonder Drug* shows that early modern European science was not only far from hegemonic but also was, at times, impotent in the face of the heterogeneous social, cultural, and natural worlds that comprised European empires. Rather than a narrative of European science and empire dominating the peoples and places of the globe hand in hand, the case of quina offers a reminder that the entanglements of science and empire were not necessarily destined to succeed. Consequently, this episode challenges the dominant view that has emphasized the successful and symbiotic relationship between early modern European sciences and empires.

In the past few decades, many historical studies have taken as their main objective to demonstrate that European sciences and empires developed an effective and mutually beneficial relationship.[38] Such scholarship has played an important role in undermining the myth of science as the disinterested pursuit of knowledge that had little to do with imperialism; it has also exploded the myth of the "civilizing mission" of European empires—an ideological framework that cast science, technology, and medicine as Europe's gift to the "savage" peoples of the world.[39] The majority of this scholarship emphasizes the success stories in which imperial connections provided the practitioners of various sciences—from cosmography to natural history—with access to natural phenomena on a global scale as well as access to the resources and authority of colonial government. In return, the sciences provided the knowledge to facilitate the imperial and economic expansion of Europe as well as new scientific (usually racial) theories to justify European hegemony over peoples and places around the globe.[40] Much of this work has admirably demonstrated the ways in which the sciences were implicated in Europe's imperial enterprises and vice versa. However, *The Andean Wonder Drug* suggests that such narratives give a distorted impression of the efficacy of European science and of the synergy between science and empire, especially in the early modern world, a period when imperial states were relatively weak and when many people did not take the authority of science for granted as their counterparts in later centuries would.[41]

In addition to calling attention to the fragility of early modern European science, *The Andean Wonder Drug* also demonstrates the importance of the Spanish Atlantic to the histories of science and the Atlantic World. Until relatively recently, the Iberian World (Spain, Portugal, and their colonies) has received little attention in these areas of historical inquiry.[42] One reason is the persistent perception of the early modern Iberian World as an uncivilized and backward region that remained mired in ignorance, superstition, and brutality, and contributed little to the emergence of the modern world. This perception traces back to the seventeenth and eighteenth centuries when Northern Europeans, mostly in Protestant lands, used the Iberian World as the foil and antihero in historical narratives that placed the development of science, enlightenment, and modernity in other parts of Europe and the Americas.[43] Similarly, modern histories of science and of the Atlantic World have, until recently, focused much more on the Northern Atlantic, especially Britain and its empire. While other regions of the Atlantic World, such as the British Empire, can provide useful points of comparison for analogous developments in the Spanish Empire or the Iberian World as demonstrated by J. H. Elliott's magisterial *Empires of the Atlantic,* it is imperative to avoid the tendency to use the histories of the British Empire or other regions of the Atlantic World as yardsticks, benchmarks, or ideal types against which the Iberian World is measured.[44]

By highlighting the epistemic culture of Spanish colonial government, this book demonstrates that the relationship between science and empire in the Spanish Atlantic was not derivative or deficient relative to similar activities in other regions of the Atlantic World. Instead, it shows that things in the Spanish Empire were simply different. After all, every European enterprise in the Atlantic World—commercial or imperial—wrestled with the challenges of acquiring and accumulating the knowledge and expertise necessary to exploit natural resources, coordinate trade, and administer colonial territories across vast expanses of land and sea. Is it any wonder then that a distinctive epistemic culture developed in the Spanish Atlantic World? And just as there is no reason to expect that the integration of science and empire would take the same form in different parts of the Atlantic World, there is no reason to expect that Spanish efforts to wield European botany would achieve the same results as in other imperial contexts, especially at a time when there were few models to emulate and when it was far from clear that science could be an effective tool of empire (something we can only appreciate in hindsight). Moreover, as the this book shows, one of the distinguishing features of science and empire in the Spanish Atlantic was that the two enterprises were more intimately intertwined than elsewhere in the Atlantic World.

Nature and the Politics of Knowledge in the Spanish Empire

What makes the case of quina especially vexing is that studies have demonstrated that the shortcomings of botany and natural history as imperial sciences in the Spanish Atlantic World cannot be attributed to a lack of interest and resources or a lack of engagement with the intellectual currents of the Scientific Revolution and the Enlightenment.[45] Officials, reformers, and elites in Spain and Spanish America supported scientific endeavors and embraced the utility of science to government as much as their contemporaries in other regions of the Atlantic World.[46] In the late eighteenth century, the Spanish Crown and its advisers sought to challenge contemporary perceptions of Spain as an uncivilized nation mired in ignorance and superstition, a view promulgated by Enlightenment thinkers beyond the Pyrenees. In the second half of the eighteenth century, royal patronage supported the establishment of a Royal Botanical Garden (1755) and a Royal Cabinet of Natural History (1772) in Madrid as well as nearly sixty scientific expeditions to Spanish America between 1759 and 1808.[47] Consequently, in the early nineteenth century, Alexander von Humboldt (1769–1859), the renowned Prussian scientific traveler, observed, "No other European state has invested a greater sum to advance the knowledge of plants than the Spanish government."[48]

To Humboldt, one result was clear: science benefited from empire. The reverse, however, was not necessarily the case. The new confluence of science and empire that took shape in the late eighteenth-century Spanish Atlantic World offered few, if any, of the practical or economic benefits that botanists promised and imperial reformers envisioned. Much of the vast corpus of images and text produced by Spanish naturalists did not make it into print.[49] Moreover, Spanish botanists identified many new genera and species of plants including some species of cinnamon and tea that they hoped would substitute for similar products from the East, but it proved quite difficult to put this knowledge into action. By the end of the eighteenth century, the grandiose new vision of science and empire promulgated by botanists, imperial reformers, and the Crown was revealed as little more than rhetoric.[50]

The case of quina offers new insight into this historical mystery by elucidating the various material, social and intellectual challenges associated with converting nature into an imperial natural resource. In the case of quina, a variety of obstacles threatened and thwarted the Crown's efforts to control the bark. Cinchona trees became harder to find. The bark degraded and lost its medical efficacy while in transit. Merchants committed fraud and adulterated quina shipments by mixing other barks with their cinchona bark. They also traded the bark as contraband to foreign merchants. To make matters worse, there were many different kinds of bark whose medical efficacy varied depending on the variety of cinchona tree from which it was harvested or whether the bark came from the trunk of the tree or its branches. Ultimately, effective exploitation of quina required more than a set of policies that asserted government control of its production and distribution. It also required knowledge—specifically reliable knowledge of how to identify, distinguish, and evaluate the different varieties of cinchona tree and different kinds of quinas.

Because the cinchona tree and its bark were objects of interest to many groups in Spanish colonial society and the Atlantic World, *The Andean Wonder Drug* shows how the ineffectiveness of European science was, in part, a result of the active role that different groups in Spanish colonial society took in challenging and undermining the new vision of science and empire promulgated by the Spanish Crown and imperial reformers. As noted above, firsthand experience was highly valued with the episteme culture of the colonial government. Consequently, the Crown often ordered officials in South America to seek information and advice from *hombres péritos* (experts), who had direct experience of the matter in question.[51] When it came to natural resources like quina, officials throughout the colonial government consulted indiscriminately with indigenous healers, missionaries, bark collectors, merchants, and local officials as well as physicians, pharmacists, and botanists. As a result, the documentary record of the royal reserve of quina contains an impressive range of voices and perspectives. Anyone with relevant experience and expertise was a potential adviser to the empire. It is a testament to a surprising amount of promiscuity and pragmatism that existed in the epistemic culture of the colonial government. At the same time, the various groups, on whom imperial officials relied for information and knowledge, represented distinct and, at times, discordant political and economic interests. In this way, the colonial government provided an institutional context in which colonial subjects not only challenged imperial reform as well as European science but also coopted the mechanisms of colonial governance for their own ends. In other words, knowledge of the natural world was not prior to but a product of the tensions and competing interests that pervaded the Spanish colonial government.

These features of the case of quina are significant because they provide new perspective on the environmental history of colonial Latin America.[52] As noted by Mark Carey, the history of environment in colonial Latin America has largely been a tragic tale of the environmental degradation and destruction due to the allegedly "outside" forces of capitalism and colonialism that took hold in the wake of the arrival of Europeans.[53] The case of quina challenges such narratives by showing that it was the colonial government that first recognized the destructive environmental effects of bark harvesting and attempted to take steps to promote the conservation of cinchona trees.[54] It also shows how local interest groups took an active role in the exploitation of this natural resource not out of ignorance but out of a clear sense of their own priorities and local interests. This book further highlights the inadequacy of framing the analysis of complex phenomena like environmental degradation and resource extraction in terms of overly simplified oppositions of local versus imperial, indigenous versus foreign, or Creole versus European.

Organization of the Book

This book is divided into two parts that are organized chronologically and thematically. Part I explores the complexities of knowing nature in the early modern Atlantic World as manifested in the transformation of quina from a local Andean remedy into a botanical commodity and an imperial natural resource in the Spanish Atlantic World from the mid-seventeenth to the mid-eighteenth centuries. The chapters in this part show how each dimension of the bark's identity— as medicament, as commodity, as natural resource— was not necessarily inherent to the bark itself but was, in some sense, a consequence of the bark's integration into Andean, Atlantic, and imperial networks the supported the movement of people, objects, texts, and images. The first chapter suggests that the persistent association of the bark with the Andean province of Loja was not just an accident of history or biogeography but also a reflection of the region's long tradition as a center of medical knowledge and practice as well as the ingenuity and expertise of indigenous healers, who employed a sophisticated understanding of the nature, disease and the body to cinchona bark. An understanding of the centrality of the Andean World and the sophistication of Andean medicine provides the foundation for making sense of later challenges faced by Spanish pharmacists and botanists in their efforts to assert the superiority of their knowledge of the cinchona tree and its bark.

By the mid-eighteenth century, when the Spanish Crown formally declared its interest in the bark as an imperial natural resource, quina had become a product as much of the Atlantic World as the Andean World. Chapter 2 goes on to show how epidemiological, environmental, and economic developments in the wider Atlantic World facilitated the bark's transformation from a local remedy into a global botanical commodity. The consequences of this transformation were significant. First, to meet increasing demand, quina had to be produced on an unprecedented scal— a development that was, on the one hand, an economic boon to the quina-producing regions of South America and that, on the other hand, put increasing pressure on the cinchona forests of the Andes. Second, the transformation of quina into a botanical commodity gave rise to new understandings of the bark as physicians and pharmacists in

Europe tried to explain its medical effects, as merchants and healers beyond the Andean World developed their own techniques for assessing its medical efficacy, and as travelers from Europe sought new information on the trees and region that gave rise to this early modern wonder drug. Just as in the case of the Andean World, ideas about the cinchona tree and its bark that circulated along Atlantic trade networks provided another vantage point from which the members of Spanish colonial society could challenge the authority of European science.

It was only in the wake of its transformation into a botanical commodity that quina became an object of empire with the establishment of a royal reserve of quina in Loja in 1751. As officials in Spain realized that they needed to know a lot more about quina and the quina trade, the Crown mobilized the individuals and institutions of Spain's colonial government to provide that knowledge. Chapter 3 then examines this process and describes the distinctive epistemic culture of the colonial government—the set of values and techniques employed by officials throughout the Spanish Empire to make natural resources, like quina, intelligible and, more importantly, amenable to imperial administration. Collectively, the chapters in part I elucidate the indigenous, Atlantic, and imperial networks that facilitated the circulation and production of the bark itself and knowledge about the bark as well. In short, these chapters show how quina was an object of Andean medicine, Atlantic commerce, and imperial policy at the same time as it was an object of European science and medicine. Appreciation of the bark's multiple identities in the Atlantic World is important because its simultaneous existence as a meaningful object to different communities of knowers and users influenced the possibilities for Spanish pharmacists and botanists to know the bark and to assert their authority speak for the cinchona tree and its bark later in the eighteenth century.

Part II describes the conflicts that emerged in the late eighteenth century as the Spanish Crown tried to use the European sciences of pharmacy and botany more exclusively to assert greater control over the cinchona tree and its bark. These episodes provide useful insight into the politics of knowledge of the natural world and into the fragility of European science in the contexts of colonial society and colonial governance. The first chapter in part II examines the relationship between the Royal Pharmacy in Madrid, which received and examined much of the quina from South America, and the royal reserve in Loja, which produced the annual quina shipments for the Crown. In particular, it explores a disagreement that erupted between pharmacists in Madrid and bark collectors in Loja over which kind of quina was the best. This disagreement was a pivotal moment as bark collectors emerged victorious and reasserted the importance of local knowledge to the Spanish Empire relative to the claims of Spanish pharmacists. Chapter 5 goes on to explore one of the main consequences of this episode: the expansion of the community of experts in Spain involved with the testing and examination of cinchona bark. With the authority of the Royal Pharmacy in question, officials in Madrid in the 1770s and 1780s began looking to other medical experts and botanists to provide the authority and certainty that the Royal Pharmacy apparently lacked. Such evidence shows how the practical problems associated with quina led to the further intertwining of science, especially botany, and empire. It also shows how earlier disagreements over the identity of quina encouraged the Crown to alter its tactics in its efforts to assert greater control over cinchona trees and their bark.

The final two chapters of the book look specifically at the experience of botanists acting as agents of empire. In the final decade of the eighteenth century, the Crown's interest in quina provided an opportunity for Spanish botanists to make good on the promises of their predecessors that their science was useful to the empire. In 1789, the Crown appointed a "botanist-chemist" as the new director of the royal reserve of quina in Loja. Such an appointment was virtually unprecedented in the history not just of Spanish imperial governance but European imperial governance in the early modern Atlantic World. Yet, as chapter 6 shows, this botanist-chemist soon discovered that the utility and authority of European science had limited efficacy in the face of the larger challenges of colonial society. Science, on its own, was not enough to guarantee effective imperial control of the cinchona tree and its bark. Finally, chapter 7 offers new insight into a major debate among Spanish naturalists over the classification of quina. While there were real intellectual questions and issues of scientific methodology at stake, this debate was also fueled by competing visions of the imperial order. This episode shows how intimately intertwined science and empire had become in the Spanish Atlantic World in the twilight of Spanish imperial rule. Taken together, the chapters in part II demonstrate the ways in which preexisting networks of knowledge production and the epistemic culture of the colonial government could actually undermine the effectiveness of European science as a tool of empire. Rather than suggesting some kind of deficiency in Spanish science or imperial rule, these episodes offer a more important insight: that there was no natural affinity between science and empire. They also provide an appreciation for how much work it took to transform European sciences into the sciences of empire.

The Objects and Politics of Historical Knowledge

To some readers, it might seem unusual to write the history of science and empire from the perspective of an object like tree bark.[55] A more obvious approach might have been to focus on a scientific discipline, institution, or individual. But such approaches often provide only part of the story, especially because such approaches often start with an implicit definition of what counts as "science" or as "scientific." By working from the perspective of an itinerant object defined by many bodies of knowledge beyond European science, we reduce the risk of reinscribing an imperial politics of knowledge in which European ways of knowing nature are implicitly labeled "science," while other (i.e., non-European) ways of knowing are labeled "non-science" or "superstition."[56] The documentary traces of the movements of cinchona bark in the early modern Atlantic World make visible a heterogeneous network of people, places, texts, images, and objects that supported the production of knowledge about nature. Quina was not just a botanical specimen; it was also a medicament, a commodity, and an object of imperial policy. One of the main advantages of this approach is that it sets European science, without privileging it, among the many different ways of knowing nature in the early modern Atlantic World.[57] Such an approach also makes clear the ways in which the authority of European science was still uncertain and its practical application largely untested.

Knowledge of nature was an artifact of European imperialism as much as it was an artifact of any of the other worldviews and enterprises that existed in the early modern Atlantic World. Consequently, the politics of knowledge remain a challenge for historians, who work with the archival traces of an empire's knowledge.[58] The question of who gets to speak for nature in historical narratives is just as important as it was in the debates that took place in the eighteenth-century Atlantic World. The central conundrum is how to recognize the asymmetries and inequalities produced by European imperialism while being sensitive to the agency of those whose knowledge and culture was marginalized or assimilated by imperial regimes. Like many other early modern and modern European empires, the effective mobilization and exploitation of natural resources of colonized territories was as central to the Spanish imperial enterprise in the Americas as was the mobilization and exploitation of human resources, especially indigenous peoples and African slaves. Knowledge of these resources was essential to the survival and success of the colonial project. Some of this knowledge was knowledge that Europeans brought with them across the Atlantic Ocean (knowledge that itself was the product of cross-cultural encounters in the Afro-Eurasian World in the preceding centuries). In the late fifteenth and early sixteenth centuries, for example, systematic knowledge of the stars and seas made the European voyages of exploration possible and facilitated the encounters between Europeans and the indigenous peoples on the Atlantic islands and coasts of the Americas. When it came to the sustained exploration, colonization, and exploitation of the Americas in the sixteenth century and beyond, Spanish colonialism—like that of contemporary European states—relied on and assimilated the knowledge of Amerindians and Africans. After the period of the earliest encounters and exchanges between Africans, Amerindians, and Europeans in the Americas and elsewhere, much of the "science" in the Atlantic World was the result of the interactions between the efforts of Africans, Amerindians, Europeans to produce systematic knowledge about the natural world.[59]

In many ways, quina—as an object—is a good metaphor for European science in colonial contexts of the Atlantic World. Both were powerful but fragile. And it required a significant amount of labor from different groups of people to ensure that both retained their efficacy across long distances. Cinchona bark had the power to restore health, but as early modern healers knew, its medical virtues were fickle. Sometimes the bark worked, and at other times it did not. The same was true for science. In many cases, European sciences and their representatives proved powerful agents of empire. In other instances, this was not the case. Just as the forces of nature (humidity) and society (fraud) could undermine the integrity of quina, myriad forces could undermine the authority and efficacy of European science in colonial contexts. Both a fuller understanding of science and empire and a more global history of science emerge from exploring the shortcomings, in addition to the successes, of European science in the eighteenth century, a transitional period to the modern era in which European sciences and empires would achieve unprecedented global reach and influence.

PART I

Andean, Atlantic, and Imperial Networks of Knowledge

One

Quina as a Medicament from the Andean World

On 3 February 1737, Fernando de la Vega (b. circa 1672) welcomed an unusual visitor to his modest home in Malacatos, a small village near Loja, the capital of a province by the same name in the southern sierra of the viceroyalty of New Granada.[1] This visitor was Charles Marie de la Condamine (1701–1774), a scientific traveler from France. La Condamine was a member of a joint French-Spanish expedition organized by the Royal Academy of Sciences in Paris and sent to the Audiencia of Quito in 1735. The official purpose of the expedition was to measure a degree of latitude at the equator in order to settle a dispute among European natural philosophers about the shape of the Earth.[2]

But it was not the equator that brought this scientific traveler from Europe to Malacatos. La Condamine had come on unofficial orders from the French Crown to study Loja's cinchona trees. "The best, or at least the most famous, *quina*," observed La Condamine in a report to the Royal Academy of Sciences, "comes from the mountain known as Caxanuma that is two and a half leagues to the south of Loja."[3] The only problem was that La Condamine had no idea what the trees actually looked like. He had never been to South America let alone laid eyes on the "quinquina" tree as he called it. Moreover, the imprecise images and vague written descriptions of the tree available in early eighteenth-century Europe would have proven useless in the face of the verdant and diverse forests of the Andes.[4] So the Frenchman came to Fernando de la Vega—a man who had worked with cinchona trees and their bark for much of his life.

Vega's experience was a common one. Even in the early 1700s, after centuries of contact and colonization in the Americas, scientific travelers from Europe still arrived knowing relatively little about American flora and fauna.[5] As a result, these travelers depended on local contacts' knowledge of local landscapes and natural phenomena. The situation is especially striking in the case of the cinchona tree considering that Europeans had been using its bark therapeutically for nearly a century by the late 1730s when La Condamine visited Malacatos.

One irony of such encounters is that, in travel accounts and scientific treatises printed upon their return to Europe, travelers like La Condamine often cast themselves as the sources of knowledge and enlightenment in these unfamiliar lands. By marginalizing, if not omitting, the contributions of their local contacts, European scientific travelers claimed all the credit for the knowledge of natural worlds beyond Europe. La Condamine was no different. He was not about to share the glory of producing the first botanical description of the cinchona tree with Vega. After all, the man in Malacatos had no formal training in the study of the natural world nor was he a member of the Royal Academy of Sciences in Paris as La Condamine was. While the Frenchman styled himself a scientific hero, his host in Loja became an anonymous assistant.[6]

In the end, La Condamine's decision not to mention of Vega in his report on the "quinquina" tree would have significant consequences. Since its publication in the *Mémoires* of the Royal Academy of Sciences in Paris, La Condamine's study has been commemorated numerous times as one of the first "scientific" accounts of the cinchona tree.[7] In 1753, Swedish naturalist Linnaeus (1707–1778), one of the architects of modern botanical taxonomy, used the Frenchman's report as the basis for the description and classification of the genus *Cinchona,* which Linnaeus added to a new edition of his *Species Plantarum.*[8] Meanwhile, Vega has been largely forgotten even though it was probably his choice of specimens for La Condamine that made this enterprise possible. In the nineteenth and twentieth centuries as Western medicine came to recognize quinine, the antimalarial alkaloid derived from cinchona bark, as a modern wonder drug, histories of the drug's discovery and development continued the pattern of exclusion but did so on a much grander scale by ignoring the entire Andean World. After teaching European colonists about the bark's medicinal properties in the early seventeenth century, the people of the Andean World virtually disappear from the story—as if Andean healers stopped using the bark and innovating in its therapeutic application once they shared their knowledge with Europeans.[9]

This chapter (re)establishes, at the outset, the importance of the Andean context to the history of cinchona bark and emphasizes the role of the Andean World as one of the centers of knowledge production in the Americas and the Atlantic World. Although exact details of how Europeans first learned about the medicinal properties of quina are unknown, the earliest European reports agree that this transfer of knowledge took place in the Loja region (figure 1).[10] Recent scholarship has shown that place matters when it comes to knowing the natural world.[11] In the modern science, certain places and spaces, such as the field or the laboratory, bestow epistemic authority and credibility to the knowledge produced at those sites. Building on the insights of this literature, this chapter argues that the Andean context mattered and that the association of Loja with quina was not merely a coincidence but evidence of the importance of Andean networks of knowledge production and of Andean bodies of medical and botanical knowledge. After all, the various species of cinchona tree could be found throughout the Andean forests from Bogotá to La Paz and even in the forests of the Amazon basin in Brazil (figure 1).[12] But Loja remained the only Andean region consistently associated with quina and retained a reputation for producing the best bark well into the nineteenth century.

Perhaps, the cinchona trees in Loja had some special quality or maybe it was the unique characteristics of the local environment that made quina from Loja so potent. It is difficult to say, especially because no one in the early modern Atlantic World engaged in a systematic study of the medical efficacy of bark from Loja.[13] Undoubtedly nature played a role, but it is important to recognize that it also took considerable skill and knowledge to know where to find the trees, to identify which trees or parts of the tree had the best bark, and to harvest and prepare the bark in a way that would best preserve its medical virtues. Such knowledge was not easy to come by. It required a community of specialists—Andean healers—dedicated to the accumulation and transmission of such knowledge and techniques over time and across space. Like many other botanical medicaments quina was in the first instance as much an artifact of sophisticated Andean medical theory and practice as it was a product of the natural world. This chapter thus describes an Andean culture of medical knowledge and practice in order to show that the *curanderos* (indigenous healers) of the Andes, as much as European naturalists, were representatives of a dynamic tradition of specialized knowledge about the natural world. Greater appreciation of the bark as a natural object and a cultural artifact provides one way to reinsert indigenous healers and the Andean World back into the global history of science and medicine. In addition, the presence of an existing tradition of medical knowledge and community of specialists with knowledge of plants would, later, expose the fragility of European science and serve as an obstacle in the efforts of Spanish pharmacists and botanists to assert the superiority of their knowledge of plants such as the cinchona tree and its bark.

The Healers of the Andes and Their Medical Cosmology

Curanderos of the Andes were likely the first group with specialized knowledge of healing and the natural world to use cinchona bark therapeutically.[14] With the outbreaks of malaria in the Andes that occurred in the wake of the Spanish conquest of the region in the 1540s, these healers would have found themselves along one of the many epidemiological frontiers created by Spanish colonization. As a result, they were the first community of healers in the early modern world to have regular experience with malarial fevers and reliable access to cinchona bark. Under such conditions, Andean curanderos were likely the first to recognize the bark's efficacy in treating the intermittent fevers, one of the most prevalent symptoms associated with malaria. At the same time, it is important to recognize that these healers did not have access to the explanations of malaria and the therapeutic efficacy of cinchona bark offered by modern medicine.

How and why did Andean healers ever think to apply this particular tree bark to the new cases of malarial fevers? After all being located in one of the most biologically diverse regions of the world, Andean healers had access to an extraordinary variety of medicinal plants. To answer this question, it is important to look at the practices and medical cosmology of Andean healers. The discovery of the antimalarial properties of the bark can be made intelligible in terms of Andean theories of health and healing, suggesting that the discovery did not occur by chance. Evidence also suggests that the continued use of the bark by Andean healers should be attributed as much to their sophisticated medical knowledge and practice as to practical necessity of having to assist their communities in dealing with the deadly outbreaks of disease that occurred in association with Spanish colonization.

It is difficult to know what exactly Andean medicine was like before the arrival of the Spanish because of the scarcity of sources and material artifacts.[15] Archaeological evidence and modern anthropological studies from southern Ecuador and northern Peru do suggest that curanderos possess "rich shamanic lore" rooted in a long tradition of healing practice and specialized medical theory.[16] Some of the earliest evidence comes from the ceramics depicting scenes of healers and healing of the Moche civilization (100–800 CE), one of the earliest complex societies in the Andes.[17] Though it is important to recognize Andean ethnomedicine as an evolving body of practice and knowledge, comparison of the use of particular kinds of "ceramics, staffs, [and] stones" in the healing rituals of today's curanderos with archaeological evidence reveals continuities between medical practices past and present.[18]

Recent anthropological studies of the practitioners of traditional medicine or ethnomedicine in modern Andean communities provide additional insight. One finding shows that these communities recognize different types of healers that specialize in distinctive ways of healing. For example, a study of Northern Peru has identified four main types of healers: *curanderos mayores* (master healers), *curanderos menores* (novice healers), *parteras* (midwives), and *hueseros* (bonesetters). Another study of the broader Andean World suggests a wider variety of healers, including *brujos* (witches), *curanderos generales* (general healers), *curandero-hierbateros* (herbalists), *curanderos del espanto* (healers specializing in the treatment of *espanto* or terror), *sobadores* (bonesetters), and *parteras* (midwives).[19] Since members of these groups produce the next generation of healers through apprenticeship, these specializations have persisted over time.[20]

Most Andean healers command great respect in their local communities and attract clientele from far away. Regarding the function and importance of healers in the indigenous communities of South America, anthropologist Michael Taussig has observed: "folk healing and magic in today's Third World are not bizarre and meaningless discourses, but specialized forms of folk art and wisdom addressing the perplexity and mystery of daily life in the modern world."[21] Similarly, as indigenous communities faced colonization first by the Inca and later by the Spanish, local healers may have used their "specialized" wisdom to help individuals and their communities address the challenges of daily life under colonial rule.[22]

To further appreciate the role of Andean healers as specialists and experts in their communities, let us consider their medical cosmology. Two key concepts in Andean understandings of health, disease, and the body are balance and reciprocity. One version of the concept of balance is expressed in the idea that all things in the world can be divided into two broad categories: hot and cold.[23] Various indigenous groups in the Andes apply this system to a broad variety of phenomena including "geographical features, food sources, the body and the diseases that affect it."[24] This hot-cold system of classification along with the worship of mountains, water spirits, and the spirits of plants appears to be a distinctive feature of Andean medicine that, in all likelihood, predate European colonization.[25] At the same time, this theory of health and disease had much in common with that of early modern European healers, who, drawing on the works of Hippocrates and Galen, also employed a humoral theory in which health, illness, places, and plants were classified according to the opposition of qualities of hot and cold and also wet and dry.[26] It may be one reason why European colonists in the early modern Andean World were so amenable to the medical and botanical knowledge of Andean healers. Current medical practice among Andean healers reflects the influence of a historical and ongoing process of hybridization of Andean and European medicine since the colonial period.

Because of their applicability to all kinds of phenomena, the categories of hot and cold provide a framework that informs the daily activities of indigenous peoples in southern Ecuador and northern Peru. Under this system, health is a state that occurs when the hot and cold phenomena influencing the body are in balance. Maintenance of health is, thus, a complex task due to the diversity of phenomena that can exert an influence on the body. These include hills, lagoons, plants, foodstuffs, parts of the body, winds, medicaments, colors, and illnesses. As a result, healers must take into account the local environments, diets, and physical state of their patients as well as the hot or cold influences of these factors before treating an illness. In addition, if the healer decides on a course of treatment that involves the use of medicinal plants, then plants with the right hot or cold qualities must be selected to restore balance and health.[27]

In addition to this system of hot and cold, Andean conceptions of health, disease, and the body were also informed by ideas about the relationship between the individual, society, and the sacred. Health was a state of balance not just between hot and cold phenomena but also between social, natural, and supernatural forces. Disease could also be the result of an imbalance between these forces. According to Irene Silverblatt, the Quechua term *ayni,* which means both balance and reciprocity, best expresses the ideal of "universal equilibrium" that was the basis of health in the Andean cosmology.[28] When this balance broke down, the result was disease. The illness caused by such disequilibrium could afflict a whole society in addition to any one individual. As a concept, ayni also implies the need for human intervention to maintain the balance between individuals, society, and the sacred. Enter the healer.

According to this worldview, Andean healers needed to know how to mediate the supernatural and the sacred as much as they needed an understanding of the physiological symptoms of disease and the physical properties of the plants or other natural substances used to treat disease. Consequently, many Andean healers were simultaneously "expert herbalists and masters of ritual necessary to restore cultural order and thereby cure disease."[29] In the course of treating disease, these curanderos relied on a variety of practices— "divination, knowledge of the medicinal properties of plants, a sort of confession and ritual cleansing, and the ability to direct rites that accompanied offerings to sacred deities."[30] In this way, Andean healers engaged in magical as well as material practices in the course of diagnosing and treating disease unlike modern physicians that deal strictly with the treatment of physical symptoms. It is difficult, if not impossible, to understand the practices of curanderos and their role in Andean societies without some appreciation of their medical cosmology and the broader Andean worldview.

A good example of the intertwining of the material, the social and the sacred in theory and practice can be found in the Kallawaya, an important group of itinerant indigenous healers in the Bolivian Andes that have been the subject of several studies. Instead of a strong emphasis on balance, the cosmology of the Kallawaya emphasizes flow. According to their worldview, nature is understood as a "system in a cycle" such that the goal of healing is to keep a patient's humors flowing.[31] In particular, the Kallawaya possess a hydraulic theory of medicine "in which liquids are concentrated or distilled by centripetal forces and dispersed to the periphery by centrifugal forces."[32] In this view, health depends on the cyclical motion of fluids—water, air, blood, and food. The body distills these fluids into both secondary fluids (mucus, bile, sweat, urine, gas, milk, and semen) and semifluids (feces and fat). Kallawaya do classify these fluids according to two sets of opposing characteristics—hot/cold and wet/dry; however, they understand these qualities primarily in terms of their effect on the viscosity of the fluids and the cycling of the hydraulic system in the body.

The medical cosmology of the Kallawaya is a good example of how healers serve as mediators of social interactions while addressing the material causes of disease. Their understanding of the body is linked to their understanding of the fundamental unit of Andean social organization: the *ayllu*.[33] An ayllu is an extended group of households linked by social and economic ties that are organized vertically on a mountain with one community in the highlands, another community in the central lands, and a third community in the lowlands. The Kallawaya understand the ayllu as a body and they think of its different parts in anthropomorphic terms.[34] Like the human body, healers in the Bolivian Andes also understand their ayllus and the mountains on which they live as a cyclical system. Just as the health of the physical body depends on the circulation of fluids in the body, the health of the social body of the ayllu depends on circulation of goods and people that bound the subunits of the ayllu into a single community.

Some forms of Andean healing try to keep humors in balance or in circulation within the body, whereas other forms of Andean healing focus more on the supernatural causes of illness. Within the medical cosmology of curanderos in southern Ecuador and northern Peru, disease can have a natural cause, a supernatural cause or, in some cases, both. Supernatural causes of disease include the malevolent influences of the spirits of winds (*los vientos*), of hills (*los cerros*), and of lagoons (*las lagunas*).[35] It is likely that healers in the northern

Andes share with their Kallawaya colleagues in the south an understanding of health as a state of equilibrium or circulation that could be influenced by the animated spirits of winds, mountains, and lagoons.

Other supernatural causes of disease include the machinations of a malevolent shaman as well as violations of divine law, a reflection of the influence of Catholicism on the cosmology of today's Andean healers. For Andean healers, such supernatural causes exist alongside a variety of natural or material causes of disease that, in addition to the imbalance of hot and cold humors, can include socioeconomic conditions, the influence of the ambient environment, and even the phases of moon.[36] All these features provide insight not just into the important and varied roles that healers played in Andean communities but also into the sophisticated medical cosmology that informed their healing practices.

This evidence of the complexity of the Andean medical cosmology contrasts with the recent trend in historical scholarship to emphasize the empirical knowledge of the indigenous informants of European scientific travelers in the Atlantic World.[37] Several studies have suggested that Europeans came to value the empiricism that was so characteristic of the "new science" of seventeenth-century Europe because of their experiences trading and colonizing unfamiliar regions around the globe. By the eighteenth century, European naturalists came to value the contributions of their Native American and African assistants because they considered these groups to have direct experience with the natural world. Although European emphasis on empiricism could give indigenous and African knowledge considerable authority, it is important not to reduce these bodies of knowledge to simply a set of empirical observations. Andean healers were not merely empiricists or pragmatists but representatives of a bodies of knowledge just as dynamic and complex as the sciences and medicine of early modern Europe.

Andean Healing along a Sixteenth-Century Epidemiological Frontier

Many of the earliest European accounts of cinchona bark printed in the late sixteenth and early seventeenth centuries recognized the utility of Andean medical knowledge and credited "Indians" with the discovery of the bark's medicinal properties. More important, several early reports associated the bark specifically with the northern regions of the Viceroyalty of Peru that later became the Audiencia of Quito. For example, in his 1571 treatise on medicinal substances from the Americas, Spanish physician Nicolas Monardes (1493–1588) described a medicinal tree bark from Guayaquil, the main port city serving the Loja region. Historian of medicine Fernando Crespo Ortiz has convincingly identified this bark as quina.[38] Similarly, in his *Moral Chronicle of the Order of Saint Augustine in Peru* published in Barcelona in 1639, Antonio de la Calancha (1584–1654), an Augustinian missionary, wrote, "Peru produces a tree that is called the tree of fevers in the region of Loja." He also noted that the "barks" of this tree were the color of cinnamon and reported that a drink was made using pulverized bark weighing the equivalent of two *reales*. This concoction, he continued, "gets rid of fevers and *tercianas* [intermittent fevers that occurred every three days]."[39] Through such reports, European physicians and pharmacists were able to emulate the therapeutic techniques of Andean healers.[40]

Anthropological and historical evidence suggests that the association of the bark with Loja was no coincidence. On the one hand, it is a reflection of the importance of Loja as part of a vital network of healing practice and medical knowledge in the colonial Andean World. On the other hand, it is a reflection of the contingencies of the Spanish conquest and colonization in that the healers in this region were some of the first to encounter the Old World diseases, including malaria, brought by European and African colonists. In the sixteenth century, healers in Loja and its neighboring regions not only found themselves along one of the main epidemiological frontiers created by early modern European imperial expansion but were also located in a region with both a long-standing tradition of medical knowledge and an extensive botanical pharmacopeia derived from the diversity of the local flora.

According to the work of Peruvian anthropologist Lupe Camino, Loja lies along a nearly continuous line of cities, towns, and settlements from Quito in Ecuador to Puerto Eten in Peru that form an "axis of health" (figure 1 inset).[41] What makes this region an "axis of health" is the presence of a shared understanding of health, disease, and healing among healers in modern Andean communities. The prevalence of a shared medical cosmology is not only evidence of a high degree of intellectual exchange among healers in the region but also suggests a shared tradition of medical knowledge and practice. In a recent study of medicinal plants used by healers in southern Ecuador and northern Peru, ethnobotanists have confirmed this characterization and describe the region as "the healing center of the old Central Andean cultural area that stretched from Ecuador to Bolivia."[42] In other words, when the Spaniards first conquered this region in the sixteenth century, they unknowingly stumbled upon an important center of Andean medical knowledge that connected an extensive network of indigenous healers.

Several factors made this region, especially Loja, a "healing center" in the Andean World. One was the local environment. Naturalists and ethnobotanists have identified the region straddling the border between Ecuador and Peru as one of the regions of the world with the highest biological diversity due to "the rapid transition between humid mountain forests of the northern

Andes to the dry, deciduous forests and deserts of the northern Peruvian lowlands."[43] Such diversity meant that healers had access to a vast array of plants for therapeutic use. Whereas healers and botanists in Europe had to travel and collect specimens of plants from around the globe to enrich their botanical gardens with diverse species, Andean healers were surrounded by a diversity of plants.[44] Consequently, the Andean World was one in which healers have developed knowledge of the medicinal use of an extensive variety of plants. In addition, access to such diversity may have fostered a culture of therapeutic innovation and pharmacological experimentation among these healers.

Another factor that made Loja a "healing center" was local geography. According to Linda Newson, the mountain passes near Loja are at a lower altitude than anywhere else along the Andean mountain chains.[45] This feature of the local topography allowed for "an easy exchange between the flora and fauna of the Amazon Basin and the Pacific lowlands." [46] Such conditions would have also promoted the exchange of knowledge of medicinal plants between the curanderos of Loja and those in adjacent areas, including the coastal regions to the west and the jungles to the east. In other words, Loja's curanderos lived and worked in a botanically diverse region and cultural crossroads, where interactions with itinerant healers and the local environment encourage the production and circulation of medical knowledge.[47] Consequently, when malaria and other Old World diseases started to decimate their communities, Andean healers had the knowledge and access to materials necessary to try to alleviate the negative health effects of Spanish conquest and colonization.

In the early sixteenth century, it was European commercial and colonial expansion that put the nascent Atlantic World into contact with the established Andean World, which, at the time, was experiencing yet another phase of Inca expansion.[48] In conjunction with exploration of the west coast of South America in the 1520s and the conquests of Peru in the 1530s, European colonists and enslaved Africans introduced many new diseases to the indigenous populations of the Andes as they did elsewhere in the Americas.[49]

One of these was the blood-borne disease malaria.[50] Because indigenous populations suffered outbreaks of many Old World diseases simultaneously, it is difficult to pinpoint exactly when malaria arrived in South America.[51] Nonetheless, we do know that the disease would have found favorable environmental conditions in the coastal areas and lowlands of colonial Ecuador. These regions would have served as breeding grounds for various species of Anopheline mosquito, the main vector of malaria aside from humans and animals. Andean healers found themselves situated along an epidemiological frontier in which there were repeated outbreaks of disease as Old World pathogens encountered the previously unexposed populations of the New World. Moreover, Andean healers would have been some of the first medical practitioners

TABLE 1. Geography of Hot and Cold along the Andean Axis of Health

Country	Department	Locality	Characteristic
Ecuador		Quito	Cold
		Zamora	Hot
		Loja	**Hot**
Peru	Piura	Talaneo	Cold
		Huaringas	Cold
		Guar-Guar	Cold
		Huancabamba Valle	Intermediate
		Cerro Pariacaca	Hot
		Cerro Witiligum	Cold
		Sondor-Sondorillo	Hot
		Huarmaca	Cold
	Lambayeque	Salas-Penachi	Intermediate
		Desierto de Olmos	Hot
		Monsefú Mochumi	Intermediate
		Eten	Hot

Source: Lupe Camino, *Cerros, plantas y lagunas poderosas: La medicina al norte del Perú* (Piura: Cipca, 1992), 43.

to have both regular experience with malaria and reliable access to cinchona bark. By the time the bark was first imported to Europe in the 1630s, these healers would have had more than a century to pit their local pharmacopoeia against these new diseases that ravaged their communities. In other words, Andean healers had more than enough time to figure out that this bark was effective against the intermittent fevers caused by malaria.

How did these healers even think to apply cinchona bark to cases of intermittent fever? It may have been through trial and error, but it is hard to imagine these healers applying medicaments at random, especially because evidence suggests that their healing practices were informed by a sophisticated medical cosmology. As it turns out, according to the logic of the Andean medical cosmology, it would have made sense in theory to use cinchona bark to treat fever.

If Andean healers of the sixteenth and seventeenth century employed medical cosmology similar to the one employed by their modern counterparts, then the system of hot and cold would have provided one justification. In particular, we know that today's healers classify different regions along the "axis of health" that runs north to south from southern Ecuador to northern Peru as being either hot, cold, or intermediate as shown in table 1.[52] According to this system, the hot or cold character of plants stands in an inverse relationship to the hot or cold character of the region where they grow. In other words, healers along the axis of health from Quito, Ecuador to Eten, Peru characterize plants gathered in "hot" regions as "cold," while they characterize plants gathered in "cold" regions as "hot."[53] According to this logic, cinchona bark from Loja, a "hot" region, would be characterized as "cold" plant, making it an appropriate treatment for a "hot" disease like fever.[54]

Here, it is worth pausing to make a comparison with early modern European medicine. As we have seen, according to the logic of their system of hot and cold, Andean healers had justification for using cinchona bark to treat fever. By contrast, many physicians and pharmacists in late seventeenth-century Europe employed their own humoral theory to reject the use of the bark in cases of fever. According to the Greco-Roman humoral theory that informed the European medical cosmology, cinchona bark was a "hot" remedy on account of its bitter taste. As a result, some European physicians considered it inappropriate and unethical to use the bark to treat fever because it would lead to a further imbalance of the hot and cold humors.[55] Thankfully for those suffering from intermittent fever in early modern Europe, the bark worked in enough cases that therapeutic use continued in the face of theoretical resistance; thus allowing this embodiment of Andean medical knowledge to triumph over European ignorance.

Andean Healing under Spanish Colonial Rule

Printed sources from the early centuries of Spanish colonial rule provide evidence that contemporaries valued the knowledge of Andean healers and even recognized certain regions of the Andean World as important centers of healing practice and medical knowledge. Many of these sources come from seventeenth-century missionaries, one group of European colonists that not only had extensive contact with indigenous healers but also published accounts of those encounters. Many of these missionaries were part of the campaigns in the seventeenth-century to extirpate "idolatry" in the Viceroyalty of Peru, an effort by the Spanish colonial government to investigate and eliminate any remnants of native Andean religion (considered idolatry in Spanish eyes).[56] Paradoxically, many missionaries recognized the power and knowledge of Andean curanderos, even as they attributed the efficacy of Andean healing to the Devil. In his 1621 handbook on identifying and eliminating non-Christian practices in Peru, *The Extirpation of Idolatry in Peru,* the Jesuit Pablo José de Arriaga (1564–1622) characterized several classes of indigenous healers as "Ministers of Idolatry."[57] A decade later in his chronicle of the missionary efforts of the Augustinians in Peru, Antonio de la Calancha (1584–1654) characterized indigenous healers similarly by describing them as *hechizeros* (sorcerers).[58]

Many of these Spanish sources made a distinction between hechizeros and herbolarios (herbalists). Such a distinction itself is evidence of the importance that European colonists assigned to plant knowledge by separating it from other forms of Andean healing practice and knowledge that were often considered forms of sorcery among European missionaries. Herbalists fared quite well in the eyes of seventeenth-century Europe authors, including even Arriaga, who recognized the utility and authority of indigenous knowledge of the healing properties of local flora. "Many of the sorcerers," explained Arraiga, "are *Ambicayos,* or healers, as they are called."[59] Unfortunately, from Arriaga's perspective, many of these ambicayos included "superstitious and idolatrous practices" in their treatments. Of course, what Arriaga saw as idolatry was a vital part of the function of healers operating according to a medical cosmology in which disease could have both supernatural and natural causes. At same time, Arriaga recognized the utility of "their knowledge and use of certain *herbs* and other simples." As a result, he encouraged priests to "instruct" the ambicayos to heal solely through the use of medicaments rather than engaging in practices that were "superstitious and evil."[60]

La Calancha, an Augustinian missionary was, by contrast, a little more circumspect in his endorsement of the plant knowledge of indigenous healers. Rather than acknowledging healers directly, La Calancha praised what he called "the pharmacy of the Indians."[61] Peru, he observed, "produces a vast number of medicinal herbs and beneficial roots that cure Spaniards in places where there are no pharmacies."[62] Notice that he emphasizes the place—Peru—as the purveyor of medicinal plants, rather than healers themselves. He offers no recognition that Spanish colonists in most cases would have needed the help of an Andean curandero to help them identify the "medicinal herbs and beneficial roots" that the land produced. La Calancha also noted that people "in cities where there are Physicians" were known to use the local medicinal herbs of Peru as part of their treatments.[63] It is likely that Andean healers administered these botanical medicaments as well, or, at the very least, assisted Spanish physicians and pharmacists in finding and preparing them.

Missionaries like Arriaga and La Calancha, engaged in the various campaigns in seventeenth-century Peru to extirpate idolatry, could praise indigenous healers' plant knowledge in part because plants were not inherently sacred objects from a European perspective, so the knowledge of their medicinal effects could be treated as secular knowledge to be extracted from the larger cosmology (erroneous from the missionary's perspective) that undergirded Andean medical theory and practice.[64] Missionaries thus played an important role in the decontextualization and secularization of indigenous healing knowledge—a role that has heretofore been underappreciated.[65]

Though the emphasis on indigenous plant knowledge derived in part from missionaries' desire to exploit useful medical knowledge even as they extirpated superstition and idolatry, their emphasis on indigenous knowledge of medicinal plants was likely the product of the influence of Andean culture as well. Several colonial sources note that the Inca also valued curanderos, especially those that had knowledge of medicinal plants. In his seventeenth-century chronicle of Incan history before and after the Spanish conquest, the Inca Garcilaso de la Vega (1539–1616) noted the "hervolarios" or herbalists were especially well regarded in Inca society. "[Great herbalists]," wrote Garcilaso de la Vega, "were very famous in the time of the Incas. . . . They possessed knowledge of the virtue of many herbs and taught [this knowledge] to their children, who became Physicians."[66] The reputation and fame of these "great herbalists" was reflected in their clientele that included not only "the [Inca] Kings and their relatives" and "the *curacas* [native lords]," but also "the relatives of the herbalists."[67] Garcilaso de la Vega emphasized that indigenous herbalists used their treatments on their own families as well as Inca rulers and elites in order to establish the trustworthiness and authority of their knowledge and their remedies.[68] In addition to the endorsement of the Inca elites, a healer's willingness to use his remedies on his own family members was a powerful testament to his confidence in the efficacy of his practices.

Anthropologist Joseph Bastien has suggested that Garcilaso de la Vega's "herbalists" may be a reference to the Kallawaya. As evidence of their high standing under the Inca, Bastien references an illustration from a manuscript chronicle of Inca history submitted to King Philip III by the Quechua nobleman Felipe Guaman Poma de Ayala (1550–1616) in the early seventeenth century.[69] In a chapter on the patrimony of the Inca rulers, Poma included an illustration of the Inca ruler (figure 2).[70] For Bastien, service in the Inca emperor's litter was a sign of the prestige and high social standing assigned to the Kallawaya under the Inca. While it is not unreasonable to think that the Inca emperors would have traveled with their own personal curanderos or herbolarios, there is some question as to whether the "yndios de callauaya" mentioned in the caption to Poma's illustration worked as healers. After all, the phrasing of Poma's caption could be interpreted as an indication of the geographical origins of the Inca's litter bearers. In other words, the phrase "yndios de callauaya" may simply mean "Indians from Callauaya."[71] As the editors of a modern edition of the *Huarochirí Manuscript* explain, Callauaya was the name of an "agricultural satellite village of Huarochirí" cited in legal documents in 1594.[72] Yet this reference is also ambivalent. It does not explic-

Figure 2. "Yndios de Callauaya," The Royal Library—Copenhagen, GKS 2232 4°: Guaman Poma, *Nueva corónica y buen gobierno* (1615), fol. 331, Drawing 333, "AN-DAS DEL *INGA, QVISPI RANPA*."

itly associate these people with a certain place and the description of the Maca Uisa's litter bearers as "the people called Calla Uaya" emphasizes their ethnic identity, perhaps more so than geographical origins.

Whether Garcilaso de la Vega's healers were the Kallawaya or not is ultimately inconsequential. Sources from both early mestizo chroniclers and Catholic extirpators of idolatry attest to the importance of indigenous healers in the Andean World as reflected both in accounts of their high status under the Inca and under the Spanish, despite fears that indigenous healers were agents of the Devil. These sixteenth- and seventeenth-century observers gave rise to a discourse of the importance of indigenous healers and the value of their knowledge.

In eighteenth-century Peru, creole botanists and physicians would continue to champion the importance of Andean healers and their knowledge of medicinal plants, even as their counterparts in Enlightenment Europe, such as Charles Marie de la Condamine, marginalized the knowledge and contributions of indigenous healers. Consider an article published by Hipólito Unanue (1755–1833), a creole naturalist and physician in Peru. His article appeared in the *Mercurio Peruano,* a learned periodical published in Lima. In the first installment of a piece entitled, "Introduction to the Scientific Description of the Plants of Peru," Unanue praised the knowledge of Andean herbalists.[73] He even went as far as to describe the "Indians" as the "Fathers and founders of the Botany of Peru."[74] Unanue cited La Calancha's account of the Incan law that required those who could not engage in agriculture or warfare to serve as herbalists. He also cited several other works of the sixteenth and seventeenth centuries that praised indigenous knowledge of medicinal plants. Drawing on a passage from the work of the Inca Garcilaso de la Vega, Unanue repeated the claim that some "Indians" made such "progress" in their "knowledge of medicinal plants" that they became physicians to the Incan emperors and the Incan elite, a sign of the authority of the medical knowledge of indigenous healers.[75] Even as the writings of seventeenth-century missionaries discounted many of the practices of Andean curanderos, Unanue cited both missionary and mestizo sources to bolster his claims about the authority of indigenous knowledge of plants.

Spanish colonization provided a new epidemiological and cultural context to which Andean healers had to adapt their ideas and practices. The extirpation of idolatry campaigns of the seventeenth century, at times, made it more difficult for these healers to openly engage in the spiritual or magical elements of their therapeutic practice. At the same time, many missionaries still recognized the value of curanderos' knowledge of medicinal plants and so they advocated the preservation of this knowledge even as they actively sought to destroy the worldview that made the therapeutic use of these plants meaningful. In the end, the Spanish—like the Inca before them—clearly valued and respected the knowledge that indigenous healers had of medicinal plants. Moreover, as evidence suggests, curanderos in the Andean World had enjoyed a good reputation in the preceding centuries under the Inca and the civilizations that preceded them.[76] Consequently, it is no surprise that the transmission of healing knowledge of cinchona bark to Europeans took place in Loja, a region located in one of the centers of the healing knowledge that Spanish colonists and European missionaries came to value.

Let us return to the meeting between Fernando de la Vega and Charles Marie de la Condamine in Malacatos in 1737. Judging by the Frenchman's 1738 report to the Royal Academy of Sciences in Paris, Vega was no one of consequence to La Condamine's enterprise. La Condamine notably fails to identify his host by name and describes Vega simply as "a man from [Loja], who lives at the foot of the mountain and makes his living by collecting the bark of the quinquina tree."[77] This description was meant to suggest that Vega was merely a *cascarillero,* a Spanish term used in South America to refer to seasonal laborers that made their living by making contracts with local landowners and merchants to collect bark from cinchona trees growing wild in the Andes. By casting his local contact as merely a laborer, La Condamine followed the trend of his contemporaries in Enlightenment Europe of marginalizing the role of Andeans in the broader enterprise of knowing New World nature.

Other evidence suggests that Fernando de la Vega was much more than a bark collector. In 1743, La Condamine made a second visit to Loja during his return trip from France. He described this visit in a travel account published in Paris in 1751. This time, La Condamine identified Vega by name. More than that, the Frenchman described how his host in Malacatos provided him with samples of the "best kind" of "quinquina." And how did Vega know which cinchona bark was the best? The most likely answer is that he knew because he lived and worked in Loja, one of the healing centers of the Andean World, and likely had ties to local networks of healers that were the custodians of a long-standing, yet dynamic, body of knowledge and practice regarding the medicinal use of local flora.

Additional evidence suggests that Fernando de la Vega was a curandero. In his 1751 travel account, La Condamine reported that on his second visit to Loja Vega gave him samples of "an extract and a salt recently made from the bark according to the procedure that he learned from Monsieur [Joseph] de Jussieu," a physician and naturalist that was also part of French-Spanish expedition to Quito.[78] This seemingly minor detail raises an important question: Why in the world was a bark collector making extract and salts from his quina? As a cascarillero, Vega would have needed the bark to meet his contracted obligations to the merchants or landowners that hired him to collect the bark in the first place. So why waste it on making extracts and salts? The most likely explanation is that he was a curandero.

The only other piece of evidence regarding Fernando de la Vega confirms this conclusion. In 1752, he hosted another visitor to Loja. Like La Condamine, Miguel de Santisteban had come in search of the cinchona tree, but he was not a botanist. He was a government official sent by the viceroy of New Granada from Santa Fé de Bogotá to collect information on the quina trade.[79] During an interview with Santisteban, Vega revealed that he regularly made a variety of infusions, extracts, and salts from different parts of the tree. The official from Bogotá dutifully recorded this information as well as the curandero's descriptions of how he used these concoctions to treat the various ailments of those who sought his medical counsel.[80] This image of the man who lived at the foot of Caxanuma Mountain contrasts strikingly with La Condamine's portrayal of Vega as an anonymous laborer.

This new perspective on Fernando de la Vega has broader implications for our understanding of the history of cinchona bark and the history of science in the Atlantic World. First, it encourages us to recognize that La Condamine was partaking of an established tradition among imperialists and colonizers in the Andes (Inca as much as Spanish) of relying on local contacts and their knowledge to make sense of Andean nature. Of course, La Condamine would not have understood his interaction with Vega in this way, but the pattern is notable and further suggests the importance of Loja as an Andean center of healing knowledge. Second, since Fernando de la Vega was a curandero, his encounter with the scientific traveler from France is better understood as a meeting between two representatives of specialized bodies of knowledge about the natural world: La Condamine as a representative of European natural history and Vega as a representative of Andean medicine. In addition, this episode shows that even in the early eighteenth century, after several centuries of contact and colonization, European science and its representatives continued to benefit from the knowledge of Andean healers.

More important, by highlighting Loja's long-standing role as a center of medical knowledge in the Andean World, this chapter suggests how people throughout the Atlantic World suffering from intermittent fevers also benefited from Andean medicine into the eighteenth century. After all, it was primarily indigenous laborers, many of whom may have been curanderos like Vega, that extracted this medicinal bark from the Andean forests so that it could be traded and consumed within and beyond the Andean World. As we follow the transformation of this Andean medicament into a global medical commodity and an imperial natural resource, it is important to keep in mind the centrality of the Andean World not only to the production of the bark but also to the production of knowledge of the bark. In the case of quina, Loja remained as much a center of knowledge production in the eighteenth century as any scientific society or botanical garden in the imperial capitals of Europe. This long-standing and robust network of knowledge among Andean healers would continue to pose a significant challenge to Spanish pharmacists and botanists that assumed the superiority of their understanding of the natural world and underestimated the sophistication and authority of Andean botanical and medical knowledge.

Finally, Vega's story is not unique but emblematic of the experiences of many indigenous, African, and mixed race peoples in the colonial Americas that worked with and for European colonizers and scientific travelers, whether by choice or by force.[81] Most records of science and discovery published in Enlightenment Europe bear little trace of their contributions, especially as Europeans often privileged their own ways of knowing the natural world while coopting or marginalizing the ways of knowing the natural world espoused by other peoples around the globe. Moreover, as the *philosophes* of Enlightenment Europe increasingly defined their knowledge of nature as "science," everyone else's knowledge of nature became "superstition," especially in the case of healers like Fernando de la Vega, whose practice was often a mixture of spiritual techniques and natural substances.[82] European imperial expansion thus gave rise to the modern distinction between Western "science" and the "primitive" knowledge of non-Western societies and cultures.[83]

Recent historical studies have done much to bring such omissions to light and to demonstrate the importance of a variety of marginalized groups in colonial American societies, such as Amerindians, enslaved Africans, and women to various scientific enterprises in the Atlantic World.[84] This work, in conjunction with studies on other regions of the globe, has shown that early modern science was as much the product of the cross-cultural interactions provoked by European commercial and imperial expansion, as it was the product of a handful of enterprising minds in Europe.[85] Even as they sought to marginalize indigenous, African, mixed race, and creole peoples in colonial societies, Europeans remained dependent on these peoples, whom they recognized as having experiential knowledge with local plants, animals, and minerals.

At the same time that his experience reflects this marginalization of indigenous knowledge by Europeans, evidence of Fernando de la Vega's role as curandero also offers new insight into the dynamics of such encounters. This episode serves as a reminder that many of the indigenous peoples and Africans in the Americas that shared their knowledge with Europeans should not be reduced simply to their role as informants. To do so risks endorsing the perspective of European colonizers and scientific travelers who only saw their local contacts in the Americas as sources of information and purveyors of empirical facts. In the Americas, many Europeans adopted this perspective with regard to indigenous peoples, because they saw America's native inhabitants as being "closer to nature." In the eyes of many Europeans, indigenous knowledge was not a product of culture but came directly from nature itself. For example, La Condamine reported that the indigenous Andeans had learned about the medicinal properties of cinchona bark by observing "lions" drinking water from a pond in which some cinchona trees had fallen.

In order to dispense with such historical myths of knowledge production, it is vital to take account of the social and cultural contexts of indigenous knowledge as much as European knowledge. By appreciating the authority that knowledge of Andean healers, especially their knowledge of medicinal plants, commanded from Spanish colonists as well as indigenous communities, we have a better understanding of the challenge that European science would face later in the eighteenth century as Spanish pharmacists and botanists increasingly asserted the truth and authority of their knowledge of cinchona bark over that of the local knowledge of the cinchona tree and its bark.

The case of Fernando de la Vega is a good entry point and that is why we started in Loja. As a curandero, Vega was a product of a dynamic tradition of medical theory and healing practices that had existed for thousands of years in the Andean World.[86] In comparison to European intellectual traditions, it is much more difficult to trace the historical development of Andean medicine for one simple reason: a paucity of sources. Nonetheless, recent work by anthropologists, ethnohistorians, and ethnobotanists provides insight into Andean ways of knowing and manipulating nature. Such evidence encourages us to recognize that local knowledge in the Andean World could be just as meaningful and sophisticated as the knowledge and ways of knowing espoused by European scientific travelers. In the case of quina, the ways in which the bark was a product of the Andean culture as much as Andean nature cannot be overlooked, especially as the identity and knowledge of the cinchona tree and its bark was not entirely determined by Europeans. Throughout the colonial period, there were myriad European encounters with Andean medicine and its practitioners. From this perspective, Vega's encounters with La Condamine during his visits to Loja were not unique but part of a larger pattern of increased commerce and exchange of information between the Andean and Atlantic worlds in the late seventeenth and early eighteenth centuries. To further appreciate the challenges that European sciences faced in the colonial contexts of the early modern world, some understanding of existing bodies of knowledge and networks of knowledge production is necessary.

Two

Quina as a Product of
the Atlantic World

In 1707, Antonie van Leeuwenhoek (1632–1723), a Dutch naturalist and early master of microscopy, published an article on cinchona bark titled "Microscopical Observations on the Cortex Peruvianus" in the *Philosophical Transactions* of the Royal Society of London.[1] After recounting the therapeutic use of the bark by a "Doctor of Physick at Middleburgh," Leeuwenhoek described the bark's physical properties as well as the bark's "particles" as observed under the microscope. With his "microscopical observations," the Dutch naturalist hoped to shed new light on the mystery of the bark's efficacy in treating intermittent fevers. This issue had baffled physicians in Europe for decades. According to the dominant understanding of medicinal plants in Europe, the bitter taste of the bark meant that it should have a warming effect on the body.[2] Yet, according to European medical theory, fever was a disease that required the application of cooling drugs to counteract the excess heat generated by an imbalance in the four bodily humors (blood, phlegm, black bile, and yellow bile).[3] In theory, quina should not have worked. But, in practice, it did (most of the time at least).[4] Leeuwenhoek hoped to explain this anomaly by making the bark's therapeutic efficacy intelligible in terms of the physical and chemical properties of its "particles."

Leeuwenhoek's article is a striking example of how cinchona bark had become an object of philosophical and medical interest in late seventeenth- and early eighteenth-century Europe. His article is also emblematic of the ties between the philosophical study of nature in Europe and the emerging global trade in botanical medicaments and other natural commodities.[5] Consider the illustrations that accompanied the article (figure 3). At the time, most Europeans would have encountered quina as it appeared in the images at the bottom of Leeuwenhoek's illustration: a dried and curled piece of bark. By portraying quina as it commonly appeared in markets and apothecary shops, this illustration offered a palpable reminder that it was the commercial networks of the Atlantic World and its burgeoning drugs trade that connected the forests of the Andes to the markets and pharmacies of Amsterdam, where Leeuwenhoek likely acquired his bark specimens.[6] The images of microscopic particles at the top of Leeuwenhoek's illustration evoke the transformation of this Andean medicament into an object of European science, but the images at the bottom evoke the bark's transformation into a botanical commodity of the Atlantic World. In addition, the blank white background of the images is emblematic of the ignorance or indifference of Europeans toward the human and natural contexts of the Andean World that gave rise to the bark in the first place.

The transformation of cinchona bark into a major commodity and natural resource involved much more than its physical characteristics and medicinal properties. Using evidence of the spread of malaria and the growth of the quina trade, this chapter argues that, by the early eighteenth century, quina had become as much a product of the Atlantic World as it was a product of the Andean World. Andean healers were the earliest users of the bark and the first to recognize its efficacy in treating intermittent fevers; however, it was the expansion of the Atlantic World that facilitated the spread of the bark beyond the Andean World. Moreover, at the same time as trade networks provided the infrastructure that brought this Andean remedy to fever sufferers in Europe, Africa, and other parts of the Americas, major developments in the Atlantic World in late seventeenth century actually created more demand for the bark by facilitating the spread of malaria, the disease against which quina was most effective.

In 1751, when the Crown declared quina to be "an object worthy of interest, curiosity and attention" in a royal order, the bark was being produced, traded, and consumed at unprecedented levels.[7] As shown in this chapter, it was only after private traders had established the bark as a valuable commodity that the Spanish Crown came to recognize quina as an object of imperial interest. In addition to increasing the economic value of the bark, the growing trade in quina had another important consequence: it also increased the circulation and exchange of knowledge about the bark. The Spanish colonial government officials—in their quest to understand and control quina—often

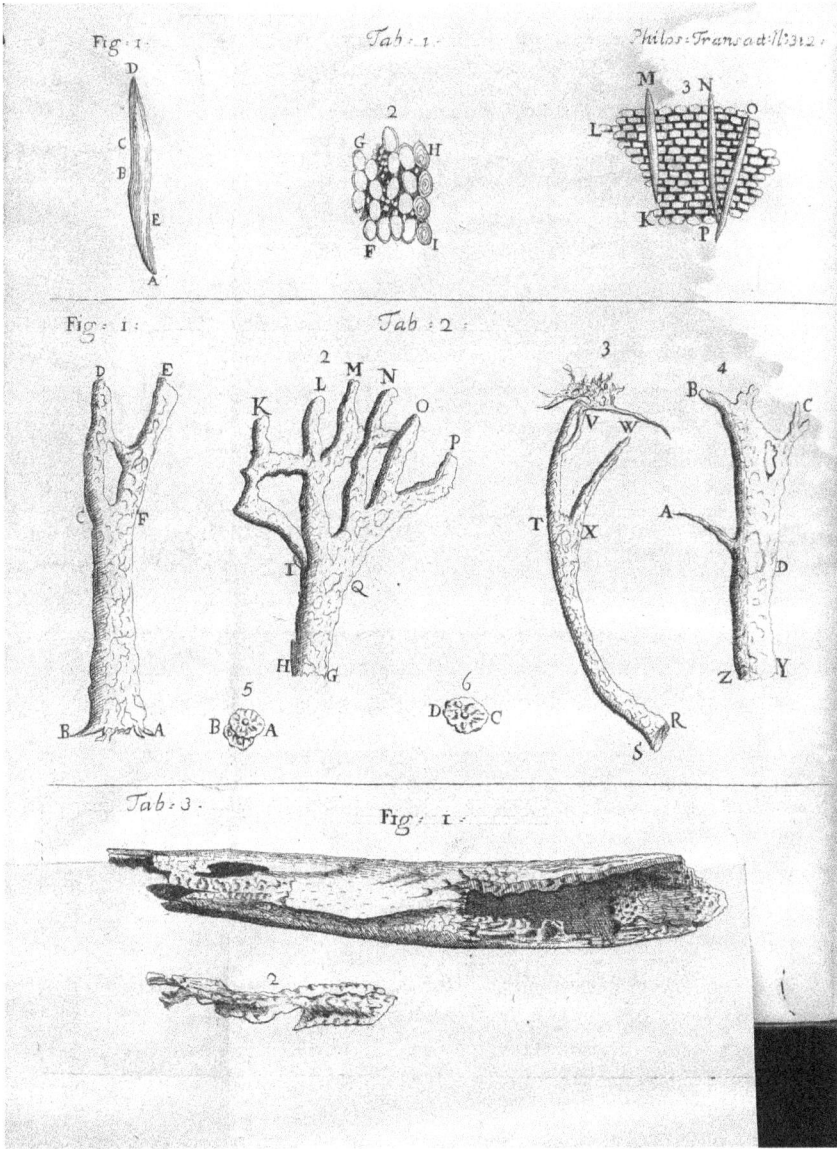

Figure 3. Antonie von Leeuwenhoek "Microscopical Observations on the Cortex Pervianus" (1707). General Collection, Beinecke Rare Book and Manuscript Library, Yale University.

relied on the commercial knowledge of the Atlantic World as much as the indigenous knowledge of the Andean World. As a result, some of the challenges that the Spanish Crown later faced in its efforts to control cinchona bark were a product of the dynamics of the quina trade that developed in the late seventeenth-century Atlantic World and persisted for the rest of the colonial period. Informal transatlantic and transimperial networks played a vital role not only in transforming the bark into a global medical commodity but also in spreading knowledge and transforming perceptions of the bark.

Greater appreciation of quina as a product of the Atlantic and Andean worlds also provides insight into the fragility of European science in the Spanish Atlantic World. In the 1770s, when the Spanish Crown first turned to pharmacists and botanists for their assistance in determing which kinds of quina should be collected by the royal reserve and used by the Royal Pharmacy, these specialists encountered a natural object about which there were already communities connected to the trade networks of the Atlantic World that had established methods for identifying and evaluating different kinds of quina. It is important to recognize the significance of these networks of knowledge because different groups in Spanish colonial society would use their connections to the quina trade to claim the authority to speak about the cinchona tree and its bark in order to undermine the efforts of Spanish pharmacists and botanists to exert greater control over quina in the name of empire. In their efforts to assert the superiority of their knowledge, Spanish pharmacists and botanists could not rely on their association with the Crown nor could they rely on convincing people of the truth of their claims about cinchona bark. They also needed to displace these established ways of knowing the cinchona tree and its bark that had a significant presence not just in colonial society but also in the colonial government.

As it entered the long-distance trade networks of the Atlantic World, the bark would retain its associations with the Andean World but it also acquired a new identity as a commodity; fittingly, new names for the bark were coined. In Spanish South America, the bark came to be known as *cascarilla*. In Europe, quina was the most popular among many new monikers including Peruvian Bark and Jesuits' Bark. The name "quina" may have been based on an original Andean term, but we will never know for sure because the Andean name was never recorded and soon forgotten—a fitting illustration of the significance of the bark's Atlantic crossings. If the bark was originally *from* the Andean World, after the 1630s it became a product *of* the Atlantic World—quite literally. Several major developments made in the Atlantic World transformed this Andean tree bark into a new entity that would become known as "quina."

The Feverish Atlantic: Malaria in Europe, Africa, and the Americas

Disease played an important role in the bark's transformation from a local Andean medicament into a global medical commodity. In the early sixteenth century, the simultaneous expansion of the Inca and Spanish empires brought the Atlantic and Andean worlds into contact.[8] In conjunction with exploration of the Isthmus of Panama and the west coast of South America in the 1520s and the conquests of Peru in the 1530s, European colonists and Africans (enslaved and free) introduced many new diseases to the indigenous populations of the Andes as they did elsewhere in the Americas.[9] One of these almost certainly would have been the blood-borne disease malaria. Because indigenous populations suffered outbreaks of many Old World diseases simultaneously, it is difficult to pinpoint exactly when malaria arrived in South America.[10] Nonetheless, it is likely that the coasts and lowlands of colonial Ecuador provided the right environmental conditions for the disease to flourish, especially because these regions served as breeding grounds for several American species of *Anopheles* mosquito, the main vector of the malarial parasite aside from humans. It was in the epidemiological context of the Spanish conquest that cinchona bark took on new importance. As it turned out, the bark proved an effective treatment for the intermittent fevers associated with malaria.[11]

Malaria was the link that connected the fate of quina to two major developments in the late seventeenth-century Atlantic World: the expansion of plantation agriculture in the Caribbean and the intensification of the Atlantic slave trade. To understand this connection, let us first review the etiology and epidemiology of the disease. In everyday speech, the term "malaria" is generally used to refer to a distinctive disease condition in humans. However, common usage of the term can create some confusion because the disease is not caused by a single species of microorganism, as is the case with many viral and bacterial diseases. Instead, there are four identifiable kinds of malaria caused by four different species of a protozoan parasite known as plasmodium: *Plasmodium falciparum, Plasmodium ovale, Plasmodium malariae,* and *Plasmodium vivax.*[12] Biologically, all four plasmodia are similar with complex, multistage lifecycles in which different phases occur in the gut of female mosquitos in the genus *Anopheles* and in the circulatory system of humans.

The disease consequences of the four malarial parasites are also similar— fever, chills, sweating, aching, and in some severe cases anemia. Intermittent fever, when patients suffer recurring bouts of fever at regular intervals, is the most iconic symptom of malaria. The periodicity of the fever depends on

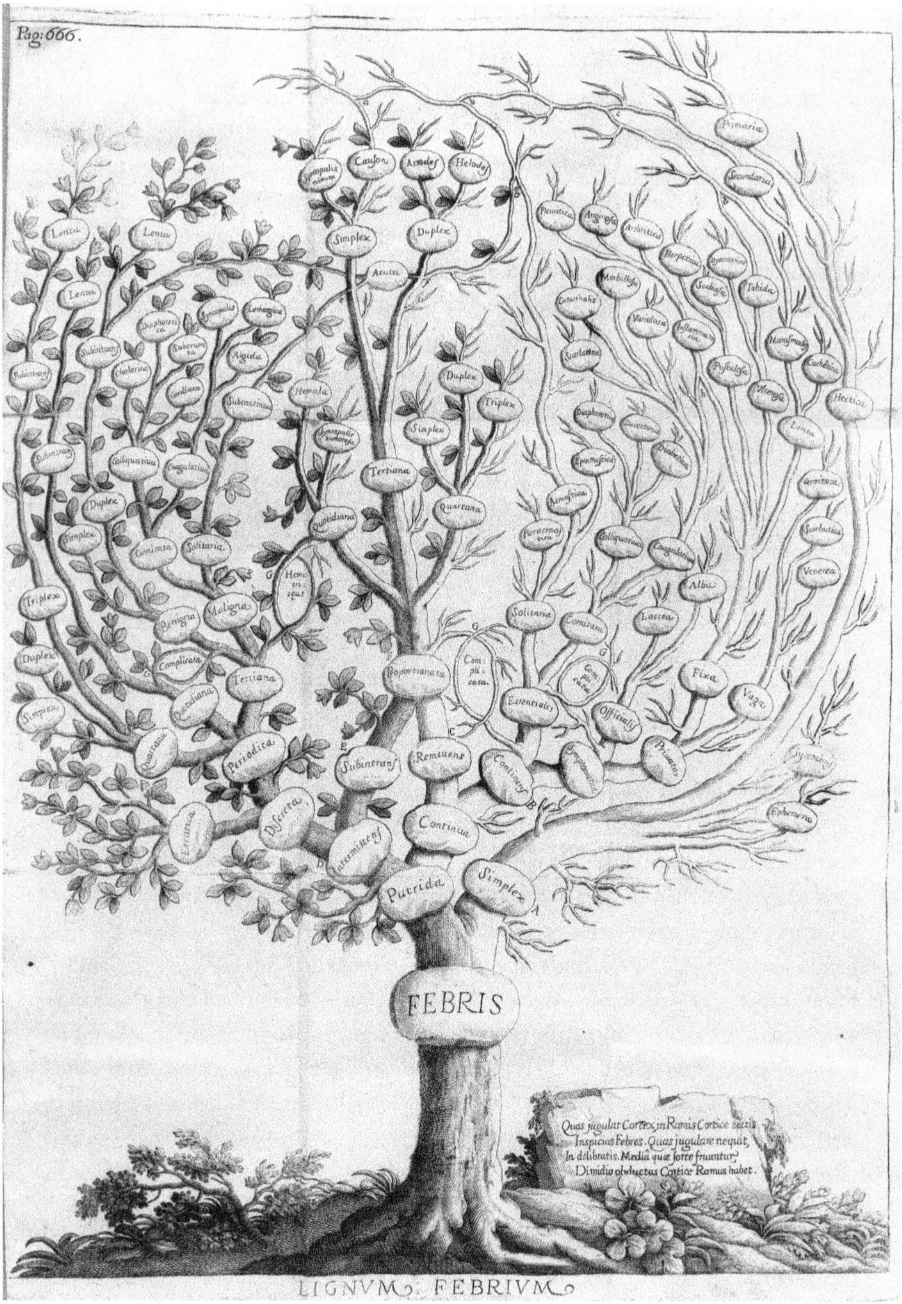

Figure 4. "Tree of Fevers" from Francesco Torti, *Therapeutice Specialis* (1712). Wellcome Library, London.

the species of plasmodium. Malaria sufferers infected with *P. falciparum*, *P. ovale*, or *P. vivax* experience a bout of fever every forty-eight hours, while those infected with *P. malariae* experience a bout of fever every seventy-two hours. Today malaria is a treatable and, even, curable disease; yet some forms can be lethal. The majority of deaths from malaria worldwide result from cases caused by *P. falciparum*.[13]

Although healers in the early modern Atlantic World did not understand malaria and its etiology as we do today, they were keen observers of the body and its changes. In Europe, fever had been a major focus of study and medical treatment for centuries preceding the introduction of cinchona bark to Spain and Italy in the 1630s. By the early modern period, European physicians had developed a complex classification of fevers based on centuries of medical observation. In the early eighteenth century, the Italian physician Francesco Torti represented the medical classification of fevers in the form of a tree, a common practice since the Middle Ages (figure 4).[14] Torti's image, known appropriately as the "Fever Tree," included all the different kinds of fever that European physicians had observed. Focusing on the branches of the tree, we note that one of the most complex branches of the tree corresponds to intermittent fevers; leaves on select branches of this part of the tree indicate those fevers that could be treated with cinchona bark.[15] Ultimately, Torti's image is emblematic of the sophisticated understanding of fever employed by European physicians, many of whom conceived of fever as a disease in itself rather than a symptom of an underlying condition.

It was mainly through the observation and classification of fevers that malaria was intelligible to healers in Europe and other parts of the Atlantic World. Those informed by European medical theories recognized two common varieties of intermittent fever: tertian fever, in which bouts of fever occur every forty-eight hours, and quartan fever, in which bouts of fever occur every seventy-two hours. They also recognized the variable mortality rates of some of these fevers as reflected in the distinction between a benign tertian fever and a malignant tertian fever.[16] Although a direct correlation of early modern and modern terms is difficult, table 2 suggests how early modern European terms for different kinds of intermittent fever correspond to the symptoms produced by the different species of plasmodium recognized by modern medicine. It is important to note that retroactive identification of malarial cases based on early modern sources can be difficult because the presence or periodicity of intermittent fever is not a reliable basis for diagnosis. On the one hand, many other diseases can have intermittent fever as a symptom. On the other hand, not all cases of malaria manifest tertian or quartan fevers. Although some cases of *P. falciparum* result in irregular fevers or fevers with a twenty-four hour periodicity (quotidian fever), it is also possible to be infected

TABLE 2. Modern Malarial Parasites and Early Modern Fevers

Type of Plasmodium	Fever Periodicity	Early Modern Name
P. vivax	every 48 hours	Benign tertian fever
P. falciparum*	every 48 hours	Malignant tertian fever
P. malariae	every 72 hours	Quartan fever

* Most cases of malignant tertian fever were likely *P. falciparum* since that is now recognized as the most lethal form of malaria. However, malarial cases caused by *P. falciparum* can result in an irregular fever or even a quotidian fever, in which bouts of fever occur every twenty-four hours, instead of a tertian fever.

by more than one species of plasmodium at the same time resulting in a seemingly erratic fever with no identifiable periodicity.[17]

The etiology of malaria provided by modern medicine is useful for highlighting the biological and environmental factors that helped to spread the disease in the early modern Atlantic. Malaria appears to be one of the oldest human diseases and remains one of the most lethal.[18] Humans acquired many of their most common diseases as a result of associating with animals first domesticated about 10,000 years ago, but the origins of malaria trace back to as early as 100,000 years ago, when malarial parasites, carried by mosquitos, first infected the ancestors to modern humans.[19] Over the tens of thousands of years that followed, malaria flourished among early humans as they increased in number, settled down in ever more densely populated communities, and transformed the landscape through agriculture, a process that often created the ideal environmental conditions for malaria-carrying mosquitos to thrive.

When humans first departed Africa and colonized the Eurasian continent, malaria traveled with them. The disease went almost everywhere humans did during the first globalization of human species except for the Pacific Islands and the Americas. During the last Ice Age, when humans crossed the land bridge to the Americas, they left malaria behind because mosquitos and malarial parasites could not survive in the colder climate and harsh conditions at the northern latitudes.[20] Malaria only became a truly "global" disease in the sixteenth century when it finally reached the Americas by crossing the Atlantic as a stowaway in the bloodstreams of Europeans and Africans who migrated voluntarily and involuntarily.

Historian James Webb has identified three main "meta-conditions" that shaped the historical geography and epidemiology of malaria during its spread from Afro-Eurasia to the Americas after 1492.[21] The first meta-condition is that malaria, which has a complex lifecycle involving both human and mosquito hosts, requires dense human populations in order to survive and thrive, like many other human diseases. The second meta-condition is the expansion of "disease-experienced populations" into new regions. The Europeans and Africans that came to the Americas all came from populations where malaria was endemic and had been for thousands of years. In the Americas, these disease-experienced groups encountered indigenous populations that had no prior exposure to malaria and many of the other disease endemic to Afro-Eurasia. As a result, for indigenous peoples of the Americas, the encounter with Europeans and Africans was a biological one resulting in population decline and, in some areas, demographic collapse due to epidemics.[22]

Distinctive zones of endemic infection are the third meta-condition. These zones were areas where certain types of malaria were predominant while others were not. This meta-condition played an especially important role in shaping the historical epidemiology of malaria in the Atlantic World. All of the malarial parasites readily crossed the Atlantic after 1492 except for *P. ovale,* which remained (and remains) confined to tropical Africa for reasons that are not yet entirely understood. Historical epidemiologists have identified three different zones of malarial infection that existed on the Afro-Eurasia continent before 1492: a vivax zone in Northern Europe, a mixed zone of vivax, falciparum, and malariae around the Mediterranean basin, and a falciparum zone in tropical Africa. As a result of the regional distribution of the different types of malaria, it is likely that northern Europeans were most responsible for bringing *P. vivax* malaria to the Americas, while most of the *P. falciparum* malaria came from Africans. It is not surprising then that new zones of intense falciparum infections developed in those regions in the colonial Americas where African slaves made up a majority of the population, such as the sugar-producing islands of the Caribbean and the regions of intensive sugar production in Brazil.[23]

Environmental conditions and the history of acquired immunities of human populations were two key factors that defined the geography of malarial infections in the Atlantic World. Malaria caused by *P. vivax* flourished in Northern Europe and the colder climates of the Atlantic World due to an extended incubation period that enabled the parasite to wait out the cold winters and reemerge during the summer months.[24] Whereas *P. vivax* had an advantage in the climate of Northern Europe due to the peculiarities of its lifecycle, acquired immunity to *P. vivax* among the human populations in tropical Africa gave rise to a zone where infections from the falciparum parasite were predominant. In tropical Africa the mild climate and year-round mosquito activity met the biological and environmental requirements of malarial parasites. Yet, after thousands of generations of exposure to the vivax parasite, probably the oldest of the four main types of malaria, human populations in tropical Africa had developed a genetic mutation known as the "Duffy negativity," a condition in which a change to the antigen receptor on red blood cells prevents the vivax plasmodium from gaining entrance to the cell and feasting on its hemoglobin.[25] *P. falciparum* was unaffected by this particular form of acquired immunity. Even though *P. falciparum* was a more lethal form of malaria, African populations had some resistance to this type of malaria due to the prevalence of a recessive gene for sickle-cell anemia, which causes a deformation in some red blood cells that makes it difficult for malarial parasites to thrive. The presence of the sickle-cell trait also made West Africans useful carriers of the disease from the parasite's perspective because they could still spread it without manifesting or suffering from any of the symptoms.

Colonization, migration, and the Atlantic slave trade ensured a steady supply of Europeans and Africans as potential carriers of malaria; yet other factors also fostered the spread of malaria and its transformation into an endemic disease in many regions of the Americas. One key factor was the presence of several native species of *Anopheles* mosquito in the Americas. In addition to serving as hosts for part of the lifecycle of the malarial plasmodia, these mosquitos also spread the disease over the short distances between the human communities in the colonial Americas. Another important factor was the introduction and spread of plantation agriculture, a defining feature of many societies in the colonial Americas that played a vital role in the emerging Atlantic economy. As the primary source of demand for slave labor in the Americas, plantations were the main reason for the continued enslavement and forced migration of West Africans throughout the seventeenth and eighteenth centuries.

In addition to ensuring a steady stream of malaria carriers, plantations also created the environmental conditions for mosquitos to thrive. First, many plantations were home to open containers of standing water—an excellent breeding ground for many species of *Anopheles* mosquito. Second, the use of domesticated animals on plantations provided yet another population of hosts for the malarial parasites.[26] Consequently, when the malarial parasites arrived in the bloodstream of African slaves and European colonists, plantations provided many of the conditions for these parasites to thrive.

In the late seventeenth century, the Atlantic World witnessed a significant increase in slave trade in association with the spread and intensification of plantation agriculture especially in the Caribbean. These two key features of the Atlantic World formed a feedback loop that fostered the spread and prevalence of malaria throughout the colonial Americas. It was also in the late seventeenth century that healers beyond South America began to learn about a new botanical medicament that was effective against the intermittent fevers. First imported to Europe in small quantities by Jesuit missionaries and Spanish officials returning from Peru in the 1630s, news of the bark's healing powers spread rapidly. By the end of the seventeenth century, quina had become one of the most important medical commodities in the Atlantic World in terms of both volume and value of the trade.

Expansion of the Quina Trade in the Atlantic World, c. 1650–1750

Data on the imports of cinchona bark to London and Cádiz, two important European port cities on the Atlantic, show a significant increase in the cinchona bark trade in the late seventeenth century. The reasons for the increase are far from clear. Undoubtedly, because malarial fevers afflicted many populations in early modern Europe, some of the increase can be attributed simply to the growth of the trade to meet existing demand. As more healers in Europe learned about quina and became convinced of its medical utility, merchants would have received more requests for the medicament. At the same time, some of the increase can be attributed to the spread of malaria into new regions in the Americas. It is no coincidence that this increase in the trade in cinchona bark correlates with the intensification of the slave trade and plantation agriculture that began in the late seventeenth century and continued for much of the eighteenth century.

These data are significant not simply because they demonstrate the economic importance of cinchona bark in the early modern Atlantic World. They are also significant because they provide new perspective on the transformation of this Andean tree bark into the medicament and botanical commodity known as "quina." Ever since the seventeenth century, there has been debate about who deserves the credit for the discovery of the bark. Some have attributed the discovery to the Andean healers, who were probably using the bark well before their encounter with Europeans.[27] Others have attributed the discovery to European missionaries and physicians because they were, allegedly, the first to recognize and understand the bark's efficacy in the treatment of intermittent fevers. Yet, the evidence from the drugs trade suggests that "quina" is better understood not as a discovery to be attributed to one individual or one group but as a product of the interactions between Native Americans, Europeans, and Africans in the Atlantic World. After all, it was in the context of the expansion of the Atlantic World that new applications of the bark were developed. In addition, as we will see, the trade networks of the Atlantic World not only spread the bark to expanding markets of malaria sufferers but also fostered the circulation and exchange of information about the cinchona tree and its bark between bark collectors, merchants, and healers on all three continents.

A recent study of the drugs trade in London provides some of the best evidence for the emergence of a mass market for cinchona bark in the late seventeenth and early eighteenth centuries. In addition to serving as the "epicenter" of the drugs trade in England, the case of London is also suggestive of general trends in the drugs trade of the Atlantic World because records indicate that merchants in London exported drugs to ports throughout Europe and the British colonies at a high rate. Overall, the sources, including Port Books and the annual ledgers of the Inspector General of Customs, show a major increase in the imports of commodities classified as "drugs" to London from 1567 to 1774.[28] However, when considering the general outlines of this trade, it is important to keep in mind that "drugs" was a category that in early modern Europe referred to a much broader collection of phenomena than just substances used for medicinal purposes.[29]

Quina, known as the Jesuits' Bark in England, occupies a prominent place in the records of the English drugs trade in the seventeenth and eighteenth centuries. In addition, trends in the importation of the bark are notably different than the general trends in the import of other drugs used primarily for medicinal purposes. First, cinchona bark was a late arrival relative to other materia medica from the Americas. Other American botanical medicaments, such as guaiacum and sarsaparilla, began trickling into Europe in the sixteenth century, whereas the first record of shipments of cinchona bark arriving in London dates to 1685, when 2,778 pounds were imported over the course of the year. Because records of drug imports to London are incomplete for the seventeenth century, it is difficult to determine exactly when merchants began importing a significant volume of Jesuits' Bark. There is no mention of the bark in the 1660 Rate Book for London, which suggests that significant imports only began between 1660 and 1685.[30] Nonetheless, despite being a late arrival to the London drugs trade relative to the other medicinal drugs in the seventeenth century, the trade in cinchona bark boomed, and, overall, imports of the bark accounted for 40 percent of all drugs imported to London between 1567 and 1774, which is remarkable considering that this period includes data on the drugs trade decades before the first imports of cinchona bark.[31]

The trends in the imports of Jesuits' Bark continued to diverge from the overall trends in drug imports to London during the seventeenth and eighteenth centuries. When considering all commodities classified as "drugs," the most significant increase in the volume of drug imports occurred in the seventeenth century. As for drugs used mainly for medicinal purposes, the volume of imports declined in the early eighteenth century only to recover by midcentury and stagnate thereafter.[32] By contrast, the most significant increase in imports of cinchona bark to London occurred in the first half of the eighteenth century. Whereas the annual volume of imported bark averaged 2,045 pounds between 1699 and 1701, the average annual volume increased to more than 98,000 pounds between 1752 and 1754. At the same time, the whole-sale price of the bark decreased from 60.8 dollars per pound between 1696 and 1706 to 26.5 dollars per pound between 1796 and 1799. The value of the bark in London seems to have remained relatively constant in the first half of the eighteenth century. Among the ten leading "medical drugs" as ranked by value of imports, Jesuits' Bark ranks first between 1722 and 1724 and second between 1752 and 1754. Ultimately, these data show that the trade in the bark "boomed" in the eighteenth century even as the trade in most other medicinal drugs grew much more slowly over a similar period.[33] At the very least, the data show that English merchants were trading significant volumes of the bark—a subject of much concern to Spanish officials, who thought that only Spain and its merchants should reap the benefits of the trading in the goods that were unique to Spanish America.

Records of imports to the Spanish port city of Cádiz provide additional insight into the growth of the quina trade. From the early sixteenth century until early eighteenth century, one Spanish port had a monopoly, in theory, on all trade with Spanish America—Seville. Charles V first bestowed this honor on Seville, a city that straddled the Guadalaquivir River about fifty miles from the Atlantic Ocean, when he established the Casa de Contratación (House of Trade). The House of Trade was to oversee all matters, including navigation, related to colonial trade. Over time, ships arriving from the Americas found it increasingly difficult to reach Seville as the Guadalquivir River began to silt up. As a result, in 1717, in the wake of the War of Spanish Succession, the Spanish Crown designated Cádiz, a city on the Atlantic Ocean, to replace Seville as the only port city to handle the trade with Spanish America. Cádiz held the formal monopoly on Spanish colonial trade until 1778, when the Crown began to allow additional port cities on both sides of the Atlantic to participate in the imperial trade. Due to the prevalence of corruption and contraband, the official import records of Cádiz provide an incomplete picture of the volume of trade between Spain and its American territories for the fifty-year period. Nonetheless, these data remain one of the best sources for assessing the general trends in the Spanish imperial economy in the Atlantic World.[34]

Archival records of imports to Cádiz from 1717 to 1778 include those goods that arrived via the annual convoy, known as the *flota,* and those goods that arrived via an emerging system of individual, registered ships.[35] Due to the loss of records and the interruption of the convoy system because of the various wars of the eighteenth century, these records include data on the arrival of only thirteen annual convoys to Cádiz in the periods from 1718 to 1737 and 1758 to 1778.[36] Similarly, records of imports via "registered" and "mixed" ships are available annually or nearly annually for the years 1717 to 1739 and 1747 to 1778. Nonetheless, these data show a significant increase in the import of cinchona bark—listed as cascarilla—to Cádiz from the early eighteenth century to mid-eighteenth century, just as in London.

The clearest evidence of the increase in imports comes from comparing the annual volume of cinchona bark import for three periods in which there are seven consecutive years of import data. In the period from 1718 and 1724, merchants in Cádiz reported importing a total of 103,375 pounds (4,135 *arrobas*) of cinchona bark with an average annual import volume of approximately 14,775 pounds (591 arrobas).[37] These amounts can be compared to those of the seven-year period from 1752 to 1758. During this period, merchants in Cádiz reported importing an astounding total of 2,225,487 pounds (89,019 arrobas) of cinchona bark with an average annual import volume of approximately 317,925 pounds (12,717 arrobas). In other words, in the three decades between the early 1720s and mid-1750s, imports of quina to Cádiz increased more than twentyfold.[38]

In the next seven years after 1758 (1759–1765), the data show a decline, with merchants in Cádiz reporting a total of 646,219 pounds of imported quina (table 3). In this period, the average annual volume of imported quina was only 92,317 pounds. Even though this amount was a fraction of that of the early 1750s, it still represented a sixfold increase over the average annual import volume in the early 1720s. The trend in increased imports of cinchona bark continued into the 1770s. From 1767 to 1770 and from 1772 to 1777, merchants in Cádiz reported importing a total of 2,565,344 pounds (102,613.75 arrobas) of cinchona bark. During these years, Cádiz merchants had an average annual import volume of approximately 256,535 pounds of bark.[39]

In the eighteenth century, both London and Cádiz experienced a significant increase in the import of cinchona bark. The comparison of the two sets of import data suggests that the volume of the quina trade in Cádiz was much greater than that of London. Whereas merchants in Cádiz imported an average of approximately 242,000 pounds of cinchona bark annually between 1752 and 1754, merchants in London imported an average of 98,000 pounds of bark annually during the same period. These numbers make sense given that, in theory, Cádiz was the port through which all of Europe's cinchona bark was to be imported. In practice, much of this bark was likely funneled to foreign merchants, many of whom had effectively infiltrated the monopoly that Cádiz had on the trade with Spanish America.[40] Such activities explain how officials in Madrid could have had difficulty finding cinchona bark to purchase even though merchants in Cádiz were importing significant quantities of the bark. Moreover, the evidence suggests a significant increase in the volume of the

TABLE 3. Imports of Cinchona Bark (Cascarilla) to Cádiz on Annual Convoys and Registered Ships (1717–1778)

Date Range	Total Bark (pounds)	Annual Average (pounds)
1718–1724	103,375.00	~14,775.00
1752–1758	2,225,487.50	~317,925.00
1759–1765	646,218.75	~92,317.00
1767–1770, 1772–1777	2,565,343.75	~256,535.00

Source: Antonio García-Baquero González, *Cadíz y el Atlántico: El comercio colonial español bajo el monopolio gaditano* (Cádiz: Diputación Provincial de Cádiz, [1976] 1988), vol. 2, tables 19–21, 131–152.

quina trade in the early eighteenth century leading up to the Spanish Crown's decision in 1751 to intervene directly in the production and distribution of the bark by establishing a royal reserve of cinchona trees in Loja.

Economic Development and the Quina Industry in the Audiencia of Quito

Undoubtedly, the Atlantic World played an important role in cinchona bark's transformation into a major medical commodity. But the increases in demand and the volume of the trade are only part of the story. As demand increased, producers of the bark in the Andes had to keep pace. Coincidentally, several developments in the regional economy of the Audiencia of Quito fostered the expansion of the quina industry.

One of the limiting factors in the production of quina throughout the eighteenth century was labor. Harvesting and preparing the bark for commercial and medicinal use was more complicated than it might seem at first. As a result, landowners and merchants in the quina-producing regions of South America needed skilled laborers to collect bark for Atlantic markets. The first challenge was finding trees from which to harvest the bark. Before the mid-nineteenth century, all cinchona bark was collected from trees growing wild in the verdant forests of the Andes.[41] If the trees did not die from having their bark removed (as many did), bark collectors in South America, known as *cascarilleros,* had to wait several years for the trees to regrow their bark. As the quina trade became more lucrative, extraction outpaced the natural rate of reproduction and regrowth of cinchona trees. As a result, cascarilleros pushed the frontiers of bark collection into new regions of the Andean forests almost every year. The unsustainability of bark extraction became especially apparent in the late eighteenth century when Spanish colonial officials started to receive alarming reports of the impending disappearance of cinchona trees from several different regions including Loja.[42] In short, as rising demand led to an increase in and intensification of extraction, the task of locating the trees amid the rugged terrain and lush vegetation of the Andes became more difficult.

Once harvested, the bark had to be properly dried and then packaged for transport in either leather sacks or canvas-lined crates. Because it rained much of the year and the bark needed protection from humidity, harvesting of the bark could take place only during the summer months. As a result, cascarilleros were seasonal laborers. At the beginning of the harvesting season, bark collectors would enter into a contract with a merchant or *hacendado,* who often gave European manufactured goods to the bark collectors in exchange for the delivery of a predetermined amount of bark by weight to the merchant or hacendado by the end of the harvesting season.[43] In the southern sierra of Quito, merchants and landowners would hire gangs of twenty to fifty laborers to scour the hills and forests around Loja for cinchona trees. Cascarilleros were thus a crucial part of the production process and the health of many in the Atlantic World depended on their ability to find enough bark to meet growing demand.

Unrelated developments in the colonial economy at large and the regional economy of the Audiencia of Quito led to an influx of laborers into the quina-producing regions of southern Quito in late seventeenth and early eighteenth centuries. One factor was the decline of textile industry in the northern highlands of the Audiencia of Quito.[44] Because of the laws governing Spanish imperial trade, Quito's participation in the emerging Atlantic and global economies during the colonial period was indirect as most goods had to pass through Callao, the port of Lima, the capital city of the Viceroyalty of Peru.[45] Although the region did have small deposits of precious metals, these were not enough to support a significant volume of trade and, by the eighteenth century, mines in the region had experienced a sharp decline in production.

Before the late seventeenth century, Quito's main connection to the emerging networks of transatlantic and global trade was the export of *paña azul,* a blue cloth for which the region was known. Cloth from Quito's textile industry met the demand for cheap cloth in the urban and mining centers of South America, especially in the Viceroyalty of Peru.[46] This burgeoning industry attracted laborers from throughout Quito especially the northern highlands, the epicenter of textile production. Since they comprised the majority of the population (over 60% in most regions), Amerindians were the main source of labor for Quito's *obrajes* (textile mills). From the late sixteenth century to the late seventeenth century, the Amerindian population in "the north-central sierra (from Ibarra to Riobamba)" swelled from 145,000 in 1591 to more than 270,000 a century later.[47] The increasing population of Amerindians meant cheap labor for the textile mills, which in turn meant that paña azul was comparatively cheaper than cloth from other regions. Merchants and elites in the Audiencia of Quito exchanged their cheap woolen textiles at markets in Peru and New Granada for precious metals and specie, which they used to purchase European goods and "maintain a comfortable European lifestyle."[48] Undoubtedly, some of these Amerindian migrants would have provided some of the seasonal labor needed to extract quina.[49]

In the late seventeenth century, the textile economy of the northern highlands entered a period of crisis and decline leading to a greater emphasis on other economic activities including agriculture and the collection of cinchona bark in the southern sierra. Decreased demand in Peru for Quito's textiles was, in part, the result of a decline in mining in the viceroyalty as well as the growth of textile production in provinces closer to the mining towns that were major drivers of the vice-regal economy.[50] Several natural disasters magnified the impact of these economic trends. In 1687, a series of earthquakes devastated Lima. According to Kenneth Andrien, these disasters undermined the city's "strength as a marketing center for over a decade."[51] In 1691, the Audiencia of Quito then suffered a severe drought resulting in a famine, the impact of which was magnified by a series of epidemics (measles, smallpox, and fevers) that lasted until 1695. In 1698, Quito itself suffered from an earthquake. The earthquakes undoubtedly damaged the infrastructure that supported regional trade in Quito's textiles, and the epidemics reduced the Amerindian population, which, in turn, resulted in an increase in labor costs. In other words, at precisely the moment that Quito's paña azul and other textiles began to face competition from cloth producers closer to the mines, owners of obrajes saw an increase in labor costs, which made their cloth more expensive and less competitive in regional markets.

Just as the region came out of more than a decade of drought, famine, epidemic disease, and earthquakes, Quito's textile mills faced a new source of competition: "cheap, high-quality European cloth."[52] During the War of Spanish Succession (1700–1716), the official Spanish system of annual trade convoys to Portobello and Cartagena broke down. Spanish America would have remained even more disconnected from international trade if it had not been for an increase in contraband trade. During the war, French traders supplied European goods, including textiles, to the Pacific Coast of South America both with and without the permission of the Spanish Crown.[53] Between 1700 and 1725, French legal and contraband trade with Peru "totaled nearly 100 million pesos," which accounted for "approximately 68 percent of the viceroyalty's foreign trade."[54] Quito's cloth producers could not compete with the French merchants, who offered a variety of European textiles at different price ranges to meet the variable needs of consumers in South America. Moreover, those traders engaged in contraband trade and did not pay taxes, thus they could offer the cloth at much lower prices undercutting those of their South American competitors.[55]

The decline of textile production led to general economic decline in the Audiencia of Quito, but not all regions of the Audiencia experienced these changes in the same way. As the estates of the north-central sierra, the epicenter of textile production, fell on harder times, agricultural producers in the southern sierra of the Audiencia of Quito "experienced steady growth from the 1690s." Several factors contributed to the prosperity of the southern provinces of the Audiencia related to the northern provinces. The owners of large estates continued to find robust demand for agricultural products and textiles in the regional economy, especially "the local highland markets, Guayaquil and northern Peru." Another major component of the economic prosperity of the southern sierra was the prevalence of cheap labor as indigenous Andeans migrated south in search of work in the wake of the collapsing textile industry.[56] These laborers harvested the bulk of the cinchona bark that fueled the boom in the quina trade in the early eighteenth century.

As to the local economy of the southern sierra of the Audiencia of Quito, the two provinces—Cuenca and Loja—that comprised the region shared many similarities but also had some important differences. According to data from the late eighteenth century, the cultivable land in the province of Cuenca was divided into 105 estates, 103 (98 percent) of which Andrien classifies as large estates based on the number of *conciertos* (resident laborers on Spanish estates). Estates in the province of Cuenca produced a variety of goods for the regional economy including livestock, grains, fruit, sugar cane, and cinchona bark. Collection of quina was a specialty among the *haciendas* in the southeast corner of the province. In the province of Loja, large estates were less common than in Cuenca due to the lower population density. Of 109 estates in the province, 93 (85 percent) were small haciendas that employed less than ten conciertos. As in Cuenca, these estates produced a variety of lowland products such as sugarcane, *yerba maté, añil,* and *maní* as well as highland products such as grains, beans, *ají,* fruits, and vegetables. Estates that specialized in the collection of quina were clustered in the eastern highlands of the province.[57] Nonetheless, landowners in both regions were able to take advantage of the influx of laborers in the late seventeenth and early eighteenth centuries, some of which they used to extract cinchona bark from the forests. Consequently, just as demand from Europe for quina was increasing, merchants and landowners in the Audiencia of Quito had both the motivation and the labor supply to meet these increasing demands. Moreover, the increase in both supply and demand were driven in part by the ties between the Andean World and the Atlantic World.

Commerce and Knowledge in the Early Eighteenth-Century Atlantic World

By the mid-eighteenth century, the quina industry and trade were booming. As more bark became available to more people and places throughout the Atlantic World, more information about quina became available as well. Whether through written documents or word of mouth, trade networks facilitated the movement of information alongside the movement of goods. Works published in the late 1730s about the cinchona tree and its bark show how the informal networks associated with commerce of the Atlantic World fostered the production and circulation of new knowledge of natural phenomena across long distances. It was not a coincidence that the French scientific traveler Charles Marie de la Condamine traveled to Loja in 1738, when the quina trade was in full swing.

To appreciate the epistemological consequences of the growing bark trade, let us return to the images of cinchona bark published by Antonie von Leeuwenhoek (figure 3) and contrast them with the images that accompanied La Condamine's report on the "quinquina tree" (figure 5). Whereas Leeuwenhoek offered images of bark on a blank page, La Condamine depicted the bark in the context of part of a tree branch as well as illustrations of the leaves, flowers, seeds, and fruits of the tree. In the accompanying article, he provided information on the broader geographical, historical, and economic contexts that gave rise to quina. La Condamine described the local geography and climate in which the tree grew. He also provided a historical account of the discovery of the tree and its medicinal properties, in which he claimed that South American Indians learned about the bark by watching animals ingest it. And, finally, La Condamine described the harvesting of the bark in Loja with details on the quina trade. Ultimately, the quina described by La Condamine was much more than the disconnected medical commodity of Leeuwenhoek's article; it was an integrated part of natural and human systems of production.

While the images that accompanied La Condamine's article were the first botanical illustrations of the cinchona tree published in Europe, it is important to recognize that his article was just one of several published accounts of the cinchona tree that appeared in Europe between 1735 and 1745. These publications are a testament to the increasing production and circulation of knowledge of New World nature in the eighteenth-century Atlantic World. In addition, greater attention to the conditions of their production provides in-

Figure 5. Engraving of a branch and flowers of the cinchona tree from La Condamine's "Sur l'Arbre du Quinquina" (1745). Wellcome Library, London.

sight into the role of informal networks in the production of knowledge about key natural resources such as cinchona bark.

In 1737, about six months before La Condamine's article appeared in Paris, John Gray published an article on the "Peruvian or Jesuits Bark" in the *Philosophical Transactions* of the Royal Society of London.[58] Like La Condamine's account, Gray's article provided a range of information on the cinchona tree and its bark, including a physical description of the tree, a description of the four main kinds of bark, the location and climate in which the trees grew, and a description of the process of harvesting and drying the bark. Gray's article is significant because, like La Condamine's article, it highlighted the environmental conditions and human activities that transformed the bark into a useful commodity. At the time of publication of Gray's article, even though cinchona bark had become a regular part of medical practice in Europe, physicians, pharmacists, and other healers only had limited access to reliable reports on the tree from which the bark was harvested and how the bark was produced.[59]

Unlike La Condamine, Gray had not visited any of the quina-producing regions of South America. So, where did the Englishman get his information? One of his sources may have been quina traders in South America. It is likely that Gray composed his article during his residence in Cartagena, a port city on the northern coast of New Granada. Ever since the sixteenth century, when the Spanish Crown established its system of annual trade convoys, known as the Carrera de Indias, Cartagena had been a major entrepôt in the Atlantic World. It was also one of the ports along the convoy route known as the Galeones a Tierra Firma that connected Cádiz to Peru via Cartagena and Portobello, a port on the Isthmus of Panama.[60] Cartagena in the late seventeenth and early eighteenth centuries was, thus, an ideal place for an Englishman to collect information on the natural products of South America, including cinchona bark, as most, if not all, trade goods from Peru and other parts of South America would have passed through Cartagena on their way to Europe.

Gray also gathered information about the cinchona tree from the papers of a Scottish surgeon by the name of William Arrot. Although Gray received the papers from the Scotsman directly, he provided little information about their provenance except to mention that the author was a surgeon. He indicated that Arrot had firsthand experience with the cinchona tree as a result of collecting bark "at the Place where it grows in Peru" in the provinces of Loja, Ayavaca, and Cuenca. Although all three provinces were known for producing quina, Gray noted that only bark harvested near the city of Loja was "the true, genuine and fine Jesuits Bark."[61] During his travels, Arrot spent more than a month in the province of Loja. It is likely that other European travelers had visited Loja before him. But what distinguished Arrot from the travelers that preceded him was that he was able to get his information published—with the help of John Gray, of course. Just as bark collectors and merchants extracted quina from the Andean forests of Loja, Gray extracted information from the papers of Arrot and communicated this information to the Royal Society of London.

It was not only Britain and France that reaped the benefits of the emerging networks of information and knowledge in the Atlantic. So did Spain. Around the same time that La Condamine and Gray were preparing their reports on quina, Francisco Suarez de Ribera, a physician of the chamber in Spain and professor at the University of Salamanca, was preparing his own account of the "Quina tree" and its bark that would appear as a chapter in his 1738 *Clave Botanica, o Medicina Botanica Nueva y Novissima* (The botanical key, or the new and newest botanical medicine).[62] The main purpose of Suarez de Ribera's *Clave Botanica* was to identify and discuss plants that could be substituted for each other in the preparation of medicaments. After reviewing the utility of botany to medicine in terms of the identification of the medical virtues of plants, Ribera provided several chapters describing specific plants and their medical uses including a chapter devoted to "Quina-quina" that included a description of male and female cinchona trees.[63] Like Gray, Ribera had not visited the quina-producing regions of South America. In fact, he had not visited any region of the Americas. So, where did he acquire his information about the cinchona tree? For lack of firsthand experience, Ribera emphasized that he "consulted, orally and through letters, with trustworthy people from America."[64]

Collectively, the reports from La Condamine, Gray (Arrot), and Ribera illuminate several key characteristics of the role that informal information exchanges in the Atlantic World played in the production of knowledge about American medicaments. First, the published accounts of Gray (Arrot) and La Condamine show that knowledge of the cinchona tree—a natural resource unique to Spanish America—followed transimperial pathways that led to its publication in the imperial capitals of Spain's main rivals in the Atlantic World. Yet the enterprises of these travelers could not have been more different. La Condamine traveled under the banner of a state-sponsored scientific expedition with the formal recognition and assistance of the Spanish Crown. By contrast, Arrot's expedition to Loja—if we can call it that—was decidedly unofficial. And the knowledge that he gained from this experience followed a much more contingent itinerary from Loja's forests to London's *Philosophical Transactions*.[65] Similarly, Ribera relied on letters and word of mouth.

Several material and textual practices played a key role in facilitating the movement of knowledge across long distances including inscription, extraction, reiteration, and illustration. In the case of Arrot, he first transformed the cinchona tree and its natural and social context in Loja into his "papers" through inscription. Gray then transformed Arrot's "papers" and the knowledge they contained into an article through reinscription and extraction. Finally, the Royal Society transformed this information into a printed article. As in the modern game of telephone, the information captured by Arrot became and, more important, stayed mobile through a mediated process of reiteration. The same was true for La Condamine, except that the French explorer transformed the cinchona tree into both text and image. His inscriptions and illustrations then traveled to Paris where the text was read before the Royal Academy of Sciences, which then transformed La Condamine's manuscripts into printed text and images. Similarly, Ribera relied on information gleaned from letters sent by his contacts in the Americas. The practices of reiteration and reinscription played a vital role in all cases and show that the techniques and practices that kept knowledge moving were just as important as those that made it mobile in the first place.

These examples also show that making and moving knowledge in the Atlantic World was a collective enterprise that involved a range of actors from the bark collectors and indigenous healers in Loja, the original sources of much of this knowledge, to the explorers, surgeons, merchants, travelers, artists, and printers that inscribed and reinscribed this knowledge. In this way, the emergence of quina as a medical commodity resulted from transatlantic and transimperial networks that depended as much on the knowledge and experience of laborers, landowners, and merchants in the forests of Loja as on the journals and printing presses of Europe's scientific societies and their natural philosophers.

In these circuits, mediators abounded—many of them connected to trade. Just as Arrot mediated the knowledge of Loja's bark collectors, Gray mediated Arrot's knowledge and so on. Similarly, La Condamine acted as a mediator between the Royal Academy of Sciences in Paris and Fernando de la Vega, the indigenous healer, who acted as La Condamine's host and guide in Loja. Mediators also moved knowledge in the opposite direction from Europe to South America. During his stay in Quito, La Condamine corresponded with Thomas Blechynden, an official working for the British South Sea Company, who was stationed at the company's warehouse in Portobello on the Isthmus of Panama. It was a key link in the quina trade as all bark bound for Europe passed through Portobello. According to La Condamine, it was Blechynden who informed quina merchants in South America about "chemical analyses and experiences" of the bark in England that was giving rise to an increasing preference for thin bark rather than thick bark.[66] Such instances show that the movement of knowledge about quina and other natural products in the

Atlantic World is best characterized as an exchange or dialogue rather than diffusion from (European) centers to (American) peripheries.[67]

Oftentimes, such mediators took credit for the information collected from local informants. Consequently, as we recognize the role of mediators and go-betweens in the production of knowledge in the Atlantic World, it is also important to remember that these mediators did not act alone. In this case, key informants were the bark collectors and merchants that helped Europeans like La Condamine locate quina trees among the bewildering variety of plants in the tropical forests of the Andes and provided additional information that they had gained from years of experience harvesting and preparing the bark for commercial use. Such dynamics in the interactions between local experts or informants and Atlantic travelers occurred repeatedly and in many different contexts.[68]

As to the structure of this collective enterprise, we note that the communication of information and knowledge tended to follow the pathways of trade, clandestine or otherwise, and empire with each of these three published accounts appearing in the capital cities of three major European empires in the Atlantic World: London, Paris, and Madrid. Yet imperial divisions were not impermeable as attested by the very existence of eyewitness accounts from a British and French observer being published in their home countries.[69] Just as Spain had no monopoly on the quina trade, it did not have a monopoly over the production and circulation of knowledge about the quina tree and its bark. Moreover, it is important to emphasize that places like Loja that produced natural products of special commercial and medicinal interest were and could be centers of knowledge production on par with London, Paris, or Madrid, especially if viewed from the perspective of the object in question. Thus, the geography of knowledge about quina and many other American medicaments was truly transatlantic and transimperial and it included not only places like the Royal Society of London, which had a critical mass of natural philosophers, but also the forests of Loja, which had a critical mass of laborers and healers with a vast store of collective and experiential knowledge, only a fraction of which ever appeared in European publications.

One of the main effects of the increase in the quina trade and the increased circulation of knowledge about the cinchona tree was that the bark attracted the attention of the Spanish Crown and its advisers. In a sense, trade made quina visible to the empire. The case of quina is interesting, in part, because the Spanish Empire was so slow to take a direct interest in the bark. As Antonio Barrera has shown, the Spanish Crown had supported activities for the discovery and development of new medicines from the Americas since the sixteenth century. Yet somehow quina was overlooked. Instead of a conscious discovery of imperial bioprospecting, the Andean medicament that became quina is better understood as the product of a conjuncture of several large-scale developments in the Atlantic and Andean worlds. Given the relative paucity of new natural commodities resulting from explicit imperial policies to discover and develop them, the case of quina was not entirely idiosyncratic but the norm when it came to imperial development of natural resources in the early modern period. In this regard, European empires were rather conservative.

It was only in 1751—more than a century after the introduction of cinchona bark to Europe—that the Crown declared quina to be an "object of interest, curiosity and attention."[70] The phrase appeared in a royal order addressed to the viceroys of Peru and New Granada as well as the president of the Audiencia of Quito. The Spanish Crown had finally decided to intervene directly in the quina trade by converting the forests around Loja into a royal reserve that would provide annual shipments of the best bark exclusively to the Crown. Notably, out of the dozens of books, articles, and pamphlets published about quina since its introduction to Europe in the 1630s, Charles Marie de la Condamine was the only author explicitly referenced in the royal order.

What is the significance of this link between La Condamine's article on the cinchona tree and the establishment of Spain's royal reserve of quina? On the one hand, it signaled the Spanish Crown's affirmation of the utility of botany to the identification and exploitation of the natural resources of Spanish America, especially because it was La Condamine's explicit goal to provide a textual and visual account of the cinchona tree to facilitate its classification according to the conventions of Linnaean taxonomy. On the other hand, the Crown's reference to La Condamine's article is further evidence of networks of knowledge in the Atlantic World. If the discourses of botany and political economy of the early Enlightenment provided a general vision of New World nature as a source of commodities, the social and material practices that facilitated the movement of natural knowledge played a vital role in focusing the Spanish Empire's attention on the cinchona tree and its bark. Although there was not necessarily a causal link between La Condamine's description of the cinchona tree and Spain's royal monopoly, it was no accident that the Crown established its royal reserve of quina shortly after the publication of La Condamine's report on the "quinquina" tree. With the greater circulation of information and knowledge about the cinchona tree and the production of the bark, the quina of the eighteenth century became a different object than the quina of the seventeenth century. The bark became more than a novel medicament from the Andes; it became a valuable commodity of the Atlantic World.

With the establishment of the royal reserve in Loja in 1751, Spanish officials on both sides of the Atlantic quickly realized that they knew little about the cinchona tree and its bark in order to effectively regulate its production and distribution. The next chapter shows how quina's new identity as an imperial natural resource led to the mobilization of yet another network engaged in the circulation and production of knowledge about New World nature: the Spanish colonial government. This imperial network of knowledge production did not displace the existing indigenous and commercial networks of the Andean and Atlantic worlds. Instead, the Spanish colonial government sought to appropriate the knowledge of merchants, bark collectors, and indigenous healers by asking them to provide information and advice on the cinchona tree and its bark. In this way, these representatives of Andean and commercial bodies of knowledge became the Spanish Empire's first experts on this vital natural resource. Ironically, even as the Spanish Crown sought to assert greater control over the informal (and often illicit) trade in quina, its colonial government benefited immensely from the experience and knowledge of those that participated in that trade as it reached new heights in the mid-eighteenth century. By 1750, the transformation of cinchona bark from an Andean remedy to a transatlantic commodity was well established. It was then that the Crown took first steps toward transforming this valuable Andean medicament and Atlantic commodity into an imperial natural resource. Yet, when seeking knowledge about this vital natural resource, officials in Spanish colonial government relied as much on bark collectors, merchants, and local officials as on physicians, pharmacists, or naturalists. As a result, Andean and commercial ways of knowing the bark would come to have a significant presence in the colonial government's understanding and knowledge of quina—a feature of the epistemic culture of the Spanish colonial government that would highlight the fragility of European science in the decades to come.

Three

Quina as a Natural Resource
for the Spanish Empire

In June 1748, Juan Francisco Toro, a member of the Society of Jesus, a Catholic religious order, received a letter from the viceroy of Peru, José Manso de Velasco (1688–1767). The viceroy wanted Toro to comment on a recent royal request for the fangs of a caiman and the bones of a manatee, among other things. These two items were part of a list of thirty-three natural products from Peru that the Spanish Crown had ordered Manso de Velasco to collect and send to Madrid. The majority of the items were botanical products including three kinds of sandalwood, two kinds of tamarinds, a few balsams as well as nutmeg, ipecac, and "cascarilla de Loja," one of the many names for cinchona bark at the time. While the list read a bit like the desiderata for the cabinets of curiosities and museums of natural history that were popular among European elites, Toro had immediately recognized that these items had a much more mundane purpose.[1] They were materia medica to be sent to the Royal Pharmacy at the Royal Palace in Madrid for further study.[2] Although not a physician or a pharmacist, the Jesuit had earned the "admiration" of "the Pharmacy of the Kingdom of Peru" as the result of his "knowledge and intelligence" as well as his "forty years of experience" as a missionary.[3]

After writing a few comments about each medicament, Toro sent his report to Lima, where Diego de Sales, one of the viceroy's scribes, promptly made a copy. This transcription of the Jesuit's report was eventually sent to

Spain, along with a letter from the viceroy and another report on the same list of materia medica. The viceroy had solicited a second report from a professor of medicine at the University of San Marcos in Lima, Juan de Avendaño, who also served as the *protomedico* (first physician) of Peru.[4] We know that these documents successfully made the trip to Spain because they currently reside in the Archivo General de Indias in Seville, the main archive of the Spanish colonial government. They are filed in a *legajo* (a bundle of documents) with hundreds of similar reports from Spanish America submitted to Spain's Ministry of the Indies throughout the eighteenth century. When considered in the context of the colonial archive, it becomes clear that the reports of Toro and Avendaño were not unique, but part of the regular activities of Spanish colonial government.[5]

It is no secret that European empires in the Atlantic World took great interest in the botanical resources of their colonial territories. After all, in the preceding centuries, the spread of plantation agriculture and booming trades in all kinds of botanical commodities made many in Europe (and the Americas) very wealthy. At the same time, new botanical products from the Americas improved people's lives by curing their diseases, enriching their diets, and stimulating their senses.[6] Such tangible and tantalizing results proved a powerful spur for European rulers and elites to support various enterprises to survey and study the natural resources of the Americas. As recent studies have shown, the sciences of description and classification in Europe, such as natural history, took on new importance in the eighteenth century with the growing interest in finding new, useful, and profitable plants to be exploited by Europe's commercial and imperial enterprises.[7] Even as many European naturalists, such as Linnaeus, celebrated their studies of the natural world as a way to appreciate the handiwork of God, the practical utility of their work was undeniable.[8]

Spain and its empire engaged in such activities as much as the rest of Enlightenment Europe. Studies by Daniela Bleichmar, Helen Cowie, and Emily Berquist Soule have shown how natural history, especially botany, flourished in Spain and Spanish America throughout the eighteenth century.[9] Notably, the Spanish Crown, under the Bourbons, became a major patron of natural history and botany in the interest of bringing both prestige and profit to Spain and its ailing empire. At the same time, creole elites throughout Spanish America supported or directly engaged in the study of the natural world often driven by a patriotic interest to promote the utility and natural diversity of their regions to the empire and to the world.[10] Many of these activities were associated with new scientific institutions established thanks to the generosity of the Crown as well as elites and high-ranking officials in Spanish America, including the establishment of a Royal Botanical Garden in Madrid (est. 1755) and, later in the century, a series of scientific expeditions to New Spain, New Granada, Peru, and other locations throughout the Spanish Empire.

Botany became an important tool of empire in the late eighteenth century, yet it is only a piece of the broader story of knowledge and empire in the Spanish Atlantic World. Herein lies the significance of Juan Francisco Toro's 1748 report on Peru's materia medica: it points to the larger apparatus of knowledge production embedded in the Spanish colonial government. After all, the colonial government had been collecting information and producing new knowledge about the Americas in a variety of forms since the early sixteenth century.[11] As Paula De Vos has pointed out, the various scientific activities of the late eighteenth century were a relatively late addition to the Spanish Empire's other efforts to systemically collect information on American flora in the hopes of identifying new, useful, and profitable botanical resources.[12] Before botanists were deployed as agents of empire, bureaucrats—the officials running the colonial government in Spanish America—played a vital role in Spain's efforts to make its American territories knowable as well as governable.[13]

By the time that the Crown set up the royal reserve of quina in the mid-eighteenth century, the techniques of knowledge production in the colonial government were well established. Starting in the sixteenth century, officials at the Council of the Indies and the Casa de Contratación, the two institutions in Spain that oversaw commercial and colonial activities, employed a number of techniques to produce knowledge about the peoples, places, and things of Spain's overseas empire.[14] Perhaps, the most famous examples of such efforts are the various questionnaires that officials in Spain sent to their counterparts in Spanish America in the hopes that the resulting reports, known as *relaciones,* would provide an encyclopedic vision of the human and natural resources of Spain's Atlantic Empire.[15] Many other European imperial states in the early modern Atlantic World employed similar techniques. In light of such activities, it is likely that early modern European colonial governments produced as many descriptions, images, and specimens of natural phenomena as Europe's scientific institutions did. But because much of this knowledge circulated in manuscripts and was ultimately hidden away in government archives, the role of early modern colonial bureaucracies in the production of knowledge about the natural world has been overlooked until recently.[16]

Using documents from the first decade of the royal reserve of quina in Loja (c. 1751–1761), this chapter describes the epistemic culture of the Spanish colonial government.[17] It aruges that the Spanish colonial government not only possessed an identifiable epistemic culture but also that the values and techniques of this epistemic culture were remarkably consistent throughout the myriad institutions that comprised the colonial government. Existing scholarship has provided important insights into this epistemic culture by looking at the various projects whereby the colonial government attempted to produce a comprehensive survey of the natural resources in Spain's American territories—many of which were unsuccessful. By contrast, the case of quina is unique because it is one in which the colonial government directed its efforts to a comprehensive understanding of a single natural entity, albeit one that had many facets. The colonial government was generally quite successful at acquiring and producing knowledge about quina as evidenced by the tens of thousands of pages of documents strewn across colonial archives in Spain and Latin America.

One goal of this chapter is to highlight the promiscuity and political nature of the epistemic culture of the Spanish colonial government. Such characteristics not only show how this epistemic culture reflected the structure and style of Spanish colonial governance at large but also highlight the ways in which the tensions of empire could have a significant epistemological dimension. Debates about knowledge and who has the authority to speak for nature—bark collectors in South America or botanists in Spain—often functioned simultaneously as debates about the order of empire as informed by tensions between local and imperial interests. More important, this portrait of the epistemic culture of the Spanish colonial government is vital to understanding the fragility of European science in the Spanish Atlantic in the late eighteenth century. After all, the epistemic culture of the Spanish colonial government mirrored the broader political ideology of colonial governance in which the Crown was understood as the final authority in the adjudication of competing interests in colonial society and the empire at large. Consequently, when Spanish pharmacists and botanists became a more regular part of the Spanish colonial government's efforts to produce knowledge about quina and the natural world, they were joining an established epistemic culture in which their knowledge was not necessarily privileged. Rather, it was treated as one additional perspective on the phenomena in question (quina) to be collected and assessed by officials and the Crown alongside the claims of other groups such as bark collectors, merchants, missionaries, and local officials. By examining the Spanish colonial government's efforts to survey the possible sources of cinchona bark throughout South America, this chapter outlines the contours and characteristics of the colonial government's epistemic culture that would exert a significant influence on the efforts of imperial reformers and Spanish naturalists to wield European science as a new tool of empire.

Toro's report on the Crown's list of materia medica serves as a useful introduction for two reasons. On the one hand, the production and circulation of this report provides a glimpse of the colonial government's epistemic culture in action. On the other hand, Toro's report illuminates an important moment in the history of quina in the Atlantic World: its transformation into an imperial natural resource. Whereas in 1748 "cascarilla de Loja" was just one of many Peruvian medicaments of interest to the Crown, just three years later, the bark became the focus of imperial intervention. With the establishment of the royal reserve, officials in Spain suddenly became aware of how little they knew about the cinchona tree, thus the Crown mobilized the colonial government to increase the empire's knowledge of this vital natural resource. Ultimately, the early years of royal reserve show that the consequences of quina's transformation into an imperial natural resource were as much epistemological as they were economic and political.

Quina as an Imperial Natural Resource

A good example of the relationship between knowledge and empire can be found in the 1751 royal order in which the Crown first declared its intention to intervene in the production and distribution of quina. In particular, this document shows how much the colonial government's conception of the bark owed to preexisting ideas about quina, the Andean World, and commerce. More important, this document defined a paradigm for thinking about the cinchona tree and its bark that informed imperial policy for decades to come. The royal order defined quina as medicament, botanical commodity and an "object worthy of curiosity, interest, and attention."[18] At the same time, this document makes clear that the Crown and its advisers thought of quina as a finite resource rather than a renewable one. It seems a surprising idea given that quina was a botanical product and cinchona trees could, in theory, be regrown. It is even more surprising given that plantation agriculture was so prevalent in the Americas and the Atlantic World.[19]

What were the origins of the Crown's perception of cinchona bark as a natural resource that existed in limited supply? One possible source may have been an account of quina harvesting submitted to the Crown in 1748 by Antonio Ulloa (1716–1795) and Jorge Juan (1713–1773). Ulloa and Juan were two Spanish naval officers that participated in the same expedition to Quito as Charles Marie de la Condamine, the joint French and Spanish expedition to take measurements at the equator. Upon their return to Madrid, Ulloa and Juan published a travel account of their observations and experiences in South America and submitted a separate report to the Crown that was critical of Spanish colonial society and colonial governance.[20] Ulloa and Juan devoted a chapter in their report to various natural resources that could be more effectively exploited by the Spanish Empire. Within this chapter, they included a section on the "cascarilla or quina produced in the mountains of Loja." The two officers reported that there was a "harmful disorder" in the quina trade, which was the "method of extracting the bark by uprooting the tree and removing its bark." "Since no one takes care to plant more trees in its place," they continued, "there is no doubt that in time these mountains will become bare because even though the mountains are extensive, they do have an end."[21] Notably, Ulloa and Juan also suggested that the *corregidor* and *alcalde mayor* (town council) of Loja appoint a *juez* (magistrate) to oversee the replanting of the trees.

Royal officials in Madrid that read Ulloa and Juan's report, took only one message from it—that quina was a finite resource because the cinchona trees were disappearing. Yet the 1751 royal order made no mention of the possibility of replenishing Loja's forests by replanting cinchona trees. A closer reading of the order reveals why. It shows that the officials interpreted existing knowledge of quina in light of their own preconceptions about the natural resources of the Andean World, on the one hand, and their mercantilist vision of imperial trade, on the other. Such evidence shows that in the process of reinventing quina as an imperial natural resource, officials in the colonial government did not start from a blank slate.

In a section of the royal order that discussed the possibility that "foreigners" might transplant "Quina trees to territories that they posses," the influence of existing knowledge of quina on imperial policy is apparent. Here, the Crown made explicit reference to Charles Marie de la Condamine, the only published author mentioned in the royal order. The Crown noted that the French scientific traveler had already tried and failed to transplant cinchona trees from Loja to French Guiana during his return trip from Paris to Quito via the Amazon River.[22] From the Frenchman's experience, officials in Madrid concluded that there was little reason for concern about the French, Portuguese, or Dutch transplanting cinchona trees to their own territories in the Americas, but even more importantly, this evidence suggested that the trees were not amenable to cultivation. Drawing on an analogy of growing grapes in Spain, the royal order suggested that even a difference of one "league" of terrestrial distance is enough to adversely affect the quality of the cinchona bark, such that transplanted bark was likely to be medically useless.

In addition to the information circulated by travelers in the Atlantic World, two factors reinforced the perception of quina as a finite, extractable resource rather than a renewable, cultivable one. First, evidence suggests that Peru and the Andes were generally understood as spaces of extraction rather than cultivation. The Crown and its advisers were well aware that the Peruvian Andes were home to the Potosí silver mines, the most productive silver mines in the early modern world and a major source of royal income since the sixteenth century.[23] In other words, one of the most iconic and valuable products of Peru and the Andean World was an extracted resource rather than a cultivated one. Even though cinchona bark was technically a product of the

Viceroyalty of New Granada (establishd in 1717 and again in 1739; see figure 1), many in eighteenth-century Europe understood quine as a product of Peru and may have assumed that the bark was also an extracted resource rather than a cultivated one.

Even those involved in the quina trade thought of the bark as a product akin to silver. For example, in 1753, Loja residents responded to the Crown's idea of making the best quina-producing hills in the province into a royal reserve by proposing that local bark collectors could give the Crown one-fifth of their annual quina harvest. They explicitly referenced the Crown's arrangement with miners to receive one-fifth of all silver produced annually.[24] Accordingly, this analogy between plant product and precious metal may have encouraged officials in the colonial government to assume that extraction of the bark would eventually lead to the exhaustion of the forests just as mines became exhausted. As a result, it may have seemed unthinkable to officials in Spain to purposefully increase the supply of quina through the cultivation of cinchona tree. It is not surprising then that the Crown focused its earliest efforts at regulating the distribution of the bark and limiting the intervention of foreign merchants. The goal was to make Spain the primary distributor of quina to the rest of the world.

The assimilation of the case of quina to the paradigm of precious metals also highlights the influence of mercantilist thought on imperial policy. Mercantilism was a popular political and economic theory in intellectual and government circles in Spain and other parts of Europe in the seventeenth and eighteenth centuries.[25] According to mercantilist theory, the wealth of the world existed in finite supply in the form of precious metals. From this perspective, European rulers and states competed with each other to acquire and control the greatest stores of gold and silver. By maintaining a favorable balance of trade and monopolizing trade goods, rulers and states could insure an influx of precious metals via commerce. From a mercantilist perspective, a natural resource like quina that was both useful and unique could be invaluable to the Spanish Empire. Because many believed that the Andean forests of New Granada and Peru were the only places to find cinchona trees, Spain seemed to be in a position to become the world's supplier of this precious medicament. By default, Spain's merchants should have been able to maintain a favorable balance of trade in quina. As a result, the Crown's early interventions in the quina trade sought to help Spanish merchants achieve that goal.

Mercantilism, Fraud, and the Royal Reserve

The royal order of 1751 displays the influence of this mercantilist perspective in the description of the main problems in the quina trade: contraband trade and adulteration of the bark. The Crown expressed concern about contraband trade because, in theory, all imports of quina to Europe were supposed to go through Spanish ports and be handled by Spanish merchants, especially because cinchona trees were supposed to be unique to Spanish South America. Instead, the Crown found that many medicaments like quina were "unknown in the interior Provinces of Spain." The main reason for this situation, according to the royal order, was that "industrious" foreign merchants (*estrangeros*) had diverted all the highest quality medicaments to their own countries via contraband trade networks. As the royal order further explained, illicit commerce in medical drugs meant that Spanish pharmacies were either unable to find the materia medica they needed or had to stock their shelves with medicaments that were "adulterated and of the worst quality." The situation was especially dire in the case of "cascarilla or Quina bark," which the Crown described as one of the most "valuable" products from Spanish America. The order explained that much of quina exported from the "Indies" went primarily to "foreign countries" such that Spanish merchants in Cádiz could not find quina of sufficient quantity or quality to trade and distribute to the rest of Spain.[26]

In cases of adulteration, unscrupulous bark collectors or merchants would add other tree barks to their cinchona bark and sell the mixture as if it were pure quina.[27] Although such practices resulted in a hefty profit for bark collectors and merchants, adulteration undermined the medical efficacy of quina and threatened the health of unsuspecting consumers. Moreover, because the Crown and its advisers thought that quina existed in limited supply, they saw adulteration as a likely outcome as bark collectors and merchants sought to keep pace with expanding demand or to cash in on the growing market for cinchona bark. Consequently, the Crown and its advisers recognized both an economic and moral reason for intervening in the quina trade. With the royal reserve in Loja, the Crown hoped to divert more bark to Spanish markets and, at the same time, safeguard the quality of that bark.

Since the Crown worked largely from the assumption that quina was a finite natural resource, only one course of action remained in the estimation of officials in Madrid. "There seems to be no other option," the royal order states, "than for his Majesty to buy on his account all the Quina that grows in [the Audiencia of Quito] in order to store in warehouses [an amount] sufficient to supply the Provinces [of Spain] and export that which is leftover, following the example of other sovereigns."[28] It is unclear which "other sovereigns" the Crown had in mind, but many in the Spanish Empire at the time cited the example of the monopoly that they believed the Dutch had over the trade in cinnamon from the East Indies. Nonetheless, this approach reflected the Crown's explicit goal of making sure that Spain and Spanish America received their quina before foreign nations did. In this way, the royal reserve of quina in its early decades was rooted in a mercantilist mind-set and a zero-sum understanding of the distribution of valuable resources like quina. The only way to increase Spain's imports of the bark was to decrease the amount of bark acquired by foreign merchants.

As for the implementation of the monopoly, the Crown and its Ministry of the Indies recognized that any intervention was likely to have differential effects on the groups involved in the quina trade including bark collectors, merchants, and local elites.[29] Nonetheless, the Crown believed that local and imperial interests could be pursued in tandem. To this end, the royal order identified three main objectives for the royal reserve of quina: "to care for the public health of [the king's] vassals, to benefit the harvesters of Quina, and to increase the treasury at the same time."[30] These objectives cast the project as a combination of Enlightenment sensibility and mercantilist policy. On the one hand, the Crown saw the bark as a limited commodity to be exploited by Spain alone. On the other hand, the Crown saw the royal reserve as serving the public good and the interests of bark collectors. Such combinations of justifications were in some sense a pragmatic attempt to convince other interested parties to endorse the project and convince them that the Crown was serving their interests.

Hierarchy in the Epistemic Culture of Empire

At the same time that the Crown and its advisers viewed quina and the quina trade through the lens of mercantilism, they also recognized that they needed to know much more about the bark in order to effectively implement and administer the royal reserve. To this end, the Crown ordered the viceroys of Peru and New Granada as well as the president of Quito to provide information regarding "all that could influence this matter with the addition your own opinion [on the matter]."[31] In other words, the Crown invoked a claim to comprehensive knowledge of the bark at the same time that it invoked a claim to control its production and distribution. Generally, the Crown wanted trade data including where the trees grow, what prices are paid for different kinds of bark, what were the predominant trade routes and costs of transportation, how much superior quality bark was harvested at each location, and how much bark was consumed in the Spanish America annually. Finally, the Crown wanted any officials that supported the proposed royal reserve to provide suggestions on how to keep the bark from falling into the hands of foreign merchants as well as information on any "political difficulty or other difficulties" that might impede the royal reserve.[32]

This request from the Crown and the responses to it provide insight into the key characteristics of the epistemic culture of the colonial government.

First, the epistemic culture was hierarchical. The vertical hierarchy of the colonial government shaped how knowledge was produced. In general, the Crown and its advisers in Spain, including the minister of the Indies, were the ones who determined what knowledge was of interest to the empire. High-level officials accumulated knowledge of natural resources by receiving reports from subordinates and subjects scattered throughout colonial territories. Moreover, the hierarchical structure of the colonial government gave high-level officials the power and authority to solicit information and advice from different groups in colonial society. This approach of collecting testimony from a variety of experts and interested parties allowed the Crown and its representatives to maintain their position within the empire as the ones with the ultimate authority to adjudicate between competing interests. In addition, such organization meant that the Crown and officials in Spain, in theory, had the power to decide what counted as "true" knowledge of the cinchona tree and its bark.

Empiricism in the Epistemic Culture of Empire

How, then, did officials in the Americas try to convince the Crown and its advisers that their knowledge and opinions were to be trusted? A good place to look is the reports submitted by officials in South America in response to the royal order of 1751. The Crown received three main responses from subjects and officials, who held different positions within the empire: a proposal from the residents of Loja, a report from Miguel de Santisteban, an official in the employ of the viceroy of New Granada, and a report from Viceroy of Peru José Manso de Velasco. Here I focus on Santisteban's report because it is a good example of how officials established their authority to speak for nature and also because it highlights another key feature of the epistemic culture of the colonial government: empiricism.

Biographical details on Santisteban are sparse, but compared to many other officials in eighteenth-century Spanish America, we know quite a bit about him.[33] Santisteban was born to creole parents in Panama and studied at the University of San Marcos in Lima, where he received a degree in mathematics in 1720. After a distinguished military and administrative career in the Viceroyalty of Peru, Santisteban traveled to Spain in 1740, probably in hopes of furthering his career aspirations as so many creole elites from Spanish America did in the eighteenth century. In 1749, he returned to South America in the retinue of a newly appointed viceroy of New Granada, José Alfonso Pizarro, Marqués de Villar (1689–1762). Two years later Pizarro commissioned Santisteban to study quina and collect the information requested by the Crown. The viceroy also ordered Santisteban to "establish in Loja the regular shipment of Quina bark that His Majesty has ordered be made every year."[34]

Pizarro did not state his reasons for choosing Santisteban explicitly, but it is likely that he was following the Crown's repeated instructions to seek out *hombres peritos* (experts) who had firsthand experience with quina. In other words, empiricism—direct experience of the phenomenon in question—was a key value in epistemic culture of the colonial government.[35] Santisteban was a good choice in this regard because he had traveled extensively in the region including a journey from Lima to Caracas in 1740–1741. During this journey, Santisteban gained firsthand knowledge of the quina-producing regions of the Audiencia of Quito including Loja, where two of his traveling companions were treated for their fevers with cinchona bark. At the same time, Santisteban had the distinction of being a corresponding member of the Royal Academy of Sciences in Paris, probably because of his relationship with Charles Marie de la Condamine, whom Santisteban may have met in the late 1730s during the Frenchman's time in South America. Finally, as a result of his years of experience in the colonial government, Santisteban knew the practices and protocols of the colonial bureaucracy as well as the fiscal structure of the empire. And Santisteban proved an excellent choice.

In June 1753, after two years of traveling and gathering information, Santisteban submitted his report to the viceroy, who forwarded it to the minister of the Indies in Spain. The report became a key document guiding the imperial intervention in the quina trade over the next few decades.[36] Santisteban referenced a variety of sources of information in his report to the viceroy, including his firsthand experience with the cinchona tree and its bark, information collected from other people that had experience with quina and the quina trade, and, finally, printed descriptions from Europe of the tree and its bark.

In keeping with the emphasis on empiricism in the colonial government, Santisteban emphasized his firsthand experience the most. In his report, he noted that his dissatisfaction with the secondhand information available in Quito motivated him to visit the quina-producing regions himself. Informants there reported to Santisteban that the "Quina Tree" could be found in provinces other than Loja. He was not "content" with these reports so he traveled to those regions, where he "examined the tree, comparing its leaves, flowers, and fruit" with the assistance of "intelligent persons."[37] Santisteban actively engaged in the study of new kinds of quina. After listing all known locations of cinchona trees from Piura to Cuenca, he provided a list of stands of cinchona trees that he personally had discovered. When recounting his experience in the province of Chimbo on his way to Guayaquil, he wrote, "I discovered many Quina trees that in their color and styptic taste seem to me to be good although the bark is not as compact." "I have found this Tree," he continued, "in the Mountainous regions of the Road, that goes from Quito to Santa Fe in all the places whose climate is similar to that of Loja and whose elevation, by the observations of Monsieur La Condamine, are elevated to 800 *tuesas* or 2,000 *vazas castellanas*."[38] With such declarations, Santisteban emphasized his active role in generating new knowledge of cinchona trees for the Crown and the colonial government.

In addition to his empirical knowledge, Santisteban made use of printed accounts of quina.[39] The reference to La Condamine was just one of several printed descriptions of the tree from Europe. Santisteban elected not to "dwell on the description of the Quina Tree and the diversity of its species" because "Mr. De la Condamine" had already described "the distinctions of color of the bark [and] the virtue, size and texture of the leaves."[40] This omission suggests that Santisteban expected his audience—the viceroy in Bogotá and the minister of the Indies in Spain—to be familiar with La Condamine's article. When recounting the history of quina trade, Santisteban also used information from the entry for "Quinquina" in the *Dictionnaire universel françois et latin* published by the Jesuits in Trevoux, France.[41] Such references would have given Santisteban's report more authority in the eyes of the viceroy and officials in Spain. In addition, this evidence highlights Santisteban's connection to the broader circulation of knowledge in the Atlantic World.

Overall, Santisteban's report here provides useful insight into the Spanish imperial networks of knowledge in the mid-eighteenth century. The process whereby Santisteban produced his report further elucidates of the importance of motivated and diligent officials to the successful operation of the epistemic culture of the colonial government. The Crown and its viceroys relied heavily on officials like Santisteban for their local knowledge as well as their role as mediators between the colonial government and the worlds of Andean medicine and Atlantic trade. Most important, the final report produced by Santisteban challenges common characterizations of the geography of knowledge in the Atlantic World in which colonial territories are cast as peripheries and European capitals as centers of knowledge production.[42] Santisteban's experience reveals the continuing importance of the Andean World as a center of knowledge production into the eighteenth century.

Having established his authority to speak about quina, Santisteban concluded his report with his recommendations on the Crown's proposal. He generally supported the Crown's plan to intervene in the quina trade and was one of the first to refer to the project as an *estanco* or royal reserve. Echoing the language of the 1751 royal order, he described the royal reserve as "useful to public health, to Royal interest, and to the municipality of Loja." He argued that because cinchona trees grow "wild and uncultivated in Royal Mountains," the king is "the owner of them." He thought government intervention could reduce fraud and suggested that the Crown set up a "factory" with two "officials" with "knowledge of [the bark]." These officials would examine the quina brought for sale and eliminate any "foreign barks" that bark collectors and merchants had mixed with the quina. Such efforts were a necessary remedy to the "greed of merchants," who "converted [quina] to a poison by adulterating it with other barks." "With this precaution," wrote Santisteban, "we will have access to pure quina that will produce those marvelous effects which were admired in the early years after its discovery." Finally, Santisteban characterized the need for royal intervention as a moral imperative as much as economic one. "Since *Quina* is a tree with which Divine Providence has increased the riches of this new world," he wrote, "it does not conform to the rules of politics that foreigners, especially the English, have made this product one of the sources of their commerce."[43]

Local Knowledge and Imperial Governance

The hierarchy of the epistemic culture of the colonial government was counterbalanced by the high value many officials placed on firsthand experience, especially when it came to knowledge of natural resources like quina. In cases such as the royal reserve, local knowledge took on new importance in the context of imperial reform during the reign of Charles III (r. 1759–1788) in which reformers sought to centralize colonial governance and to strengthen royal power.[44] In this context, many officials in South America appreciated the way in which local knowledge and experience provided a means to resist these tendencies. After all, the merchants, missionaries, bark collectors, and even local officials involved in the quina trade could exert significance influence over what the colonial government did or did not know about the bark.

Respondents to the Crown's proposed royal reserve argued that local knowledge and expertise should be integral to the project. We have already seen Santisteban's recommendation that the Crown needed individuals with "knowledge of the bark" to run the royal reserve. In a report, submitted in November 1753 from Lima, Viceroy Manso de Velasco echoed this recommendation.[45] Manso de Velasco emphasized that it was vital for the Crown to employ the services of "intelligent people" from the region because they were the only ones who could make sense of the enormous variety of cinchona barks extracted from the forests each year. To this end, the viceroy provided the Crown with a clear picture of how the production of quina was a skilled trade requiring knowledge and expertise.

Manso de Velasco recognized that quina from "the Mountain of Caxanuma in the Province of Loja" was of the "best quality, according to experience," but he also emphasized the ways in which human activities affected the bark.[46] Manso de Velasco described the various techniques for harvesting and preparing the bark and how errors in these processes could undermine the medical efficacy of the final product. He explained that exposing the bark to humidity "weakens [the bark] and makes its virtue evaporate." He then explained how bark collectors dried the bark in parallel rows on "platforms of reeds, which are called Barbacoas" immediately after stripping the bark from the tree in order to guard against the detrimental effects of humidity. "The principle indicator of its quality," he explained, "consists in the intense activity and vigorous force of its bitterness," which good bark collectors could preserve throughout the drying process.[47] Mansco de Velsaco described properly prepared bark as having "the same shape as cinnamon" with a rough texture, a light, burnt color to its interior surface and a dark color to its exterior. Yet while the viceroy could describe these qualities in words, written description was not sufficient to produce high quality. Expert knowledge of quina was a kind of connoisseurship that was acquired through experiencing the physical characteristics—taste, odor, and texture—of many different varieties of bark. When confronted with a new or unknown kinds of quina, experts or connoisseurs were able to compare the new variety to their experiences with other quinas. Many in the Andean World, including healers, bark collectors, and merchants, had precisely this kind of experience, which is why they were so useful to Crown's efforts to intervene in the quina trade.

As an example, Manso de Velasco pointed to the practices of merchants in Payta, a port on the northern coast of the Viceroyalty of Peru (figure 1). Payta was an important entrepôt where various bark collectors and merchants from inland regions brought their bark to sell and export. Payta merchants separated the incoming quantities of bark according to its physical characteristics and the part of the tree from which it came (the branches or the trunk). These merchants also cleaned the bark in order to eliminate "dust and useless leaves and other barks which fraudulently are mixed in."[48] Manso de Velasco explained how this sorting process affected the quality and value of the bark. Pricing of the bark in Payta was, ultimately, based on the purity of a given quantity of quina. He noted that one real per pound was paid for bark in "which only a third was useable," and two reales per pound for bark "which did not have too much illegitimate material." Bark with even less fraudulent material could be sold for three reales per pound.[49] Purity, as defined by the relative proportion of good or usable bark to useless or bad material, was what determined value of bark shipments at Payta; yet it was knowledgeable individuals that assessed the purity. In this way, knowledge was translated into value—and it was this knowledge that Crown needed to harness.

In his recommendations to the Crown on its interventions in the quina trade, Manso de Velasco used the activities at Payta as a model. Although the viceroy did express some reservations about "putting an end to the free trade" in the bark, he recognized that that products "subject to the Royal Hand are imported at better proportions [i.e., of better quality] and with lower risks of loss." Knowledge was key. The viceroy thought that the Crown could improve the trade by paying a "reasonable price" for the bark and by "naming able subjects to promote the harvesting [of quina] in the places which produce it." Manso de Velasco also recommended that the Crown appoint an expert to Payta in order to make sure that the bark that received the royal imprimatur was only bark "of the best quality." He speculated that the Crown's bark would ultimately fetch "six reales per pound because it had been selected very carefully by very expert individuals."[50] In other words, royal bark would prove the most potent and most valuable not jut because it came from a particular place but because it had been subject to the scrutiny of "expert individuals." Economic incentive and expertise could, thus, work together to improve the quality of quina in the name of the empire.

These remarks illuminate contemporary conceptions of the relationship between knowledge (in the form of expertise), economic value, and empire in the earliest years of the royal reserve of quina. "Expert individuals" provided the means to counter the forces of variation and adulteration that threatened the reputation of quina as a medicament and undermined the quina trade.[51] Both Manso de Velasco and Santisteban agreed that expertise was integral to the royal reserve and their reports further suggested that experts should be appointed as inspectors to serve as agents of the colonial government. Finally, their reports themselves were an example of how "expert individuals" with local knowledge could be useful to the empire.

Bark Samples, Expertise, and the Politics of Knowledge

On 30 August 1751, the Crown sent a second order to the viceroys of Peru and New Granada asking them to load "three or four hundred pounds of the most select and efficacious quina" on every ship headed for Spain from South America. The purpose of this enterprise was to build up the supply of bark in Madrid, while also providing the Royal Pharmacy with the material to determine which barks were best by conducting "observations and experiments" on the various kinds of bark submitted.[52] This request and the responses to it show that the Crown's interest in quina led the colonial government to mobilize a growing community of experts. At the same time, the process of soliciting and submitting bark samples highlights the politics of knowledge within the epistemic culture of the colonial government as different groups of experts claimed the right to speak for nature in their assessments of the bark.

Let us begin in Madrid. Just as officials in South America would come to rely on local experts to assess the quality of quina shipments produced and exported under the auspices of the royal reserve, officials in Spain relied on their own local experts, the royal pharmacists, to assess the quality of incoming samples and shipments of the bark. With the development of the royal reserve, the Royal Pharmacy became more directly involved with imperial affairs. At the same time, by having officials send samples to Madrid, the Crown indicated that only the Royal Pharmacy had the authority to make the final determination on the utility and quality of different kinds of quina. The viceroys may send what they considered to be the "most select and efficacious" bark, but only the Royal Pharmacy could determine whether this designation was apt. In this way, the Crown tried to impose a hierarchy on production of knowledge about quina and to limit the influence of local experts in South America.

As for the viceroy of New Granada, he complied with the Crown's request and apparently endorsed the authority of the Royal Pharmacy. In June 1753, he sent a letter to Spain informing the minister of the Indies of the imminent arrival of samples of bark from the provinces of Chimbo, Alausi, and Cuenca that Miguel de Santisteban had collected.[53] Viceroy Pizarro explained that he was sending the samples so that "the most skillful Botanists of the court" could examine their "virtues." The viceroy added that he hoped the examinations in Spain would "dispel the preoccupation with [quina] from the Mountains of Caxanuma near Loja."[54] This last comment illustrates how the sending of bark samples could serve local and imperial interests simultaneously. In Madrid, the Crown and its Royal Pharmacists hoped to receive a variety of barks so that the Crown could assert control over the best bark. In New Granada, Pizarro hoped that the Royal Pharmacy's recognition of the utility of quina from regions other than Loja would represent an economic boon to other provinces in his viceroyalty.

In December 1753, the viceroy of Peru sent his own bark sample from the Jungas region that bordered Portuguese territory.[55] Although he was aware that the Royal Pharmacy would review the sample, Viceroy Manso de Velasco had his local experts check the sample first. In the letter that accompanied the sample, he explained that Dr. Cosme Bueno (1711–1798), a Spanish physician and professor at the University of San Marcos in Lima, had had "good success" with the bark and observed that it produced "the same effects as that from Loja."[56] He also submitted extracts of reports from the president of the Audiencia of La Plata, the corregidor of La Paz, and the corregidor of Zica Zica. All three had experience with the quina from the Jungas region, which was supplied to them by Jesuit missionaries. Surprisingly, the president of La Plata reported that "an expert, who knows about these things" had "experimented" with this quina and found "that it is not like [the bark] from Loja"—a result that contradicted the findings of Cosme Bueno in Lima and, later, the Royal Pharmacy in Madrid.[57] Nonetheless, in July 1755, the viceroy of Peru received an order from the minister of the Indies instructing him to let the quina from Jungas enter into "commerce" so that the "regions in which this product is collected" could take advantage of the medicinal qualities of the bark and, possibly even, make some profit.[58]

The movement of Manso de Velasco's quina sample from the forests of La Plata to the pharmacy in Madrid illuminates the shared epistemic culture that pervaded the colonial government. Just as the Crown relied on its royal pharmacists to evaluate and assess samples of the bark, Manso de Velasco mobilized his own network of knowledgeable individuals to assess quina through empirical observations and therapeutic tests. More important, this example shows how the Crown's efforts to survey the possible sources of quina in South America gave rise to communities of experts on both sides of the Atlantic that provided knowledge and advice to officials in the colonial government. Yet, as the case of the Jungas quina shows, the different groups that advised the Crown did not always agree. In such instances, the final decision ostensibly fell to the Royal Pharmacy and officials in Spain. Ultimately, these examples provide insight into how officials navigated the hierarchy, empiricism and politics of the epistemic culture of the colonial government.

With the establishment of the royal reserve in 1751, quina became an imperial natural resource. It also became an object of the epistemic culture of the Spanish colonial government. When they realized that they needed to know more about the cinchona tree and its bark, the Crown and its advisers turned first to officials in the colonial government rather than to botanists, even though quina was clearly a botanical product. In doing so, the Crown was following the well-established practices of knowledge production already embedded in the colonial government, practices that dated back to the early sixteenth century. Consequently, this preexisting epistemic culture of empire would play a significant role in shaping the integration of European science, notably botany, into the Spanish imperial enterprise. As shown in the chapters that follow, Spanish pharmacists and botanist would encounter significant difficulties in asserting the preeminence of their expertise. And that was because the epistemic culture of the colonial government was one in which local officials, merchants, bark collectors, and even indigenous healers were able to claim the authority to speak for nature. In short, we cannot understand the history of science and empire in the Spanish Atlantic World without understanding the epistemic culture of the colonial government.

This chapter has highlighted the main characteristics of this epistemic culture. Hierarchy was an important part of this epistemic culture since the Crown and officials in Spain were often the ones that determined what information and knowledge was of interest to the empire. This hierarchical structure also meant that the Crown and officials in Spain expected officials and informants in South America to share their knowledge unconditionally.

Knowledge, just like cinchona bark, was a resource that the Crown hoped to control. At the same time, this epistemic culture was one that placed a high value on empiricism (broadly defined). Individuals with firsthand knowledge or experience of the phenomena in question were highly valued as informants to the colonial government. Finally, this epistemic culture was also political in that problems of knowledge were often directly connected to questions of imperial policy. In the case of the royal reserve, solicitations for information about the cinchona tree and the bark were often accompanied by solicitations for recommendations and opinions on how to best manage this natural resource. Moreover, because imperial officials consulted a variety of informants, the reports sent to the viceroys and minister of the Indies often reflected the range of political and economic interests of different groups in South America.

When the Crown defined quina as an imperial natural resource, this action eventually led to creation of an extensive apparatus of expertise throughout the empire. A good example of the developing relationship between expert and empire comes from the case of an adulterated shipment of bark that arrived in Madrid in 1756 or 1757. In this instance, the Crown asked José Ortega, a prominent pharmacist in Madrid, to review the shipment and identify the cause of the bark's adulteration.[59] Far from being surprised, Ortega noted in his report on the shipment, "The adulteration of Quina is an old [practice] in America, and for this reason this celebrated Medicament lost its credit."[60] As to the source of adulteration, Ortega pointed to an English factory in Portobelo, where he speculated that Thomas Blechiden, a merchant for the South Sea Company, was siphoning good quality quina from the Spanish trade network.[61] In this way, Ortega embraced the mercantilist proclivities of many Spanish officials by suggesting that the problem was one of distribution rather than production.

Ortega's consultation on quina is also significant because it shows how not all experts that advised the colonial government were the same. Experts in South America, like Miguel de Santisteban, often had direct experience with the cinchona tree in its natural habitat of the Andean forests, but lacked formal training in the study of nature. By contrast, experts in Spain, like José Ortega and the royal pharmacists, often had formal training in the study of plants but lacked direct experience of them in their natural habitats, especially in cases like quina. Through his consultations for the Royal Pharmacy and connections to the scientific community in Madrid, Ortega established himself as an authority on the natural world whose status as an expert derived not from direct experience with American nature, but from his training as a pharmacist and botanist, his knowledge of texts about American plants, and his greater proximity to the royal court relative to experts in South America.

It is not surprising then that Ortega did not let his audience with the Crown pass without promoting the utility of botany and medicine to the Spanish Empire. In this way, he planted one of the seeds for a relationship between science and empire that would come to fruition two decades later. As part of his report, he recounted the story of Joseph de Jussieu, a French physician and naturalist who traveled with Charles Marie de la Condamine to Quito in 1735.[62] According to Ortega, Jussieu had visited Loja at the viceroy's invitation not only to study the cinchona tree but also to instruct local merchants and bark collectors on how to distinguish "true Bark from false."[63] In short, it was a potent example of how science had already (if inadvertently) been useful to the Spanish Empire.

Despite Ortega's best efforts, the Crown would not act on his vision of science and empire until the 1770s when two royal botanical expeditions were sent to South America and botanists in Madrid came to have a greater role in the assessment of bark samples. In the meantime, botanists and other experts in Spain shared the space of expertise that surrounded the colonial government with many other individuals that served as local advisers to officials on both sides of the Atlantic. As the royal reserve developed, though, some members of these groups would test the limits of their authority within this community and the imperial bureaucracy. The next chapter explores this dynamic as manifested in the interactions between the Royal Pharmacy and the bark collectors of the royal reserve as they tried to determine which quina was the best and, more importantly, to determine, who got to decide which bark was the best. Ultimately, the epistemic culture of the colonial government provided the institutional context and figurative space that facilitated and mediated encounters between representatives of different bodies of knowledge—scientific, commercial, and indigenous—about quina and other natural phenomena. Although the practitioners of pharmacy, medicine, and botany were often assimilated quite easily to this epistemic culture, the authority and superiority of their knowledge was not necessarily taken for granted and it was at those moments that the fragility of European science became visible.

PART II

The Rule of the Local and the Rise of the Botanists

Four

Loja's Bark Collectors, the King's Pharmacists, and the Search for the Best Bark

In January 1773, the Royal Pharmacy in Madrid received a shipment of quina from the royal reserve in Loja. It had been twenty years since Miguel de Santisteban had organized the first shipment of quina from Loja to Madrid. In the intervening decades, the royal reserve had encountered significant obstacles in its efforts to supply the Spanish Crown with cinchona bark. The outbreak of the Seven Years War (1756–1763) in the Atlantic World had led to sporadic shipments. Meanwhile, the bark that did cross the Atlantic to Madrid was generally poor quality. In 1768, José Diguja, the president of Quito, acting on orders from the viceroy of New Granada, revamped the organization of the royal reserve in Loja by appointing Pedro de Valdivieso, an experienced bark trader from a prominent creole merchant family in Loja, to serve as the new director. The shipment that arrived in Madrid in 1773 was only the second one produced under Valdivieso's directorship.

The royal pharmacists' first task was to assess the quality of the quina in the shipment. One concern was that the bark might have suffered physical degradation, which invariably meant loss of medical efficacy, during its long journey from Loja. Another concern was fraud. In the regular quina trade, bark collectors and traders had a reputation for mixing their cinchona bark with other tree barks in order to increase the volume of their harvests and shipments. The pharmacists had to be sure that this quina was of the highest quality as it was to be used on the members of the royal family, given as gifts to foreign and domestic notables, and distributed to royal and military hospitals in the name of the Crown. Ultimately, the inspectors at the Royal Pharmacy determined that the entire shipment—over 15,000 pounds of bark—was useless for royal purposes.[1]

An investigation ensued and officials in Madrid decided that the main problem in this case was a problem of knowledge.[2] Valdivieso and his bark collectors in Loja simply did not know how to identify and properly harvest quina of the highest quality. The royal pharmacists offered to solve this problem by sharing their knowledge with the bark collectors. "Since the King is the Lord and Owner of this Plant," the royal pharmacists explained, "no species of quina other than that which is the most noble by its Nature, should be sent to His Majesty."[3] To this end, they developed a set of instructions on how to "collect, dry, package and transport quina."[4] These instructions were sent to Loja's bark collectors along with a sample of what the royal pharmacists considered the best bark. With this act, the royal pharmacists asserted the superiority of their knowledge and expertise over that of bark collectors in Loja. Just as the Crown claimed the right to receive annual shipments of quina from Loja, the Royal Pharmacy now claimed the right to determine what counted as the best bark.

The interactions between Royal Pharmacy and the royal reserve before and after the contentious 1773 shipment illuminate the politics of knowledge, specifically knowledge of the natural world, in the Spanish Empire that emerged in late eighteenth century. This episode shows how officials and pharmacists in Madrid embraced an imperialist epistemology that took hold in Spain and much of Europe during the Enlightenment.[5] According to this epistemology, knowledge was the product of a hierarchical division of labor between colonial peripheries, in this case Loja, and imperial centers, in this case Madrid. Officials and informants in colonial territories were to provide empirical observations and specimens of natural phenomena, while officials and experts in Europe were to analyze these materials to produce true knowledge or, in some cases, to distinguish the useful knowledge from the irrelevant and the irrational. This imperial epistemology devalued local knowledge in favor of European scientific and medical knowledge that claimed to be universal. In theory, knowledge was to radiate throughout empires from European imperial centers, like Madrid, eradicating ignorance and illuminating the peoples and places in colonial peripheries. It was this imperial vision of the production and application of knowledge that informed the Royal Pharmacy's decision to send instructions and a bark sample to Loja in responses to the 1773 shipment. Moreover, as the Royal Pharmacy accumulated bark samples from various regions of Spanish South America, the royal pharmacists acquired the synoptic view of the empire's quina resources and, with this perspective, they claimed the knowledge and authority to determine which barks were best.

This episode also shows how Director Valdivieso and bark collectors of the royal reserve in Loja challenged the authority of the Royal Pharmacy and put the underlying imperial epistemology into question. Valdivieso and his bark collectors did so by asserting the superiority of their local knowledge and experience. Their efforts highlighted the fragility of the claims of Spanish pharmacists, who faced the daunting task of reforming collection practices in Loja through the mechanism of sending instructions and a bark sample without sending one of their own pharmacists to oversee the implementation of the reforms. Recent historical scholarship has shown that such challenges to European knowledge from creole elites were common in the eighteenth-century Americas from Thomas Jefferson in Virginia to José Alzate in Mexico and Hipólito Unanue in Peru.[6] What is unique about this episode is that the contestation of European science and the associated imperial epistemology took place not in the context of printed texts but in the institutional context of colonial governance. Such evidence shows how the tensions of empire, between local and imperial interests, had a significant epistemological component as pharmacists and bark collectors debated not only which bark was best but also whose knowledge of quina was best.

This episode also demonstrates one way in which the archival records of colonial governments can enrich our understanding of the social history of knowledge in the early modern world by showing how local officials (Valdivieso) and even skilled laborers (bark collectors) could engage in the process of knowledge production as much as pharmacists and *philosophes*.

In contrast to preceding chapters that have explored Andean, commercial, and imperial networks of knowledge production individually, this chapter explores the tensions that existed and emerged between these different ways of knowing quina as they interacted in the context of the epistemic culture of the Spanish colonial government. It argues that the relationship between the Royal Pharmacy in Madrid and the royal reserve in Loja was not one between European center and colonial periphery, as many imperialists imagined, but between two centers of knowledge production—Andean forest and European pharmacy.[7] The groups at each of these centers claimed the authority to speak about nature in different ways. For the royal pharmacists, their association with the Crown and location in Madrid, the imperial capital, provided them with unique access to samples of quina and other botanical materia medica collected throughout South America via the imperial bureaucracy. For the bark collectors of Loja, their authority derived from several different sources: firsthand experience with the tree in its native habitat and a tradition of healing and plant knowledge in the Andean highlands, as well as new methods of testing different kinds of bark and references to printed texts from Europe. What connected these two sites—forest and pharmacy—was the colonial government, and it was via the bureaucratic networks of the colonial government that documents and bark samples traveled across the Atlantic.

The Royal Pharmacy as an Imperial Institution

In order to appreciate the nature of the conflict that arose in the wake of the 1773 shipment, we must first understand the role of the Royal Pharmacy within the broader Spanish imperial enterprise. At the start of the eighteenth century, the Royal Pharmacy would have seemed an unlikely candidate to become an imperial institution. After all, the pharmacy's primary functions were to produce medicaments for the "use and enjoyment of the King and his family" and to distribute medicaments to the other members of Royal Household as well as to "the poor and persons in need."[8] Located at the Royal Palace in Madrid, the pharmacy was not an autonomous royal institution such as the Royal Botanical Garden (est. 1755) or the Royal Cabinet of Natural History (est. 1772), but part of the Royal Household.[9] The royal pharmacists answered directly to the chamberlain of the Royal Household, a largely honorific post, who administered the pharmacy in consultation with the most senior chamber physician in the Royal Household.[10] As a result, the chamberlain served as an important mediator between the Royal Pharmacy and the Spanish colonial government.

In the early decades of the eighteenth century, the Crown expanded the purview of the Royal Pharmacy to include the study of the medicinal properties of plants, animals, and minerals from the Americas. Archival records of the Royal Pharmacy and the Spanish colonial government provide ample confirmation of the pharmacy's new role. For example, the "Miscellaneous" section of the Archivo General de Indias in Seville, the main colonial archive in Spain, contains a *legajo* (a large bundle of documents) labeled "Submissions of medicinal species to the Royal Pharmacy."[11] This legajo contains records of hundreds, if not thousands, of reports and samples of medicaments submitted by officials in Spanish America to the Royal Pharmacy between 1736 and 1784. Such records provide clear testimony of the Royal Pharmacy's direct ties to the Spanish imperial enterprise throughout the eighteenth century.

The Royal Pharmacy's transformation into an imperial institution occurred in the context of imperial reform in which American nature took on new importance as a source of natural resources to be exploited exclusively by Spain.[12] In the early 1740s, José Campillo y Cosío (1695–1743), a Spanish political economist, government official and reformer, articulated a vision of how the Crown might organize the study of natural phenomena from Spanish America. Campillo y Cosío suggested that officials in the America "ought to send samples to Spain of anything contained in these vast countries, be it herb, bush, root, tree, fruit, resin, mineral, rock, etc., that is known to have some special virtue for health, taste or other uses through established tradition and confirmation by experience." He added that "skillful Chemists" should examine these samples in order to "identify any object that would prove to be as useful to the Monarchy, as spices are to the Republic of Holland."[13] Campillo's plan made science integral to the Spanish Empire's efforts to profit from the natural resources of its American territories. Notably, systematic study of these natural resources was to take place in Spain not Spanish America. As a result just as in the mercantilist vision of how to organize the imperial economy, officials in Spanish America were to provide Spanish "chemists" with the raw materials from which to produce new knowledge of the natural world. In the eighteenth century, the Royal Pharmacy was one of the institutions that embraced this new vision of science and empire.

The establishment of the royal reserve of quina in Loja in 1751 only strengthened the Royal Pharmacy's role as an institution of imperial science. After the Crown ordered officials in South America to send information and samples of all known varieties of quina throughout the empire, most of this material ended up at the Royal Pharmacy alongside the annual bark shipments that started to arrive from Loja. As it acquired more and more samples of quina and other potential materia medica from the Americas, the pharmacy came to function as a "center of calculation," an institution that acquires knowledge and power through a unique accumulation of texts, objects, images, and other "data" on natural phenomena from a vast, oftentimes global, network of correspondents.[14] In this way, the Royal Pharmacy was similar to many other scientific institutions in eighteenth-century Europe, such as museums of natural history and botanical gardens, where ties to commercial and imperial enterprises led to an influx of information on natural phenomena from around the globe.[15] And just as these scientific institutions could use their panoramic collections of natural phenomena to assert the authority to make universal claims about the whole of nature, the Royal Pharmacy increasingly could use its panoramic collection of barks to assert its authority to make claims about quina.

What was the significance of the Royal Pharmacy's assessments of bark samples in the broader context of Spanish imperial rule? First, the royal pharmacists played a vital role in the Crown's efforts to exploit cinchona bark because they were the ones that inspected all bark shipments and determined if the bark was of sufficient quality for royal purposes. At the same time, the Royal Pharmacy's evaluations of the bark had much broader significance. After all, the royal pharmacists were in the unique position to compare and contrast samples from new sources of quina with the samples in their existing collection of cinchona bark. By setting samples side by side, the pharmacists could decide whether the new quina was equivalent to quina from Loja, the gold standard in the bark trade. Many officials in South America understood this fact and recognized the utility of a favorable review from the Royal Pharmacy. For example, in June 1753, the viceroy of New Granada sent samples of quina from Cuenca, a region just north of Loja, and expressed hope that examination of the bark's "virtues" in Madrid would "dispel the preference for [quina] from the Mountains of Caxanuma near Loja."[16] Although the royal pharmacists were not the only ones in the Atlantic World judging different kinds of quina, their assessment mattered more than most. As an institution of imperial science with close ties to the Crown, the Royal Pharmacy exerted considerable influence over the value assigned to different kinds of quina in the Spanish Atlantic World. Just imagine what it would have meant to a merchant or a bark collector to say that their quina was the same quina used by the king.

The Royal Pharmacy may have had considerable influence, but it was not all-powerful. For example, the royal pharmacists depended on bark collectors and local officials in the quina-producing regions of New Granada and Peru to supply samples of their barks. Just as these samples contributed significantly to the Royal Pharmacy's knowledge of cinchona bark, the mobilization of officials in Spain's American territories was crucial. Unlike other Spanish institutions of imperial science, such as the various botanical expeditions sent to Spanish America later in the eighteenth century, the Royal Pharmacy did not send its own agents to collect and observe American nature in situ.[17] Instead, the pharmacists waited for quina and other medicaments from Spanish America to come to them via the colonial bureaucracy. Moreover, when it came to making changes in the harvesting, processing, or packaging of the bark at the royal reserve in Loja, the royal pharmacists could make recommendations but they relied on the cooperation of bark collectors and local officials to implement the changes. Although the royal pharmacists generally viewed bark collectors as their subordinates, they had little direct control over the royal reserve and its activities.

The Production of Knowledge at the Royal Reserve in Loja

Officials and bark collectors at the royal reserve in Loja recognized the utility of working with the Royal Pharmacy in Madrid. In most cases, it was in their interest to do so. Nonetheless, the royal reserve in Loja remained an important center for knowledge about the cinchona tree and its bark. The early career of Pedro de Valdivieso y Torres, director of the royal reserve from 1768 to 1784, shows how. In addition, his early career also shows that the relationship between the royal reserve and the Royal Pharmacy could be a collaborative one and was not, by default, antagonistic.

As the longest serving director of the royal reserve, Valdivieso became a key figure in the institution and shaped its development in important ways. His involvement with the royal reserve began in 1768, when President Diguja appointed him to the newly created position of "Magistrate of the Forests" (*Juez Privativo de los Montes*) in Loja.[18] The magistrate's primary function was to oversee the production of the annual shipments of quina for the Royal Pharmacy. The creation of this position was a major change from past practice, in which the *corregidor* (royal governor) of Loja administered the royal monopoly. Diguja took this action after one corregidor, Manuel Daza y Fominaya (r. 1756–1770), proved especially inept at making sure the monopoly's shipments contained only the best bark.[19] Valdivieso's appointment represented another shift from past practice. Whereas the corregidores of Loja were royally appointed officials that came from Spain, Valdivieso was a local. He had grown up in Loja and was a member of a prominent creole merchant family involved in the quina trade among other commercial activities. Even though he attained the post as a result of his predecessor's incompetence, Valdivieso's appointment as Magistrate of the Forests represented at least some affirmation of the value of local knowledge by the Spanish colonial government.

Valdivieso understood the importance of knowledge to his success with the royal reserve. It is not surprising then that one of his first acts as Magistrate of the Forests was to send eight "explorers," bark collectors whom he described as "most experienced and knowledgeable," into the forests of Loja to survey the existing stands of cinchona trees.[20] In this way, Valdivieso emulated the efforts of the Crown and the Ministry of the Indies to survey of all sources of quina throughout the empire. Upon their return, the explorers were unanimous in their reports of the scarcity of cinchona trees and of evidence that overharvesting was to blame with several reporting that many stands of cinchona trees had been "clear cut."[21] By appropriating the techniques of the epistemic culture of the colonial government, Valdivieso had gained valuable knowledge. He acted on this knowledge almost immediately by taking two courses of action. First, in December 1768, just two months after taking over the royal reserve, Valdivieso issued a moratorium on all harvesting of quina from Loja's forests.[22] Next, he began searching for alternate sources of bark.

Valdivieso's survey of Loja's forests shows how the royal reserve could function as a "center of calculation." It was a place where knowledge of the natural world accumulated just at it did at the Royal Pharmacy, albeit on a more regional scale. While the epistemic culture of colonial government certainly facilitated such activities, we must also remember that Loja had been an important center of medical knowledge in the Andean World long before it had become part of the Spanish Empire. As a result, Loja's centrality to the production of cinchona bark and knowledge about the bark was as much a reflection of preexisting patterns of knowledge production as it was a reflection of imperial interests.

THE CASE OF COSTRÓN

Another episode from Valdivieso's first year as Magistrate of the Forests further highlights Loja's role as a center of knowledge production. It has to do with the search for alternate sources of cinchona bark. In February 1769, President Diguja wrote to Valdivieso seeking his opinion on *costrón* (quina made from thick branches). Even though bark traders generally disparaged this type of bark, Diguja wanted to know if Valdivieso would consider costrón suitable for the royal reserve.[23] The president explained that he had received reports of indigenous healers using costrón therapeutically and he wanted Valdivieso to investigate the matter further. If his inquiries showed that this type of bark might be sufficient for royal purposes, Valdivieso was to send samples to Madrid "so that the Royal Pharmacy may examine [the bark] and determine whether its effects are advantageous."[24]

Valdivieso responded to the president's request a month later. From his post at the royal reserve in Loja, the Magistrate of the Forests brought three different sources of information to bear on the question of costrón: indigenous practice, printed texts from Europe, and direct observation. Valdivieso first rejected the president's suggestion that indigenous healers used costrón, explaining that indigenous people "ignore the efficacy of this prodigious plant" (a common misperception at the time).[25] He then noted that European authors reported using the bark with good effects. In particular, Valdivieso referenced the work of the "illustrious" Benito Jerónimo Feijóo (1676–1764), a Benedictine monk and major figure in the Spanish Enlightenment. When it came time to perform therapeutic tests with costrón, it was on Feijóo's "authority" that Valdivieso "made use of this specific with all confidence."[26] While Feijóo did discuss quina in two of the letters in his *Cartas Eruditas y Curiosas* (Curious and learned letters), the Spanish monk did not indicate whether the quina in question was costrón or not. So, why did Valdivieso reference Feijóo's work? It may have simply been a shrewd move to demonstrate his familiarity with published European accounts of quina as a means to bolster his authority in the eyes of his potential readers—not just the president of Quito but also possibly officials and pharmacists in Madrid.

In the remainder of the report, Valdivieso addressed President Diguja's specific inquiry about costrón by trying to answer a more general question: Does the thickness of the bark have an effect on its medical efficacy? Valdivieso noted that merchants generally valued thinner barks more than thicker barks. But did the preference of merchants reflect the truth of matter? Was thin bark better? Valdivieso thought not. He pointed out that, a half century earlier, bark collectors and merchants had spurned thin barks from the branches in favor of quina that came from the "body" of the tree.[27] Around 1750, Valdivieso continued, merchants and pharmacists developed a preference for *canutillo* (thin bark from smaller branches) not because they had discovered that it was better but because the "corpulent trees" that produced thicker barks had been destroyed by "continual extraction."[28] In Valdivieso's view, such reversals showed that trends in the quina trade were no guide to the truth.

Experience was the only true guide and, according to his own experience administering the bark therapeutically for over twenty years, Valdivieso concluded that bark thickness had little effect on medical efficacy. Though he did successfully treat three "sick people" in Quito with costrón, Valdivieso also reported that, in his experience, patients obtained "prompt relief" of their symptoms from bark that "was neither very thick nor very thin"[29] As further evidence, Valdivieso referred to the practices of other healers. "Much variation exists," he wrote, "among the Physicians of Europe in the method, quantity, and schedule of administering the bark."[30] A similar situation existed in Loja. "Some," he explained, "put it into an infusion of quince syrup; others in virgin honey with some sour orange; others in *aguardiente*; others in wine and others in water."[31] In light of such variation in therapeutic practice, it seemed unlikely that one could even determine if the thickness of the bark had an effect on efficacy. Ultimately, Valdivieso endorsed the use of costrón for the royal reserve. "I am convinced," he wrote to President Diguja, "that the kinds [of bark] requested by Your Excellency will be approved by the Physicians of His Majesty."[32]

Valdivieso employed one additional technique to make his case: bark samples. Along with his report to the president, he submitted a small box containing three types of quina from the same "species" of tree: costrón or bark "from the trunk of a mature tree"; bark "from the thick branches of a mature tree or the trunk of medium maturity"; and samples of "very young plants" that did not produce bark but were often ground up and mixed with wood from cinchona trees. Valdivieso provided these samples in order to demonstrate a further point: although it mattered little from which part of the tree the bark was taken, the differences in the efficacy of different cinchona bark arose primarily from "the greater or lesser maturity of the plant."[33] Another goal was simply "to better inform" the president "about the classes of quina and the state of the Forests."[34] This box of samples was an embodiment of the knowledge and expertise of Valdivieso and his bark collectors.

Valdivieso sent another box of bark samples to the Royal Pharmacy, apparently on his own initiative. Box sixty-four of his first shipment to the Royal Pharmacy, which departed Loja in October 1769, contained samples of three other kinds of quina: *resaque* (bark that regrew in places where cinchona trees had been stripped of their bark), *cortezon* (another name for costrón) from trees that produced yellow quina and red quina, and a novel variety of cinchona bark known as *crespilla*.[35] For each of these, Valdivieso wanted the pharmacists to determine if they were of sufficient quality for royal purposes. He also hoped that testing of crespilla at the Royal Pharmacy would help to determine if this bark "loses its activity upon crossing the Line [i.e., the equator]."[36] Finally, he included samples of powdered quina packaged in tubes of bamboo (*canutos de Guadua*) to protect against humidity, and asked the royal pharmacists to determine if this bark had "equal activity" to bark that arrived whole, as most bark did.[37]

There are no records neither of the Royal Pharmacy's assessment of Valdivieso's box of samples nor of the Royal Pharmacy's final determination on the utility of costrón for the royal monopoly of quina.[38] Nonetheless, all these activities highlight the sophistication with which Valdivieso approached the study of the cinchona tree and its bark. It was a level of sophistication rivaling that of the Royal Pharmacy. Valdivieso relied on a variety of sources of information to strengthen the authority of his claims about cinchona bark, including firsthand experience, testimony from "experienced and knowledgeable" bark collectors, references to printed medical and philosophical texts from Spain, and the submission of bark samples. At first glance, it seems remarkable that Valdivieso's practices mirror those of the Royal Pharmacy so closely, until we recognize that both these inistitutions—the royal reserve in Loja and the Royal Pharmacy in Madrid—were connected to the epistemic culture of the colonial government, which provided a common set of values and techniques not just for producing knowledge but also for establishing the authority of that knowledge. Of course, Valdivieso had the good fortune of being located at one of the epicenters of bark production as well as a center of Andean medical and botanical knowledge. At the same time, he proved remarkably adept at understanding and navigating the politics of knowledge in the epistemic culture of empire. By 1773, when the Royal Pharmacy decided to send its instructions and bark sample to Loja, Valdivieso was already established as an expert in cinchona bark. Moreover, when he received these materials from the Royal Pharmacy, he did not see them as commandments to be blindly followed, but as evidence to be evaluated critically.

The Contentious Shipment of 1773

These earlier interactions provide the context for understanding the Royal Pharmacy's efforts to exert greater control over the royal monopoly of quina in the wake of the 1773 shipment, in which pharmacists found that none of the bark was suitable for royal purposes. In the preceding years, both forest and pharmacy had emerged as important centers of knowledge production and had proven their utility to Spain's imperial enterprise. Nonetheless, when ninety-four boxes arrived in Madrid and none of the bark inside was deemed suitable for royal purposes, royal officials knew something was wrong. This shipment turned out to be quite contentious and it proved an important test of the relationship between the Royal Pharmacy and the royal reserve. It was also a test of whether the authority and expertise of the local or the imperial would prevail. Ultimately, the rift between bark collectors in Loja and pharmacists in Madrid shows that while quina traveled quite easily across the Atlantic, knowledge of the bark did not. In the forest and in the pharmacy, quina was understood in remarkably different ways according to the broader contexts in which these two institutions—the royal reserve and the Royal Pharmacy—operated.

AN INSPECTION OF BARK IN MADRID

Existing documents give few specific details of the methods and standards whereby the Royal Pharmacy evaluated quina. Records of the examination of the 1773 shipment show that the royal pharmacists placed great emphasis on the physical characteristics of the bark as an indicator of its medical efficacy. Consider, for example, that bark from this shipment was tested in hospitals in May 1773 three months after the pharmacists had dubbed it useless according to its physical characteristics.[39] By that time, the pharmacy's instructions and sample were well on their way to Loja. Therapeutic trials, in fact, revealed that much of the bark was medically efficacious. However, the Crown never informed colonial officials of these results, nor did it rescind the order to collect bark according to the pharmacy's sample and instructions. Upon learning that the bark of the 1773 shipment was efficacious, the head pharmacist José Martínez Toledano remarked that the bark "should not be sent to Foreign Courts because it lacks [the proper] color and they will spurn it at a glance."[40] At the Royal Pharmacy, and in the gift economy of European nobles and royalty, quina had to be medically efficacious but it was physical characteristics—such as flavor, odor, color, and thickness—that distinguished the very best bark.

This emphasis on the bark's physical characteristics can be explained by looking more closely at how the Royal Pharmacy used quina. Upon receiving a shipment, the pharmacists separated the bark into one of four classes. These classes, from best to worst, were designated as follows: "first class" (*primera suerte*) was for the Royal Family, "second class" (*segunda suerte*) was to serve as gifts to notables and foreign dignitaries, "third class" (*tercera suerte*) was donated to Spanish hospitals, and "fourth class" (*cuarta suerte*) was either burned or sold as a dyestuff.[41] The Royal Pharmacy required superior bark that was both aesthetically pleasing and medically potent, as well as "noble by nature," because the lives of the Royal Family and the King's reputation as a gift-giver depended on the quality of the pharmacy's quina.

Although pharmacists and physicians correlated the bitterness of the bark with its medical efficacy, in practice the pharmacists did not treat these criteria as sufficient to distinguish "noble" from "common" quina. Many seem to have assumed that all quina had some medical virtue. Consequently, the relative availability of bark with certain physical characteristics became an important factor in Spanish conceptions of what separated superior from inferior bark. Many could acquire efficacious quina, especially since the King donated much of the Royal Pharmacy's bark to Spanish hospitals, but it would have been a mark of distinction to receive bark from the Crown that was pretty as well as potent. Ultimately, in this culture of the court and royal gift economy, the quality of quina derived from observable physical properties to the exclusion of other aspects.[42] In addition, rarity of physical characteristics became that which distinguished noble from common bark at the court in Madrid. In this context, royal pharmacists served, in effect, as connoisseurs of quina.[43] They had to not only have knowledge of the bark, but also have knowledge of the prevailing "tastes" for the bark among the members of European court culture and the royal gift economy.

We do not know exactly what the sample looked like that the Royal Pharmacy sent to Loja in 1773. As for the accompanying set of instructions, they focused mainly on techniques for protecting the bark from humidity and gave no verbal description of the physical characteristics of superior quality quina.[44] This suggests that the pharmacists thought that the sample spoke for itself. As we will see, this was a false assumption. The instructions do reveal that the pharmacists considered the sample an example of superior quina. They explained that bark collectors should "select Quina according to the sample that is in the accompanying box to the exclusion of any other species or quality of bark."[45] Upon receiving the pharmacy's materials late in the summer of 1773, President Diguja forwarded them to Pedro de Valdivieso in Loja. In addition, Diguja ordered Valdivieso to examine the pharmacy's sample and "throw out all other bark that has been collected which is not of the desired type."[46]

AN EXAMINATION OF BARK IN LOJA

Valdivieso received Diguja's letter as well as the pharmacy's materials on 16 September 1773.[47] The time and place could not have been more opportune. Valdivieso was at that "House of His Majesty for the receipt and packing of [quina]" in Malacatos, a small village east of Loja, where he went each year to receive quina gathered over the summer by his bark collectors in order to prepare the next shipment for the Royal Pharmacy (figure 6).[48] Consequently, when the pharmacy's sample arrived, it was reviewed not only by Valdivieso but also by twenty to thirty experts on the cinchona tree and its bark.[49] Valdivieso later testified to opening the box "in the presence of various white men and many natives, all with experience, who gathered around on the occasion of having come to deliver their respective *arrobas* of *cascarilla*."[50]

Valdivieso took testimony from seven of the examiners ranging from an indigenous laborer to a hacienda owner.[51] He probably did so because they had made the surprising discovery that the Royal Pharmacy's sample was not from Loja. All examiners noted that the durability, color, and taste of the pharmacy's sample differed from the bark from Loja as in the case of Pedro Cevallos, an indigenous laborer, who reported that he "detected a hint of acidity in the bitterness of the pharmacy's sample [while quina from Loja has] a pure and constant bitterness that lasted longer."[52] Both Cevallos and Mathias de Salazar, a resident of Loja and owner of an hacienda in Vilcabamba, noted that the Royal Pharmacy's sample was "harder" than the bark from Loja.[53] Two additional witnesses observed that a direct comparison of the sample from Madrid with fresh bark from Loja revealed several differences between the two samples.[54] Ultimately, these testimonials supported the claim that the Royal Pharmacy's sample was not from Loja.[55]

Meanwhile, several examiners, who had experience harvesting cinchona bark in the neighboring Province of Jaen (figure 1), were able to identify the pharmacy's sample as a product of Jaen.[56] For example, Antonio Blanco de Alvarado, a forty-eight-year-old resident of Loja, stated that the sample was "different from that collected in the hills of Uritusinga and Cajanuma and very similar to that collected in the Hills of the Province of Jaen."[57] Alvarado was certainly qualified to tell the difference. In addition to thirty years of experience as a bark collector, he had harvested quina in Jaen in 1770—the year that Valdivieso sent his second shipment, the "corrupted" shipment that arrived in Madrid in 1773. Two other witnesses, Alexandro Toledo and Juan de Aguirre de Dicastillo, also had experience working with *cascarilla* from Jaen and corroborated Alvarado's testimony on the origins of the Royal Pharmacy's sample.[58]

This finding presented a problem for Valdivieso. After all, the objective was to provide the Crown with the best bark but, as Valdivieso and his bark collectors knew, the best quina came from Loja. In contrast to the royal pharmacists, experts in Loja used the bark's physical characteristics as clues to its geographical origin—the main indicator of quality for those involved in the commercial quina trade. Ironically, though this sample represented the Royal Pharmacy's attempt to impose their knowledge on bark collectors, it had the

Figure 6. Hand-painted map of the Loja Province depicting bark collectors and cinchona trees labeled as "cascarilla" (c. 1769). España. Ministerio de Educación, Cultura y Deporte. Archivo General de Indias. MP-Panamá 179.

opposite effect of putting the expertise and knowledge of the pharmacists into question.

BUILDING CONSENSUS IN LOJA AND BEYOND

Although he trusted the results of the examination of the sample at Mala-catos, Valdivieso was not ready to openly refute the pharmacy. Instead, he asked the Town Council of Loja to examine the sample. Yet, unlike the bark collectors who had experience harvesting quina, the council members were less familiar with the different kinds and qualities of bark. Eager to enlist their support, Valdivieso made things easier for council members by providing samples of bark from Jaen and from Loja for comparison with the pharmacy's sample.[59] The Town Council confirmed the findings of Valdivieso and his bark collectors. A group of "experienced deputies" on the council observed "a nota-ble difference" between the quina from Loja and the Royal Pharmacy's sam-ple.[60] They also confirmed that Valdivieso's samples of bark from Loja were of "superior quality"—an affirmation of Valdivieso's ability to identify the best bark. The next day, the Town Council issued a statement that questioned the "reason or rule" used by the "Doctors and Employees of the Royal Pharmacy" in selecting their bark sample.[61]

Valdivieso next sought to convince President Diguja by sending a box of nine samples—the first five were from Jaen, the sixth was the Royal Pharmacy's sample, the seventh was bark from Cuenca, another important quina-producing region near Loja, and samples eight and nine were from Loja. In effect, Valdivieso created a miniature and mobile natural history collection of cin-chona bark. Note the strategic arrangement of the samples with the Royal Pharmacy's sample sandwiched between those from Jaen and from Loja to promote comparison. In the accompanying letter, Valdivieso wrote: "I should inform Your Lordship that obedience is indispensable to the execution of the Royal Will. However, the Instruction from the Individuals of the Royal Phar-macy for the selection of Quina lacks the full knowledge of its qualities and nature that comes from being able to examine [the tree] and from longstand-ing experience of knowing how to distinguish which of this species have been spurned for decades as much by natives as by merchants."[62] Here, Valdivieso challenged the Royal Pharmacy's authority and rejected their knowledge by implying that it was, in fact, the royal pharmacists who could not distinguish superior from inferior quina.[63]

In Quito, President Diguja assembled his own group of experts to exam-ine Valdivieso's bark samples including a former corregidor of Loja, a former governor of Jaen, the prefect of the Royal Charity Hospital in Quito, and a surgeon. This group concurred with previous examinations and described the pharmacy's sample as "that inferior quality bark which the Territories of Jaen and Guancabamba produce."[64] Upon inspecting a different box of bark from Loja sent by Valdivieso, the commissioners noted that the quina in this box "was found to be the best of that which is known to be harvested in Loja." This bark from Loja, they continued, was "different in every respect from that which was examined from the little box that was sent by the Royal Pharmacy."[65]

Diguja was convinced. On 20 December 1773, just three months after ordering Valdivieso to collect only the bark matched the pharmacy's sample, he informed the Minister of the Indies in Spain that "the Quina which has been selected by the Royal Pharmacy and [which] I have been ordered to send is [considered] entirely useless and contemptible by those who have knowledge of this specific."[66] He further explained that the pharmacy's sample was "not harvested in the Forests of Loja but in those of Jaen de Bracamoros, Piura and other [places] like these that produce quina that is not of superior quality." "Since such [bark] does not have value," Diguja further noted, "merchants make a mixture of that bark with bark from Loja with which they deceive those who have little knowledge of this specific."[67] Diguja's implication was clear. Not only did the Royal Pharmacy lack sufficient knowledge of quina to make determinations with regard to its quality, but the royal pharmacists had been duped by an imposter bark.[68] Along with his letter, Diguja sent a box of bark samples to the Minister of the Indies and encouraged him to see for himself that the pharmacy's bark was "inferior, stale, and not from Loja."[69]

Prior to 1773, the movement of cinchona bark in the Atlantic World was largely unidirectional from west to east, New World to Old, as merchants and the Spanish Crown extracted large quantities for consumption by feverish Europeans, while officials in South America sent samples of quina to the Royal Pharmacy for examination. In 1773, the Royal Pharmacy sent a bark sample against this current back to South America in an attempt to extend its knowledge and expertise to Loja and assert its authority over the royal reserve. This action was significant because it was one of the earliest attempts by the Spanish Crown to impose a standard developed in Madrid on a botanical commodity produced in Spanish America. Officials in Madrid thought that the Royal Pharmacy had the authority to develop and impose empirewide standards for cinchona bark because in the preceding decades the royal pharmacists had examined several shipments of bark from Loja as well as bark samples from cinchona trees throughout South America. In addition, these officials operated according to an imperialist epistemology in which knowledge of the natural world possessed and produced by scientific and medical practitioners in Europe was considered superior to the knowledge of nature possessed by officials, merchants, and laborers in the Americas.[70]

The only problem was that the royal pharmacists' knowledge did not travel with the bark. When it arrived in Loja, Valdivieso and his bark collectors created knowledge of the sample anew. Additional examinations in Loja and Quito confirmed that the sample was of inferior quality and *not* the best bark, as the Royal Pharmacy claimed. As a result, many officials in the colonial government, including even the president of Quito, doubted if not denied the authority and expertise of the Royal Pharmacy. Thus, Valdivieso and his bark collectors not only rejected the imperial epistemology held by pharmacists and officials in Madrid but also asserted the importance of Loja as a center of knowledge production.

What is remarkable is that pharmacists and bark collectors produced divergent results even though they were examining the same bark sample and used similar comparative techniques of analysis. This episode shows that even though pharmacists and bark collectors operated within the shared epistemic culture of the Spanish colonial government, this larger structure of knowledge production did not determine an outcome in favor of European science. Because the colonial government served as a network connecting the heterogeneous natural and social worlds of the Spanish Empire, differences in local sociocultural contexts mattered. On the one hand, the royal pharmacists assessed and selected their bark in the context of royal gift economy in Madrid. In this context, physical characteristics of the bark were the key to distinguishing superior from inferior bark. On the other hand, Valdivieso and bark collectors in Loja as well as officials in Quito assessed the bark according in the context of the quina trade in which physical characteristics were merely clues to geographical origin—the main determinant of the bark's quality. For the pharmacists, knowing quina in Madrid was one thing, acting on it in Loja was another. It was in action that the fragility of the pharmacy's authority as an institution of imperial science became evident.

More important than the Royal Pharmacy's ineptitude is Valdivieso's relative success at convincing members of the colonial government to trust his claims about the bark over those of the royal pharmacists. Here, he took advantage of two key features of the epistemic culture of empire. First, he used the emphasis on experience and empirical observation to his full advantage by circulating bark samples to other colonial officials and inviting them to see for themselves that the Royal Pharmacy's bark was not superior quality. Second, he recognized the importance of winning allies in the hierarchy of the colonial government, which he also achieved by circulating bark samples. Through his bark samples, Valdivieso established or strengthened his ties with the Town Council of Loja and the president of Quito. He was also able to convince them that he and his bark collectors were right about the Royal Pharmacy's sample. As a result, when the president of Quito sent Valdivieso's assessment of the Royal Pharmacy's sample to Spain, it arrived with the endorsement of several individuals in or associated with the institutions of imperial governance in Quito.

In addition to highlighting the dynamics of the epistemic culture of the Spanish colonial government, this episode is significant because it offers insight into the challenge of applying scientific and medical knowledge from Europe in the colonial contexts of the early modern Atlantic World. Recent studies of science and medicine in the Atlantic World have focused primarily on the production of knowledge about nature rather than its application.[71] Building on the insights of the early work of Bruno Latour, such studies have shown how European science and medicine gained the authority to make claims about the natural world through the circulation and accumulation of specimens and other data on natural phenomena on a global scale.[72] Several studies have shown how cultural assumptions and contingencies could short-circuit the production of knowledge. For example, in *Plants and Empire,* Londa Schiebinger has emphasized the ways in which the gendered interests and cultural preconceptions of (predominantly male) European naturalists inhibited the transmission of knowledge about the abortifacient properties of the peacock flower from the Caribbean to Europe.[73] In his studies of the French-Spanish expedition to Quito in the 1730s, Neil Safier has highlighted a number of cases in which the contingencies of travel and communications in the eighteenth-century Atlantic World thwarted the sending of reports and specimens or led to the omission of knowledge as information was transformed from manuscript descriptions to printed text and images.[74] While movement may have been a defining feature of the Atlantic World, it did not guarantee the production of knowledge.

Nor was the circulation of knowledge a guarantee of its application or implementation. In this case, however, it was not contingency that thwarted European knowledge. Instead, it was the presence of a preexisting tradition of medical and botanical knowledge in the Andean World. Anthropological and ethnobotanical studies have shown that Loja is situated along an "axis of health"—a region of the Andean World with a robust tradition of medical theory and practice. Evidence suggests that healers were just as active in this region throughout the eighteenth century. Valdivieso, himself, noted his efforts to therapeutically administer the bark and it is likely that several of his bark collectors were also *curanderos* (healers) because they would have had the knowledge and expertise to locate cinchona trees as well as to harvest and prepare their bark. From this perspective, the Royal Pharmacy appears as a distant periphery to an Andean center of medical knowledge. Ultimately, the circulation and study of bark samples in Spanish Atlantic World gave rise to a network of knowledge with multiple centers—including Loja and Madrid—

each with their local experts that could claim the authority to speak for the cinchona tree and its bark. In terms of understanding the limits of European science and medicine, this episode suggests that these enterprises had limited practical application beyond the museum, the library, the botanical garden, and the printed page in the late eighteenth century.

Unfortunately, there are no documents of the crown's reaction to Valdivieso's rejection of the pharmacy's sample. Records from the late 1770s and early 1780s suggest that the problem of corrupted quina shipments ceased. These records also show that Valdivieso continued to harvest quina from the forests of Loja.[75] Valdivieso had prevailed. Yet, at the end of this whole affair, it is likely that pharmacists in Madrid and bark collectors in Loja remained entrenched in their own understandings of the bark.[76] As for the king, he continued to receive his "noble" quina, just not according to the criteria of his Royal Pharmacy. As shipments and bark samples continued to arrive in Madrid, the colonial government came to rely on a larger community of medical experts for the testing of the bark. One of the major consequences of the Crown's interest in quina was the further integration of experts into the Spanish imperial enterprise. As the community of experts expanded in Madrid, a new group— botanists—asserted its authority as the empire's preeminent experts on the natural world. In the wake of the challenge from Valdivieso to the applicability of European knowledge, the Crown would turn to botanists in the hopes that they could achieve what the pharmacists could not.

Five

Botanists as the Empire's
New Experts in Madrid

In the late 1770s, officials involved with the royal reserve of quina became cognizant of a new problem: the increasing scarcity of cinchona trees in the southern sierra of the Audiencia of Quito. It was taking bark collectors longer and longer to collect quina because they had to travel further and further into the forests. Some even returned empty handed. To make matters worse, merchants and bark collectors continued to engage in fraud by mixing quina with other tree barks to meet demand for shipments in the quina trade and for the royal reserve. The increasing scarcity of cinchona trees encouraged such activities, especially as perpetrators faced few consequences.

More than ever, the Spanish colonial government needed experts with the knowledge and experience to locate new stands of cinchona trees and to determine the identity and quality of new sources of quina. But the question remained: Whom to trust with this task? In South America, some officials had begun to question the utility of recommendations from Europe, especially in the wake of Pedro de Valdivieso's experience with the Royal Pharmacy in 1773. Six years later, Miguel García de Cáceres, a customs official at Guayaquil, a port that handled most of the quina exports from the Audiencia of Quito, declared that "experts" in Spain and Europe only offered "vain science" because they erroneously used the bark's physical characteristics to determine its identity and medical efficacy."[1] He made this observation in an updated report on the quina trade submitted to an official from Spain appointed to conduct a *visita* or official review of the Audiencia of Quito. With the utility and authority of European science and medicine in question, officials in quina-producing regions continued to rely on the knowledge and experience of the local bark collectors, merchants, and officials. Such an approach seemed to offer the best prospects for meeting the challenges faced by the royal reserve and the quina trade in general.

In Madrid, the Crown and its advisers were well aware of the challenges facing the royal reserve and the quina trade, but they were unwilling to surrender entirely to local experts in Loja. Nor could they. At the very least, as a matter of quality control, the Crown continued to rely on its own local experts —the royal pharmacists—to review incoming shipments of bark from the royal reserve in Loja. Moreover, as the problem of the scarcity of cinchona trees loomed larger and larger, officials in Madrid needed their own local experts to assess samples of new sources of bark to determine if they were suitable for royal purposes. After all, many officials still believed that the ultimate determination of the identity and quality of new kinds of quina should be made not in the Andes but in Madrid, the imperial capital.

Although the Royal Pharmacy remained integral to the distribution of quina in Spain throughout the 1770s and 1780s, officials in Madrid began to invite other medical experts to give their assessment of the bark produced by the royal reserve. This process led to an expansion of the community of experts on which the Crown relied for knowledge and advice about quina and other natural phenomena. Yet, as the community of experts grew, so did discord as different groups of experts offered divergent assessments of the same bark shipments. With so much at stake in the empire's continued management of this vital natural resource, the Crown and its officials increasingly turned to a new group of experts: botanists. Spanish officials hoped that botany and its practitioners would prove useful not only in providing the certainty that the community of medical experts lacked but also in establishing the authority of European knowledge over local knowledge throughout the empire. Starting in the 1770s, botanists were integrated into the Spanish imperial enterprise as the Crown patronized expeditions and as officials in Madrid increasingly turned to botanists for advice on quina and other botanical matters of interest to the empire.

In general, the new emphasis on botany was in keeping with the Spanish Crown's and imperial reformers' broader efforts to wield science more extensively as a tool of empire in the late eighteenth century. Support for various scientific enterprises was a key element of the reform of the Spanish Empire under the Bourbons, most notably Charles III (r. 1759–1788).[2] Inspired by Enlightenment ideals, botanists and bureaucrats alike praised the utility of the natural sciences to the empire not only for identifying new and useful colonial natural resources but also for bringing prestige to the Crown and Spain.[3] Indeed, most accounts of the relationship between science and empire in Spain and other parts of Europe have tended to focus on the high-profile, patronage-intensive activities such as expeditions, the collection of natural objects for state institutions, and the production of visual and textual representations of natural worlds colonized by Europeans.[4] Such activities were undoubtedly important articulations of the nexus between European sciences and empires in this period. But they were not the only means by which scientific and imperial enterprises became intertwined in the Atlantic World.

This chapter argues that problems such as the scarcity of cinchona trees and assessing the quality of bark, as much as Enlightenment ideals, further propelled the intertwining of botany and empire in the 1770s and 1780s, especially in the wake of the difficulties encountered by the Royal Pharmacy. After all, empires not only produced representations of colonized natural landscapes, but also facilitated the production of colonial commodities. For cases prior to the nineteenth century, little scholarship exists on European attempts to instrumentalize scientific knowledge as part of specific political and economic enterprises in colonized territories.[5] The case of quina thus challenges and enriches current conceptions of imperial science as activity focused primarily on the production of exhibits, collections, images, and printed texts—all means of representing the natural world that often provided powerful justifications for European superiority and imperial rule.[6] This chapter further argues that in the Spanish Atlantic World, the case of quina was not idiosyncratic but instead should be understood as integral to the process whereby practical interests fostered the intertwining of science, especially botany, and empire. The case of quina also shows how botanists assimilated quite easily to the epistemic culture of the Spanish Empire, especially in their role as advisers to officials in the Spanish colonial government. In other words, botanists' involvement with cinchona bark in Spain provided the springboard for the further transformation of botany into a Spanish imperial science.

In order to elucidate the rise of botanists in the expanding community of experts advising the Crown, this chapter first examines the efforts of medical practitioners in Madrid to assess bark shipments from the royal reserve. Disagreements over the medical efficacy of specific samples of quina often led the Crown and its advisers to consult additional groups of experts. Such disagreements also led to the adoption of new techniques for assessing the bark at the Royal Pharmacy. The chapter also focuses on the ways in which the Crown's interest identifying new sources of quina further fostered the relationship between botanists and the colonial government on both sides of the Atlantic. This process is most evident in the careers of Casimiro Gómez

Ortega (1741–1818), who served as director of the Royal Botanical Garden in Madrid, and José Celestino Mutis (1732–1808), who served as director of the Royal Botanical Expedition in New Granada. The experiences of Ortega and Mutis show how practical concerns, as much as Enlightenment ideology, fostered the entanglement of science and empire in the Spanish Atlantic.[7]

Medical Testing in Madrid

Whereas they had served initially as the Crown's only experts on cinchona bark, the royal pharmacists began to share this role with physicians and pharmacists of the royal household, the royal hospitals, and the army in the 1770s. In other words imperial interest in quina fueled an unprecedented expansion of the community of expertise in Spain such that royal officials coordinated the activities of multiple medical institutions in pursuit of the common goal of assessing shipments and samples of cinchona bark. One motivation for this development was pragmatic. The small staff at the Royal Pharmacy simply could not handle all the duties associated with the royal reserve of quina, especially as the annual shipments increased in size and frequency and as the pharmacy received more samples of potentially new sources of the bark. In addition, in the wake of the corrupted shipment of 1773, it became regular practice to test bark from the royal reserve's shipments on patients at the royal hospitals.[8] It is not surprising then that such practices led to disagreements over the identity and efficacy of bark samples. This result, in turn, encouraged officials in Madrid to consult with an even wider community of medical experts. The juxtaposition with the royal reserve in Loja is worth noting. Although Valdivieso consolidated his authority to speak about quina in the wake of the contentious 1773 shipment, royal pharmacists in Madrid saw their authority undermined as the Crown and other officials increasingly sought input and advice from other medical practitioners in the imperial capital.

One of the earliest requests inviting other medical experts to assess quina from the royal reserve came from the Royal Pharmacy itself. In March 1773, José Martínez Toledano, head pharmacist at the Royal Pharmacy, asked physicians at one of the royal hospitals in Madrid to provide a second opinion on bark from the recently arrived shipment that the royal pharmacists had dubbed useless for royal purposes.[9] Martínez Toledano took a similar course of action with a shipment of 160 boxes of quina that arrived at the Royal Pharmacy in 1775. After one of his assistants examined the shipment and determined that the majority of the bark was useless, Martínez Toledano sent some samples to the General Hospital in Madrid "so that they could experiment [with it] and examine its effects."[10] An anonymous report noted that the tests at the General Hospital confirmed the "utility of quina that had previously been classified as useless."[11] In both of these cases, therapeutic testing at the

General Hospital and physical inspection at the Royal Pharmacy produced divergent assessments of the bark in question. These outcomes did not bode well for the royal pharmacists. The anonymous report on the results of the therapeutic testing of the 1775 shipment observed: "An opinion without any experiments [i.e., therapeutic tests] is not enough."[12] This observation is testimony to an emerging view that the Royal Pharmacy's assessments of the bark based on its physical characteristics were no longer sufficient on their own. Therapeutic testing was to become a regular practice in the assessment of the royal reserve's bark, which meant that the royal pharmacists increasingly lost their status as the Crown's preeminent experts on quina.

As the number of medical practitioners involved in assessing the bark increased, tensions and disagreements persisted in the expanding community of experts. Early on, physicians at the royal hospitals, as well as chamber physicians in the royal household, asserted their autonomy from the Royal Pharmacy. Because therapeutic testing involved administering a sample of the bark to a patient (usually a soldier or one of the poor), these physicians asserted their right to refuse to test any samples that were obviously inferior.[13] For example, in 1774, pharmacists and doctors at the General Hospital rejected a bark sample because it lacked "all qualities, requirements, substance, consistency, color and flavor to be classed as good or medium [quality]." They explained that they would not administer quina to their patients without at least some indication that there was "probability of relief."[14] Such a reaction came as a surprise to José Fernández-Miranda Ponce de León (1706–1783), the Duque de Losada, who oversaw the Royal Pharmacy in his position as chamberlain of the Royal Household. In response to the physicians' concerns, Losada noted that the king had given this same quina to "many others in the Kingdom without any complaints of bad effects."[15]

The physicians' objection fueled further investigation of the sample. The Duque de Losada sent an *arroba* (approximately twenty-five pounds) of the same bark to Joseph Lafarga, a chamber physician in the Royal Household, for further "experiments." Lafarga reported, contrary to the opinion of the physicians at the General Hospital, that the bark was of "good quality" and alleviated the effects of "regular tertian fevers."[16] These contradictory assessments drove royal officials to consult additional experts. Thus, at the same time that they became integrated into a growing state apparatus of expertise, physicians at the General Hospital and the Court helped to foster its expansion.

In 1779, physicians at the General Hospital challenged the assessment of the royal pharmacists once again. This time, however, the physicians claimed that their authority trumped that of the Royal Pharmacy. In August, the Crown ordered the Royal Pharmacy to send 400 pounds (16 arrobas) of quina to José Antonio de Rojas Ibarra y Vargas, the Conde de Mora, who was to distribute the bark to additional hospitals. This quina arrived just in time for the annual outbreak of "pernicious and malignant tertian fevers" in late summer and early fall.[17] By September, a junta of physicians at the General Hospital had determined that the bark was useless against these fevers, describing it as "stale and old and, as a result, inert and devoid of its virtue."[18] Inspection of the bark's storage boxes revealed that the word "useless" had been erased from the markings on two of the boxes. In their report, the junta questioned the credibility of the Royal Pharmacy since its pharmacists had approved of the bark.

After receiving news of the junta's unfavorable assessment of the bark, the Conde de Mora arranged to have more samples tested by a different group of medical experts with no connection to the General Hospital. He assembled an eight-member panel that consisted of four physicians and four pharmacists.[19] Though some members of the panel gave a more favorable assessment of the bark, they all agreed that it probably had little effect on the tertian fevers being treated at the General Hospital. The matter was all but decided save one dissenting voice.

At the same time that the Conde de Mora had his panel testing the bark, the Duque de Losada had arranged for another sample to be sent to Alfonso Lope y Torralva, a chamber physician in the Royal Household. Lope y Torralva conducted "experiments" with the bark and reported that it "happily cured" patients suffering from "tertian and quartan fevers some regular, some irregular and other pernicious."[20] Lope y Torralva's report was sent to the Conde de Mora and the junta at the General Hospital.

The junta rejected these findings on three grounds. First, they argued that Lope y Torralva's quina was not the same as the bark that they had been given originally. Second, the junta argued that even in a small portion of predominantly useless bark, one could find some efficacious pieces of bark. Finally, the physicians pointed out that Lope y Torralva's patients were not suffering from the same kind of fevers as the ones at the General Hospital.[21] In the end, the matter was decided in the favor of the junta, especially because this episode pitted "the report of one individual [Lope y Torralva] against the judgment of seventeen professors."[22] In February 1780, the Conde de Mora prohibited distribution of this quina—a clear victory for the physicians at the General Hospital over the royal pharmacists, who had originally approved the bark.[23]

The Duque de Losada, however, was not so easily convinced. During the winter months of 1779 and 1780, the "regular physicians" at the army hospitals were given samples of the quina that the junta at the General Hospital had rejected. After using the bark to treat cases of "quartan fevers [and other] complicated illnesses," these physicians found that, even though the bark was "not of excellent quality or of the most exalted virtue," it still could be distributed as "its quality [had been] experienced and confirmed."[24] In April 1780, Luis Blet, head pharmacist of the Royal Armies, also reported the favorable results to the Duque de Losada and noted that of 1,190 patients in army hospitals, "very few die[d]." It was a major challenge to the findings of the General Hospital's junta.[25] In addition, these results confirmed the Royal Pharmacy's determination that the bark was medically efficacious. It was a good result for the Duque de Losada, who oversaw the Royal Pharmacy and would have wanted to demonstrate to the Crown that his subordinates were executing their charge effectively.

Word had gotten out that the efficacy of the bark from the 1779 shipment was questionable. Consequently, when Pedro de Alcántara Fernández de Híjar y Abarca de Bolea, the Duque de Híjar, solicited some quina from the Royal Pharmacy, he specifically requested bark from a recently arrived shipment from Loja.[26] Híjar cited the rejection of bark from the earlier shipment by the junta at the General Hospital. In response, the Duque de Losada, who handled such requests for the Royal Pharmacy, rebuked Híjar for questioning the authority of the Royal Pharmacy. Losada further noted that a royal order had established that "the most beautiful and colorful pieces [of bark]" were reserved for the Crown and all the rest was to be given as "alms" to hospitals.[27]

This episode of the General Hospital's assessments of the royal reserve's quina reflects several important aspects of the growing community of medical experts in Spain. It provides a glimpse into the growing complexity of testing cinchona bark in Madrid as many different groups became involved. Ultimately, quina from the same shipment had been examined by the Royal Pharmacy, the physicians at the General Hospital, the Conde de Mora's eight-member panel, a chamber physician, and the physicians and pharmacists of the army. Even in Spain, the authority of the Royal Pharmacy to make judgments regarding the quality of quina—as embodied in the samples it distributed—was no longer sufficient nor did it go unchallenged. Like Pedro de Valdivieso in Loja, physicians at the General Hospital disputed the quality of the pharmacy's bark and questioned their knowledge of quina.

This episode also shows that therapeutic testing had become a regular part of the assessment of quina from the royal reserve in the late 1770s. Visible and physical inspection of the bark remained a preliminary means of assessment; however, experience or "experiments" with the bark became important as well. At the same time, this episode shows that no mechanism existed for consistently deciding which group of experts or set of techniques provided the true assessment of the bark and its medical efficacy. For lack of a means to establish consensus among medical experts, officials were free to cite the results that suited their needs. As a result, while the Duque de Losada, who supervised the Royal Pharmacy, cited the results that confirmed the efficacy of the bark, the Duque de Híjar, who had connections to the royal hospitals, cited the results that put the efficacy of the bark into question.[28] Such results show that within the epistemic culture of the colonial government, officials retained the power to interpret results and to decide what would count as true and reliable knowledge.

NEW METHODS OF ANALYSIS AT THE ROYAL PHARMACY

In the wake of these disagreements, the royal pharmacists became keenly aware of their increasingly precarious authority as the Crown's exclusive experts on quina and other materia medica. In an effort to reclaim some of their authority vis-à-vis other groups in Madrid's community of medical experts, royal pharmacists developed and adopted several new techniques for assessing incoming shipments of bark. With new techniques came new standards as evidenced by their examination of samples of quina from Bogotá in 1785.[29] Just a decade earlier, Pedro de Valdivieso had critiqued the Royal Pharmacy's methods for examining the bark, while the results of additional testing of bark in Madrid had resulted in the turn to therapeutic testing as a regular practice. Notably, in 1785, the royal pharmacists dubbed the "color, thickness and configuration" of the bark to be "accidental" and not indicative of "any essence"—a significant shift from their earlier emphasis on physical characteristics as the key to assessing the bark's medical efficacy.[30]

In place of the observation of the bark's physical characteristics, the royal pharmacists embraced several new techniques to make their testing of the bark more comprehensive and hopefully more reliable. Consider the royal pharmacists' analysis of bark samples received in 1782 from Santa Fé de Bogotá. First, the pharmacists subjected the samples to "analysis and separation of the principle components" according to "the method prescribed by Chemistry."[31] At the same time, the pharmacists engaged in a review of "all [works] by travelers and the most classic naturalists which discuss [quina]." [32] The goal was to determine if any of the bark samples from Bogotá matched the kinds of quina described in any of the published accounts. This review of the existing literature led them to conclude that one of the samples was a "species of new Quina" that was "equal to that of Quito."[33] A third technique was to distribute samples of this quina in powdered form to "twenty-two Medical Professors so that they could administer it to patients, examine its virtues and report on their results."[34] In a bid to maintain their status as the preeminent experts on quina, the royal pharmacists embraced the laboratory, the library, and the hospital as integral to their work. In addition, the pharmacists tried to make sure that all quina-related activity was coordinated through the Royal Pharmacy.

How effective was this shift to new techniques? It is difficult to say, especially because the physicians reported variable results. Twelve of the twenty-two physicians claimed that the quina from Santa Fé was as good as that from Quito or Peru, another nine physicians deemed the quina effective as a febrifuge but not quite as good as quina from Quito or Peru, and one respondent reported that the bark had no effect.[35] With these results and those from his assistants, Juan Díaz, the new head pharmacist at the Royal Pharmacy, declared all varieties of quina from Santa Fé—white, red, and yellow—to be "equal to those from Quito."[36] His decision was, ultimately, based on "the authority and experience" of the "[Medical] Professors" who tested the bark on patients.[37] Such observations from the head pharmacist ultimately signaled an important shift in the pharmacy's practices away from simple inspection of the immediately sensible, physical qualities to techniques aimed at revealing the essence of a sample.

Royal officials, too, were influenced by the new techniques and criteria employed by the Crown's learned experts. Consider the reaction of the Marqués de Valdecarzana, the new chamberlain of the Royal Household, to Juan Díaz's assessment of the bark. He noted that, while recent shipments of quina, including those from Santa Fé, appeared useless at first glance, subsequent testing demonstrated the medical efficacy of this bark. He noted, "two-thirds [of each shipment is comprised of] *cortezones, cañas duras* and some splinters from the trunk of the tree."[38] The dreadful state of the bark suggested that the "instructions given by the Ministry [of the Indies]" for collecting, examining, and transporting the bark were not being followed. Yet, this turned out to be inconsequential, especially as therapeutic testing revealed the bark to be useful, "even though [such quina] lack the good [and] accidental [characteristics] of color, odor, and flavor," in Valdecarzana's words. Ultimately, the chamberlain yielded to the "facts" of the new techniques and criteria, and concluded: "There is no bad *Quina* as long as it is true *Quina*."[39] It was a similar conclusion to one offered by Miguel García de Cáceres, in an earlier report on the quina trade that he composed in 1779. Such evidence suggests that officials on both sides of the Atlantic were becoming skeptical of the efforts to assess the bark on its physical characteristics alone.

Still, some officials, such as Valdecarzana, operated within the context of the royal gift economy. The chamberlain reminded his associates that "the [original] objective of the shipments of *Quina* to the Royal Pharmacy was so that His Majesty would have the best *Quina* from America in order to give [it] to various European Courts and to provide for the entire consumption of the Royal Pharmacy."[40] Consequently, an incongruity emerged. Physical characteristics remained determinants of the bark's value in the royal gift economy even as officials and experts rejected these characteristics as a means for assessing the bark's quality and efficacy as a medicament. Nonetheless, the shift away from assessing the bark only on its physical characteristics reflected the expanded range of experts and techniques involved in the community of experts advising the colonial government on quina and other matters.[41]

Botanists in Madrid's Community of Experts

Botanists also benefited from the expansion of the community of experts in association with the testing of quina. A good example is Casimiro Gómez Ortega (1741–1818), whose career is emblematic of the rise of botany in the Spanish imperial enterprise. It was often his consultations on quina, as much as his appeals to Enlightenment notions of the utility of science to the empire, that furthered the relationship between botany and empire in the Spanish Atlantic.

Many factors contributed to Gómez Ortega's success. He had strong connections to the pharmaceutical community in Madrid through his uncle, José Ortega, who not only owned and operated a pharmacy in Madrid but also had previously conducted examinations of quina at the Royal Pharmacy.[42] On 29 July 1772, Gómez Ortega replaced Miguel Barnades as director of the Royal Botanical Garden and, in that same year, the Crown named him "Examiner of Pharmacy" in the Royal Tribunal of the Protomedicato, the main regulatory and disciplinary board of medical practitioners in Spain.[43] This appointment gave him experience with the regulatory functions that science and government could perform together. As director of the Royal Botanical Garden, Gómez Ortega was not only the head of one of the most important scientific institutions of Enlightenment Spain, but also occupied an important node in the botanical networks of Europe and the Atlantic World. In 1777, Minister of the Indies José de Gálvez put him in charge of Spain's botanical expeditions to the Americas. In 1784, he was appointed "honorary Head Pharmacist" at the Royal Pharmacy, an auspicious indicator of the authority of botany vis-à-vis pharmacy at the time. More than just an honorary title, it provided another opportunity for involvement with the royal reserve of quina and the affairs of empire.

At first, officials in Madrid dealt with Gómez Ortega as an outside adviser to be consulted only in special cases. In 1777, the Duque de Losada had asked Gómez Ortega, along with the head pharmacist at the Royal Pharmacy, to assess some samples of cinnamon and clove from Quito at the request of Minister Gálvez. The latter had instructed Losada to arrange for experts to assess the "quality" of these samples and determine "if they will lose their sharp [flavor] if cultivated."[44] Two years later, Gómez Ortega was a member of the panel of medical experts that reexamined cinchona bark rejected by the General Hospital. In these instances, Gómez Ortega gained valuable experience in working with other experts in service to the Crown.

These early interactions with royal officials also gave Gómez Ortega the opportunity to promote the utility of his science to the empire.[45] For example, in his report on the samples of cinnamon and clove, he explained that "in order to ascertain the truth and proceed with certainty" regarding the effects of cultivating cinnamon or clove in the colonies, "a Botanical examination of a live Plant" was needed. He also generally praised the Crown for its support of the sciences, a claim that may have been an oblique reference to the imminent departure of the Royal Botanical Expedition to Peru and Chile. Gómez Ortega concluded his report with a suggestion that science would help the Spanish Empire break the Dutch monopoly on the spice trade. He proposed that "oil distilled in the Royal Chemical Laboratory from the Cinnamon [from South America]" could replace the cinnamon that Spain purchased from Dutch merchants. Finally, he suggested that the Crown could "restore . . . this important commerce to Spain by means of cultivating Cinnamon and Clove Trees from Quito."[46]

Around the same time that Gómez Ortega was asked to examine cinnamon and clove from Quito, Gálvez asked for his advice on quina. For example, in 1777, Gálvez asked him to review two samples of cinchona bark recently submitted by Sebástian López Ruiz, a creole physician in New Granada who submitted samples of newly discovered cinchona trees in the forests near Bogotá.[47] In his report, Gómez Ortega found that only one of the samples seemed to have potential and he recommended further investigation of this bark.[48]

Gómez Ortega did not stop there. He also advised Gálvez on how to organize and deploy medical experts to conduct further investigation of this new quina. In particular, he suggested that the minister send a box of the bark to "the *Junta* of the Hospitals" so that the physicians there could observe the "effects" of this quina when used on patients.[49] Another box, he continued, should be sent to "various individual Physicians who are the most experienced and accredited" for additional therapeutic testing on patients. He also suggested that the reports from these physicians could be compared with those of the "Professors of Santa Fe," whom the viceroy of New Granada had also asked to examine the quina samples. Ultimately, these recommendations show that Gómez Ortega was not content to serve as just another voice in the community of experts advising the Crown. Instead, he asserted his prominence over other medical experts by attempting to organize the further investigation of this new quina.

Gómez Ortega used the occasion of his report on quina to convince the minister of the Indies to use the resources of the colonial government in the service of botany. First, he requested that Gálvez order the viceroy of New Granada to send additional samples of the quina from Bogotá as well as samples of the "leaves, flowers, and fruits" from the trees, including "a few small trees." To this end, Gómez Ortega also offered to write and publish an *instrucción* (instruction) of four to six pages similar to another set of instructions "published on the order of the King so that Viceroys, Governors, *Corregidores* and *Alcaldes mayores* could [know how to] choose, prepare and send to Madrid products for the Cabinet of Natural History."[50] Gómez Ortega suggested that this instrucción could be sent to these same officials so that they would know how to prepare and send live plants to Madrid. The interests of botany and empire converged in such an enterprise. The Royal Botanical Garden would gain more specimens for its collection, while the Crown would have a powerful symbol of its possession of the botanical resources from throughout its vast empire. Though he could not eliminate the presence of pharmacists and physicians entirely, Gómez Ortega did his best to demonstrate botany was more useful as an imperial science than were pharmacy or medicine.

In 1779, Gómez Ortega published a short pamphlet, titled "Instruction on the safest and most economical method for transporting live plants from distant lands by sea or land."[51] It was a practical guide that emulated similar works published by Gómez Ortega's counterparts elsewhere in Europe, notably John Ellis in England and Henri-Louis Duhamel du Monceau in France. Indeed, in his 1777 report to the minister of the Indies, Gómez Ortega explained that Ellis's own publication on the transportation of live plants had facilitated the successful transplantation of coffee, tea, and other important plants to London and other parts of the British Empire.[52] Ultimately, Gómez Ortega's pamphlet was intended for a wide audience: other naturalists as well as any other kind of traveler—merchants, soldiers, missionaries, and colonial officials—that might be able to submit botanical specimens from around the globe.[53]

This short pamphlet from the director of Spain's Royal Botanical Garden was significant because it signified the new relationship between botany and empire. In particular, Gómez Ortega sought to mobilize the officials and institutions of the colonial government to collect and send plant specimens to the Royal Botanical Garden in Madrid. At the same time, by publishing his pamphlet on transporting live plants as an instrucción, Gómez Ortega gave his enterprise authority by presenting it in the format which the Crown used in royal orders to colonial officials, as in the case of the Crown's own instrucción regarding the collection of samples for the Museum of Natural History. Indeed, Gómez Ortega had experience writing such documents for the colonial government, as in the instrucción he prepared for Hipólito Ruiz and José Pavón on the occasion of their botanical expedition to Peru and Chile (1777–1788).[54]

The pamphlet characterized transplantation not just as scientific enterprise but also as the work of empires.[55] In it, Gómez Ortega noted that royal support of "botany" secured Spain's "possession of all the vegetable riches of the vast dominions of the King and the other countries of the world, especially those of Spanish America."[56] He also explained that "if our ancestors had not been so diligent in acquiring and propagating the useful plants of other countries in their country," then Spain would be lacking in "the most delicious fruits and most valuable plants."[57] As for models, Spain was to emulate "other nations" and the "Romans." Such references to the Romans, which appear throughout his introduction, supported the notion that botanical activities, such as the transplantation and cultivation of plants from abroad, were appropriate, if not essential, activities for empires. Other prominent groups that practiced transplantation included the "Saracens," who occupied the Iberian Peninsula during the Middle Ages, as well as the "Conquistadors of the East and West Indies."[58] Gómez Ortega thus linked the circulation and cultivation of plants not only with the Roman Empire, but also with the imperial glory of sixteenth-century Spain and contemporary colonial enterprises of England and France.[59]

Gómez Ortega had expressed a similar vision in an earlier letter to Minister Gálvez. He wrote: "I am of the firm persuasion that if the King, peaceful and wise, with the influence of his learned or educated Ministers, [were to] order the examination of the natural products of the Peninsula and his vast overseas Dominions, twelve naturalists with as many Chemists and Mineralogists dispersed through his States, will produce, by way of their wanderings, a utility incomparably greater than one hundred thousand men fighting to add some Provinces to the Spanish Empire."[60] Here, Gómez Ortega introduced many of the same elements that he would reiterate in his 1779 Instrucción. He cast science as a new mode and means of Spanish imperialism to supersede conquest. He also emphasized the utility of science to the state, as well as the fact that entire enterprise would emanate from the Crown. In some sense, his 1779 Instrucción was a continuation of the vision of science and empire that he had already been formulating and espousing in his correspondence with government officials. Distributed broadly throughout the empire, this pamphlet broadcasted Gómez Ortega's vision of science and empire as much to bureaucrats as to botanists.

This publication also strengthened his role as an organizer of the other experts on the natural world with whom the Crown consulted. Gómez Ortega's ascendancy vis-à-vis other experts became most visible in 1785 when the minister of the Indies, again asked the botanist to review a sample of quina. The samples came from Manuel Perfecto de San Andres, a citizen of Cuenca and a merchant in the quina trade, who had recently received the contract to supply bark for the royal reserve. The president of Quito hoped that quina from Cuenca could be used to supply the Royal Pharmacy while the forests in Loja were given a few years to recover.[61] There is no evidence that the royal pharmacists were even asked to examine the bark; Gómez Ortega alone was given the authority to accept or reject it. In April 1785, he reported to Gálvez that quina from Cuenca was of poor quality. Gálvez took action almost immediately.[62] On 10 May 1785, he sent an order to the president of Quito prohibiting any further collection of this bark for the Royal Pharmacy.[63] He also instructed the president to supply "a flowering branch and the fruit of the trees dried between sheets of paper with exact and specific information on the quality of the climate and terrain" in which the trees grew.[64] When it came to assessing new specimens and sources of the bark, this case shows how Gómez Ortega and the Royal Botanical Garden began to displace the Royal Pharmacy as the empire's primary experts on quina and other matters botanical.

Botany and Empire beyond Madrid

While botany was making inroads at the royal court and the Ministry of the Indies in Spain, a similar process was underway at the vice-regal court in New Granada as one of the region's most important naturalists—José Celestino Mutis (1732–1808)—achieved a prominent position advising the viceroy and other officials on many matters related to science and medicine, especially quina. This case provides a South American counterpoint to developments in Spain. Moreover, it shows that the rise of the botanists was a transatlantic phenomenon in the Spanish Empire.

Since his arrival in Santa Fé de Bogotá from Spain in 1761, Mutis had been petitioning the Crown and the viceroy for support of his scientific interests and activities, mainly in natural history.[65] These efforts finally came to fruition with his appointment as director of both the Royal Botanical Expedition in New Granada and as the director of a second royal reserve of quina near Bogotá in 1783. Most accounts of Mutis give primacy to his role as director of the botanical expedition. Yet, as in the case of Gómez Ortega, greater focus on Mutis's consultations on quina show that pragmatic needs also drove the intertwining of science and empire in New Granada in late eighteenth century.

Mutis's appointment as the "Quina Commissioner" in charge of the new royal reserve near Bogotá was another case of the displacement of medical expertise with botanical expertise alongside a similar process occurring in Spain at nearly the same time. In 1783, Juan Francisco Gutiérrez de Piñeres arrived in New Granada after being sent from Spain by the Crown to conduct a *visita general* (official review) of the vice-regal government. Notably, the Crown had ordered Gutiérrez de Piñeres to review the possibility of establishing a new royal reserve of quina to control the collection of bark from cinchona trees near Bogotá. As noted above, Sebastian López Ruiz, a creole physician, had claimed to have discovered these new stands of cinchona trees in 1774.[66] In 1778, the Crown appointed López Ruiz the "commissioner of quinas" to oversee the development of bark production in the region. For reasons that are unclear, Gutiérrez de Piñeres relieved López Ruiz of his duties in 1783. In that same year, Gutiérrez de Piñeres established a second royal reserve of quina near Bogotá and, on the recommendation of the archbishop-viceroy, appointed José Celestino Mutis to oversee the enterprise as the new "commissioner of the quinas." It was a political victory for Mutis, who claimed to have discovered the cinchona trees near Bogotá first and resented López Ruiz's efforts at self-aggrandizement by claiming to be the discoverer of this new source of quina.

In his capacity as director of the royal reserve, Mutis impressed upon the viceroy the necessity of botanical study of the cinchona tree in order to produce a "complete treatise" on it.[67] Marcelo Frías Núñez, one of his most recent biographers, characterizes Mutis as having a "propensity to hoard competence relevant to quina," suggesting that Mutis recognized the power and authority that he derived from claiming knowledge of this valuable natural resource.[68] Yet, rather than framing his efforts as an act of hoarding, Mutis's actions can be understood as an effort to organize a network of knowledge production with himself at the center. Just as Gómez Ortega actively emphasized the utility of botany to the empire, so did Mutis seek to make his expertise and the Royal Botanical Expedition a crucial part of the realization of the viceroy's and the Crown's practical programs in New Granada.

To bolster his authority and expertise, Mutis employed several means. First, with the viceroy's help, he solicited samples of quina from all regions of the Viceroyalty of New Granada. In this way, he emulated the Royal Pharmacy and the Royal Botanical Garden in Madrid, which had previously collected samples of quina from throughout the empire with the help of officials in the colonial government. In other words, Mutis sought to make the Royal Botanical Expedition, based in Bogotá, into a "center of calculation," a location where natural objects are collected and compared to produce new knowledge.[69] Mutis could bolster his power and authority by developing unique knowledge of quina derived from the possession of a range of samples that could be examined and compared in ways not possible in nature. Finally, Mutis worked to coopt or eliminate any other rival experts in the region. He tried both techniques in the case of his rival, Sebastian López Ruiz.[70] After initially discrediting him, Mutis later recommended that López Ruiz be given the duty of overseeing the preparation of quina shipments for the Crown.[71] Mutis had also established amiable relations with the other expert on quina in Bogotá, Miguel de Santisteban, whose direct experience with the cinchona trees in Loja would have been invaluable to Mutis, especially because Mutis had never made the journey to the quina-producing regions of the southern sierra.[72]

In October 1783, Archbishop-Viceroy Antonio Caballero y Góngora accommodated Mutis's requests when he ordered the president of Quito to send samples of cinchona trees from throughout his jurisdiction. Caballero y Góngora explained that compliance with his request was "crucial to the formation of regulations on the cutting and sending of Quina."[73] It is notable that the viceroy did not solicit reports or information on the quina from local officials—not even from Pedro de Valdivieso, a veritable czar of quina by this time. Mutis and the viceroy apparently had no interest in the knowledge of local experts in the quina-producing regions of the southern sierra in the Audiencia of Quito. Rather, the botanists in Bogotá would create knowledge of quina anew by examination and comparison of plant specimens. Such a move did not go unnoticed by the officials and experts of the quina-producing regions of the southern sierra. In November 1783, after receiving the viceroy's request for samples, Pedro de Valdivieso, still director of the royal reserve in Loja, expressed reservations to the president of Quito. "I do not doubt the intelligence of the Botanists of this Kingdom," he explained, but pointed out that the botanists lacked experience in the "actual comparison" of the bark.[74]

Valdivieso ultimately yielded to the archbishop-viceroy's request, but other local experts in the southern sierra proved more recalcitrant. Consider the case of Manuel Perfecto de Sán Andres, a bark trader in Cuenca (figure 1). In 1783, Perfecto de Sán Andres was selected as a quina supplier for the royal reserve. The Crown had approved the expansion of the royal reserve to include quina from Cuenca's forests in order to give the exhausted forests near Loja an opportunity to recover.[75] A contract to provide the annual shipments for the Royal Pharmacy would have meant steady income for Perfecto de Sán Andres.[76] As a result, this bark trader from Cuenca was eager to prove his qualifications to serve the Crown by establishing his knowledge of quina. Consequently, this case provides an example of how local experts in the traditional center of quina production (the Loja region) continued to assert their authority to speak for nature and strenuously resisted any efforts to undermine that authority.

As it turned out, Perfecto de Sán Andres's experience with the cinchona tree and knowledge of its bark was extensive including some understanding of the most recent chemical and botanical techniques of the time. Therefore, he was someone that Mutis had to take seriously. As early as November 1783,

Perfecto de Sán Andres provided the president of Quito with information on experiments that he had conducted with the "salts and extracts" of quina, and promised further results from studies he intended to conduct in the summer of 1784.[77] In May 1784, Perfecto de Sán Andres did send "skeletons of the Tree of *Quina*" to the viceroy in Bogotá.[78] Like Valdivieso, Perfecto de Sán Andres had little confidence in the botanists' ability to conduct his samples with those from the cinchona trees in Bogotá. As a result, a month later, he announced his intention to conduct his own examination to determine "if the Trees of *Quina* discovered in the Northern Forests [i.e., the forests near Bogotá] are of the same Nature as those that grow in the Southern [forests of the Audiencia of Quito]." Perfecto de Sán Andres also offered to "go to these Regions [near Bogotá] to conduct a visual examination and other studies pursuant to the resolution of such an important matter [i.e., the comparison of quinas]." "With my experience," he explained, "I am able to distinguish the qualities of Quina with more clarity than that which could be seen in [the examination] of Skeletons [dried plant specimens]."[79] It was a direct challenge to Mutis and the other botanists in Bogotá.

Mutis was indignant. He emphatically rejected Perfecto de Sán Andres's claims that his experience and knowledge trumped botanical study of cinchona specimens. In July 1784, Mutis wrote to the archbishop-viceroy to complain about the bark trader's insubordination. In his letter, he explained that "knowing a science of detailed knowledge" (i.e., botany) was "beyond [Perfecto de Sán Andres's] charge." "It would be a mistake," Mutis continued, "to hope that he could contribute to the scientific examination [of quina] on his own."[80] He reminded the viceroy that the scientific study of the cinchona tree and its bark was the sole province of the botanists of the Royal Expedition —not a bark trader in Cuenca. At the same time, Mutis recognized that Perfecto de San Andres was a real threat. First, Perfecto de Sán Andres was able to observe living specimens of cinchona trees in Cuenca (and probably Loja), while Mutis only had access to dried plant specimens or "skeletons" from these trees. More important, Perfecto de Sán Andres had questioned the utility of comparing the "skeletons" of different cinchona trees—a practice that was a fundamental technique used by botanists to identify and classify plant species.

Ultimately, Mutis prevailed over his rival in Cuenca but not without the help from the director of the Royal Botanical Garden in Madrid. In 1785, samples of quina collected by Perfecto de Sán Andres were given an unfavorable review by none other than Casimiro Gómez Ortega as well as the head pharmacist Martínez Toledano at the Royal Pharmacy.[81] Although this outcome served Mutis's interests, it made explicit the hierarchy of scientific institutions in the Spanish Empire. At this moment when the Royal Botanical Expedition was asserting itself as an imperial institution, the Royal Botanical Garden and Royal Pharmacy, through their decision in the quina from Cuenca, reasserted their authority of their assessment of natural resources throughout the empire. It was neither the first nor the last time that an implicit hierarchy of centers of imperial science would be revealed. Indeed, in 1787 and 1788, Mutis's own samples of quina would come under the critical gaze of the community of experts in Madrid, who often claimed the final say when it came to matters of the natural world.

This chapter has shown that Enlightenment ideals of the utility of science and empire were important; yet ideology alone was not enough. The context and the practicalities of empire also mattered. As officials in Spain began receiving reports of the scarcity of cinchona trees in the 1770s and 1780s, they turned to the empire's experts on quina for assistance in identifying and assessing new sources of quina. The testing of bark shipments and samples provided a concrete task that gave rise to a growing community of experts in Spain. Yet the involvement of more physicians and pharmacists revealed disagreements in the medical community over the identification and assessment of quina and, perhaps, confirmed some of the concerns raised by Pedro de Valdivieso in his 1773 dispute with the Royal Pharmacy. Nonetheless, it was this discord among medical experts that provided the opportunity for botanists in Madrid to insert themselves into the empire's growing community of experts. With the reputation of botany on the rise in the late eighteenth century, both Gómez Ortega in Madrid and Mutis in Bogotá asserted the superiority of botanical knowledge and worked to establish themselves as the preeminent experts on matters related to the natural world. The problems associated with the production and distribution of quina provided the opportunity not only for imperial officials to seek the advice of botanists but also for botanists to show that claims about the utility of their science were more than just words.

The experiences of Gómez Ortega and Mutis also provide insight into the ways in which botanists assimilated to the values and practices of the epistemic culture of the colonial government. For example, the fate of quina from Santa Fé highlights the persistence of the hierarchy in epistemic culture of empire. Even as Mutis became the prominent naturalist and director of the Royal Botanical Expedition in the Viceroyalty of New Granada, the authority of botanists and experts in Madrid still trumped all. In 1787, Mutis lost a powerful political ally with the death of José de Gálvez, Minister of the Indies, right around the time that the quina from Santa Fé was being reexamined in Madrid. In a 1789 letter to an associate in Madrid, José Celestino Mutis remarked: "Upon his death, [Gálvez] took to the grave confidential ideas in which we had agreed to make public the progress of the Royal Administration [of quina] in order to avoid the unjust complaints of the public and those interested in this commerce, who often bristle at the word Monopoly."[82] He also noted that the royal reserve in Santa Fé suffered the "disgrace that three Viceroys would change in one year and the most important issue of Quina would be put to rest as a result of the most profound lethargy."[83] Mutis's letter hinted at the influence of commercial interests in Peru that sought to undermine quina from Santa Fé in order to prevent competition with their own quina exports. Ultimately, Mutis placed much more emphasis on the effects of the changes in the Spanish colonial government as key allies and architects of the royal reserve either died or were replaced.

As botanists became integrated into the colonial government, they also became integrated into its epistemic culture and its politics. Initially, botanists sought to control the community of experts and reshape the empire's apparatus of expertise to their own advantage. However, entering into this epistemic culture meant becoming a part of the various networks of officials and other actors that influenced both the production of natural knowledge and implementation of imperial policies. Try as they might Gómez Ortega and Mutis were unable to transcend the hierarchical structure and political nature of the epistemic culture of the colonial government. True, many Bourbon reformers and Spanish botanists praised the utility of botany and natural sciences to the state. However, in this case, the practical and quotidian challenges of producing a colonial commodity, rather than enlightened ideologies, account more fully for the rise of botanists in the late eighteenth-century Spanish empire.

Many explanations have been offered for this sudden reversal in the Crown's policy on quina from Santa Fé. One explanation is to take the participants at their word: quina from Santa Fé was no good. Such an explanation assumes that there was a consensus on the medical utility of this bark; yet no such consensus existed. Even as experts in Madrid were deciding that quina from Santa Fé was useless, physicians and pharmacists in England, France, and Italy reported just the opposite. In the case of the British, it was partly in their interest to support quina from Santa Fé, as bark illegally imported from the northern regions of New Granada was their primary source.[84] Disagreement continued within the Spanish Empire as well. Throughout the 1790s, Mutis and his disciples continued to claim that quina from Santa Fé was just as good as that from Loja or Peru. In the early 1800s, this lack of consensus escalated into a full-blown debate among botanists (for more on this, see chapter 7).

In light of this evidence, changes in the Crown's policy cannot be explained by the bark's alleged poor quality. Even an expanded community of experts offered no definitive guide to imperial policy on quina. In turn, this situation left much room for ideology and interests rather than expert knowledge to influence decisions regarding to the royal reserve and imperial administration of the cinchona tree and its bark. Nonetheless, officials in Madrid continued to put their trust in the ability of botanists to bolster imperial control of American nature. In 1790, the relationship between botany and empire reached a new level of intimacy when the Crown appointed a "botanist-chemist" from Spain as the new director of the royal reserve of quina in Loja—a significant development not only in terms of imperial policy but also in terms of the wider Atlantic World in which few practitioners of the sciences were put in charge of the actual production of a botanical commodity. As the next chapter shows, the genesis and development of this new initiative in the royal reserve further highlights the fragility and limited efficacy of botany in the face of the broader tensions of empire evident in colonial Andean society as well as the epistemic culture of empire.

Six

Imperial Reform, Local Knowledge, and the Limits of Botany in the Andean World

In November 1790, Vicente Olmedo was in the Spanish port city of Cádiz. He was on his way to Loja to take up his new appointment as codirector of the royal reserve of quina. Olmedo's official title was "botanist-chemist" (*botánico-químico*)—a neologism coined by officials in Madrid.[1] It was a significant development not just in the history of Spain's royal reserve but also in the developing relationship between science and empire in the Spanish Atlantic World in the late eighteenth century. Whereas pharmacists and botanists in the preceding decades served the colonial government primarily in an advisory capacity, Olmedo was to serve directly as an agent of empire sent to implement reforms and enforce imperial standards at the royal reserve in Loja. Moreover, Olmedo's experience was unique, relative to many of his contemporaries. While most European botanists in the eighteenth-century Atlantic World focused on the production of specimens, texts, and images of plants for the purposes of identification and classification, Olmedo was responsible for actual production of a botanical commodity. Never before had the Spanish Crown sent a botanist (or a chemist) to Spanish America to intervene directly in the production of a botanical commodity.

Officials in Madrid who hatched the plan to send a botanist-chemist to Loja hoped that this reform would finally make the royal reserve into a powerful emblem of Spanish imperial rule enlightened by European science. In addition, the situation with quina had continued to worsen in the 1780s. Cinchona trees were increasingly scarce and the amount of bark in the annual shipments to the Royal Pharmacy was on the decline.[2] To make matters worse, in 1786 Spain suffered a fever epidemic that severely depleted the Crown's existing reserves of quina.[3] For the chamberlain of the Royal Household, who first recommended that the Crown send botanists directly to Loja, the solution was not more government regulation but more science. With specialized knowledge of plants, a botanist-chemist in Loja would ensure that the Royal Pharmacy received only the best bark as defined by pharmacists and botanists in Madrid. In addition, the Crown's instructions to the botanist-chemist required him to establish plantations of cinchona trees, an indication that officials in Spain no longer viewed quina as a finite resource. He was also required to introduce the production of a medicinal extract made from the bark.[4] The appointment of a botanist-chemist must have seemed quite promising. After all, botanists in Enlightenment Europe had proven their ability to collect, classify, and represent nature in the form of museum collections and published natural histories of plants. But, as Olmedo quickly learned, transforming plants into botanical specimens was one thing, transforming them into botanical commodities was another.

In addition to the rarity of an appointment such as Olmedo's, this episode is significant because archival records allow us to track the genesis, development, and implementation of these reforms. In particular, evidence shows that the decision to send a botanist-chemist was likely made under the influence and advice of Spanish botanists recently returned from an expedition to Peru. As a result, this episode highlights an emerging feedback loop between botany and empire. In addition to receiving royal patronage for their activities, botanists were beginning to have an impact on imperial policy. At the same time, a series of reports to the minister of the Indies from Olmedo on his experiences in Loja in the early 1790s elucidate the extent to which imperial reforms promulgated in Madrid could be enacted in colonial contexts. The story of Olmedo's efforts to implement these reforms is also compelling because some of the proposed reforms—namely the establishment of cinchona plantations and the introduction of the production of a quina extract—were precisely the same techniques that British and Dutch imperialists would use in the late nineteenth century to successfully exploit the cinchona tree and its bark.[5] In the Spanish case, however, these techniques proved much less efficacious because of the significant obstacles that Olmedo encountered in the colonial Andean World.

This chapter juxtaposes the enthusiasm in Madrid for botany as a tool of empire with the difficulties experienced by the botanist-chemist Vicente Olmedo in Loja in the 1790s. Such a juxtaposition shows that while botanists continued to exert influence at the court in Madrid, their influence in the broader Spanish Empire, especially South America, proved much more limited. Even though botanists proved quite successful at producing and accumulating new knowledge for the Spanish Empire, they were much less successful in their efforts to apply that knowledge to solve the more practical problems of controlling this imperial natural resource.

"Do Not Forget the Quina Tree"

In 1789, several officials in Madrid started to express great concern about the state of the empire's supply of quina. In September, the Crown ordered Judas Tadeo Fernández de Miranda y Villacís (1739–1810), who was the Marqués de Valdecarzana and the chamberlain of the Royal Household, to develop new "regulations so that the government might establish a system under which this important branch [of commerce] could be managed."[6] What had provided the impetus for this rush to reform imperial policies on quina? The most likely answer is botanists. In 1788, after leading the Royal Botanical Expedition to Peru and Chile for eleven years, the botanists Hipólito Ruiz (1754–1816) and José Pavón (1754–1844) had returned to Madrid. In addition to hundreds of crates of botanical specimens, drawings, and field notes, Ruiz and Pavón brought troubling news of the cinchona trees in South America. We do not know exactly what Ruiz and Pavón told the officials in Madrid but Ruiz's writings provide some indications. The similarities between Ruiz's observations on quina and the chamberlain's suggested reforms to the royal reserve highlights not only the emerging relationship between botany and empire but also shows that one of the consequences of botanists' involvement was that officials in Madrid no longer saw quina in the same way.

With input from botanists, officials in Madrid gained new understanding of the problems of the quina industry from Ruiz and Pavón, who had witnessed the development of the quina industry in northern Peru. At the same time, the involvement of botanists encouraged officials in Madrid to recognize the cinchona tree as a resource amenable to cultivation—a significant shift from their previous conception of quina as a finite resource. Ruiz and Pavón had arrived at the right moment. They had come back to Madrid, when the authority of botanists to speak about the natural world had reached a new height as a result of the previous efforts of Casimiro Gómez Ortega, director of the Royal Botanical Garden, to promote the utility of botany to empire.

That Ruiz and Pavón returned to Madrid with firsthand knowledge of the cinchona tree and its bark was by design on the part of Gómez Ortega. Shortly before their departure from Madrid in September 1777, Gómez Ortega gave Ruiz and Pavón a set of instructions indicating plants of special interest to the Crown and the Royal Botanical Garden. He advised his protégés to pay "special attention" to three plants: "Cinnamon from Quijos in Peru, Quina or cascarilla, in particular that from the province of Loja, [and] Icho which is abundant in Peru and serves among other uses to melt cinnabar and extract mercury" as well as any other unknown "trees and shrubs" that "could be most useful in medicine [or] for dyes and manufacturing."[7] Here, quina kept distinguished company with other key botanical products. Cultivation of cinnamon for Quijos was consider Spain's best chance of breaking the Dutch monopoly in the trade in that valuable spice.[8] Meanwhile, Icho was useful for Spain's mining operations because mercury was essential to the separation of silver from ore.[9] Later in his instructions, Gómez Ortega admonished Ruiz and Pavón: "Do not forget the Quina Tree and to arrange for the collection and transportation of its precious bark."[10] In addition to their botanical studies, Ruiz and Pavón were expected to intervene in the production of quina— further evidence of how the Crown increasingly relied on the expertise of botanists in the later eighteenth century.

The significance of Ruiz and Pavón's charge with regard to the cinchona tree and its bark should not be overlooked. First, in previous attempts to assert control from Madrid over the production of quina, the Crown had chosen only to send written instructions or bark samples to Loja.[11] In this case, the Crown was sending experts from Madrid to study the tree and directly implement royal standards for the bark. Second, they were two of only a handful of European botanists sent in the eighteenth century to colonial territories with orders to intervene in the production of a botanical commodity.[12] That Ruiz and Pavón were given this charge is also remarkable if we consider that previous appointees to oversee the royal reserve in Loja were career officials in the imperial government. Ultimately, Ruiz and Pavón were not able to make any significant changes to processes for producing and distributing quina. Indeed, they were not even able to visit Loja—the most famous quina-producing province —during their eleven-year stay in South America.[13] Nonetheless, they returned to Madrid, with experiences and information from Peru that gave them the authority to press for reform of the imperial policies relating to quina.

Although we do not know exactly what they reported to officials in Madrid, it is most likely that Ruiz and Pavón described their experiences in Huánuco, a province north of Lima in which cinchona trees were discovered just before the two Spanish botanists arrived in Peru to start their expedition. According to Ruiz, Don Francisco Renquifo, who was probably a merchant or bark collector, first identified "cascarilla trees" in the region in 1776.[14] When he found cinchona trees in Huánuco, Renquifo informed local officials and landowners that the people of Loja had a "considerable business in this bark" and suggested that residents of Huánuco "might make large sums of money" from trading in quina.[15]

Renquifo's observations were not the only connection between Loja and Huánuco. Shortly after the discovery, Don Manuel Alcaraz, a resident of Huánuco, took some bark samples to José Antonio de Lavalle y Córtes (1734– 1815), the Conde de Premio Real, in Lima.[16] Lavalle, whom Ruiz described as "a dealer in [cinchona] bark," confirmed the quality of the bark and promptly provided the capital and tools for Alcaraz and others "to gather as much bark as they could."[17] Ruiz further noted that "trained peasants from Loxa [Loja] were sent"—presumably under the auspices of Lavalle—"to teach the people of Huánuco how they collected [quina] in their region."[18] The people of Huánuco soon developed skill in the "harvesting and drying" of cinchona bark and extraction operations popped up in the forests of Cuchero, Casape, Casapillo, and Cayumba. According to Ruiz, by 1779, residents of the region were supplying 50,000–75,000 pounds of quina to Lavalle annually. In Ruiz's rendering, expertise from Loja plus merchant capital from Lima gave rise to this new extractive enterprise. Moreover, this episode shows how Loja continued to function as a center of knowledge production as skilled laborers from the region spread their knowledge of how to collect and prepare cinchona bark.

The problems of the burgeoning quina industry became apparent when Ruiz and Pavón visited Huánuco for a second time in 1787. In his journal, Ruiz reported that the "extraction [of quina] was carried out in disorder, resulting in serious destruction of the quina groves."[19] The main source of disorder was the treatment of indigenous laborers, many of whom were "penniless and continually in debt for one to two hundred bushels of bark."[20] In his journal, Ruiz described how local merchants used the system known as the *repartimiento de mercáncias* to hire laborers to collect the bark.[21] According to this system, a merchant and a laborer would enter into a contract in which the merchant gave the laborer "merchandise, such as light cloth, baize, clothing, and other effects" and sometimes money in advance. In exchange, the laborer was obligated to deliver a quantity of bark of equal value to that of the goods. When laborers failed to deliver enough bark, they went into debt.[22] To make the transaction even more advantageous to merchants, their goods were often overvalued so that merchants could demand even more bark—a situation that often resulted in indigenous laborers falling further into debt. According to Ruiz, both "judges and priests" aided merchants in the collecting the debts from indigenous laborers.[23] Clergy aided merchants by requiring laborers to "present the bushels that they need to pay their debts to various merchants" before the laborers could participate in church functions and festivals. In addition to making participation in church pageantry contingent upon delivery of cinchona bark, the clergy "or their intermediaries" were the ones who kept records of who owed what to whom. Ruiz noted that the clergy often abused the system by "never [settling] debts even though they may have been paid thrice over." In other instances, indigenous bark collectors "were sentenced and punished under orders from the judges and priests if they do not pay their debts promptly."[24] Ruiz's account is consistent with historian Luz del Alba Moya Torres's description of how merchants with well-placed familial relations exploited both the cinchona trees and indigenous labor in the Audiencia of Quito, especially in the provinces of Cuenca and Loja.[25]

Ruiz explained to the Crown that if this system of "exploitation" remained unchanged, cinchona trees were likely to disappear from the region. Bark collectors had little choice but to extract the bark quickly and, often, ineffectively as they tried to meet their quota. Imminent punishment from church and state for unpaid debts provided incentive for indigenous laborers to "gather as much bark as they could in a short time." To harvest bark quickly, bark collectors would first cut down as many cinchona trees as possible and return a few days later to remove their bark. This method resulted in much wasted bark as most trees dried up, making the bark difficult, if not impossible, to remove, before bark collectors could harvest it. "Being unable to peel off the bark easily," Ruiz observed, "[the Indians] take only a portion of the bark from some of the trunks and leave a large number, completely untouched, to rot away." Furthermore, "indiscriminate felling of both old and young trees" magnified the destructive impact of this approach. The overall outlook was grim, especially since Ruiz observed the "same wasteful exploitation" happening in several different towns and provinces. "According to a conservative estimate," he wrote, "the forests of these provinces have yielded more than 3.5 million pounds of [quina] in 8 years."[26] Exploitation of indigenous laborers, wasteful and inefficient bark harvesting, mishandling of the bark by merchants—these were the causes of scarcity that Ruiz offered to officials in Madrid.

What could the Crown do to remedy the situation? If anyone consulted Ruiz in 1788 or 1789, he may have offered the solution that he proposed in his *Quinología*, an account of the different species of cinchona tree that he published in 1792. In the section of the book dealing with the likely "exhaustion and annihilation" of cinchona trees to the "bad method" used in harvesting the bark, Ruiz recommended that the Crown encourage cultivation of the trees.[27] In particular, he suggested that if anyone should discover a stand of cinchona trees on royal lands, which included much of the unclaimed forests in Peru, the Crown should sell the land to the "discoverers" of the trees at an "equitable price."[28] These sales of quina-producing royal lands were to be made on the condition that the "discoverer" would "increase the number of [cinchona trees] with new plantings, and clear the terrain of all undergrowth and trees of a different type." To those who complained that such an enterprise was too difficult and costly, Ruiz pointed to "the Corrals and *Haciendas*, where the *Coca* bush is cultivated, were initially impenetrable forests."[29] After decades of the empire treating quina as a finite natural resource, Ruiz not only asked the Crown to recognize the trees as cultivable but to make the cultivation of trees part of the policy governing the expansion of the quina industry. In this way, botanists introduced not only new information to the colonial government but also a new way of thinking about cinchona trees and the supply of quina.

BOTANY, IMPERIAL IDEOLOGY, AND REFORM OF THE ROYAL RESERVE

One place to see the impact of Ruiz's observations and recommendations is in the reforms to the royal reserve of quina recommended by the chamberlain of the Royal Household, the Marqués de Valdecarzana. The two reports that Valdecarzana wrote for the minister of the Indies in 1789 led to several significant reforms to the royal reserve in Loja as well as changes to the Crown's general approach to the regulation of quina in the final decade of the eighteenth century. Valdecarzana's recommendations were taken seriously because as chamberlain of the Royal Household, he had direct experience with the workings of the Royal Pharmacy and the royal reserve in Loja. More than that, evidence suggests that Valdecarzana consulted with the botanists Ruiz and Pavón before writing his reports. Although there is no direct evidence of a meeting, Valdecarzana would have been aware that Ruiz and Pavón had just returned from Peru and that they brought firsthand knowledge of quina and the quina industry. Several of the recommendations from Valdecarzana's reports mirror those found in Ruiz's journals and printed works. For example, Valdecarzana recommended minimal intervention by the Crown and proposed a state-sponsored program to promote the cultivation of cinchona trees, a proposal that bears striking resemblance to the proposal in Ruiz's *Quinología*.[30] At the same time, Valdecarzana's reports show how imperial ideology played an important role in defining the relationship between botany and empire.

Valdecarzana submitted two reports to the minister of the Indies—one in August and another in September 1789.[31] His observations and recommendations were based on a number of sources. In addition to his own experiences and his likely consultations with Ruiz and Pavón, he also had access to a dossier of government documents relating to quina from the 1770s and 1780s. The minister of the Indies, Antonio Porlier, the Marqués de Bajamar (1722–1813), had arranged for this dossier to be compiled from the records in the Ministry of the Indies and the Royal Treasury. In a royal order, the Crown pressed the chamberlain to provide "the most certain regulations that might serve the government for the establishment of a system under which this important branch [of commerce] could be managed."[32]

As to the question of establishing a "system" for regulating quina, Valdecarzana rejected the idea, especially because he understood the request to mean that the Crown wanted to take total control of the production and distribution of quina through a comprehensive royal monopoly. Like many reformers in Spain and other parts of Enlightenment Europe, Valdecarzana endorsed free trade as the best system for encouraging commercial activity and the economic distribution of goods.[33] He also emphasized the negative effects that a royal monopoly would have on the quina-producing regions in the southern sierra of Quito. He noted that quina played an important economic role in these regions that had few other products to trade in order to acquire other goods that they needed or wanted.[34] A royal monopoly would require the enclosure of all cinchona forests, which, in turn, would mean the "privation of Commerce with those Provinces [Loja, Cuenca and Jaen]." Taking a page from Adam Smith, Valdecarzana emphasized that royal monopolies undermined the "reciprocal convenience" of trade in which different regions acquire "necessities" through "mercantile relations." "His Majesty has no reason," Valdecarzana concluded, "to get involved with the Royal Monopoly of Quina in America."[35] But this is not to say that the chamberlain wanted Spanish trade open to all merchants of the Atlantic World. He endorsed free trade within the Spanish Empire, but only when it proved advantageous to Spain and Spanish merchants.

In addition to a philosophical commitment to a mercantile vision of empire, Valdecarzana refused to recommend the establishment of a "system" because the colonial government did not have "perfect knowledge" of quina, and so lacked the "economic judgment which would point toward a stable system."[36] Existing government records, he noted, contained little "geographical and physical information on the enclosure and status of the Quina forests."[37] Valecarzana also made explicit reference to the divergent assessments of bark samples produced by physicians and pharmacists in Spain in the preceding decade. He pointedly asked: "What have we gotten from so many examinations of Quina from different places?"[38] Although some tests showed that certain types of quina were a "disappointment" and that the "quality" of the "true Tree" depends on its "situation, climate, and essence," Valdecarzana declared the results of these "experiments with Quina" to be "entirely superfluous to achieving the final goal of a system."[39]

The chamberlain observed that knowledge of different kinds of bark and their medical efficacy was difficult to translate into imperial policy. Such knowledge did not necessarily offer any guidance on how to produce more of the best kinds of bark, which was the main object of the system the Crown requested. Instead, all that the Ministry of the Indies could do was "clamor for good Quina from Loja, Santa Fé and Peru, while ordering that fraud be avoided and indicating the locations which produced the best [bark]."[40] Valdecarzana again emphasized the various disagreements among medical experts over different bark samples. "All Quina," he wrote, "from one location or even from the same Tree has suffered a thousand contrasting opinions among the *Juntas* of Physicians and Pharmacists of Madrid, [who do not give] one reason for their differences."[41] In the chamberlain's eyes, experts in Madrid had brought little practical benefit to the empire because they could not agree on what was known about quina.

While he rejected a royal monopoly, Valdecarzana was not opposed to the Crown asserting its right to purchase (at a fair price) the best quality bark from Loja and other regions.[42] It is his recommendations on these activities that also suggest Ruiz's influence. In contrast to the past practice of fulfilling the annual shipments of quina from the royal reserve in Loja, Valdecarzana recommended that the Crown send two representatives to South America to act as the official buyers of quina for the Crown. In particular, he recommended sending two "Botanical and Chemical Professors"—one to Loja, the other to Lima—to oversee the acquisition of the bark and the preparation of annual shipments for the Royal Pharmacy. [43] He further explained that these two "Professors" should be "the subordinates of [the minister of the Indies] and arrange their purchasing of quina according to the orders that they receive from [the minister]."[44] In this recommendation, Valdecarzana may have used Ruiz and Pavón as a model. After all, Ruiz and Pavón were two botanists sent by the Crown with specific instructions to oversee shipments of quina and this arrangement seemed to work. Yet Valdecarzana wanted these new botanists to be even more directly connected to the minister of the Indies.

Valdecarzana's recommendation to send two "Botanical and Chemical Professors" is significant for several reasons. First, it was a departure from past practice in the royal reserve. Whereas the royal reserve relied heavily on local experts, such as Pedro de Valdivieso, to collect and assess the bark, Valdecarzana's plan eliminated the role of local experts and instead put the power over the annual shipments in the hands of appointees from Spain that answered directly to the Ministry of the Indies. His recommendation was also consistent with the general spirit of the Bourbon Reforms in the late eighteenth century, in which the Crown sought to replace *creoles* with *peninsulares* (Spaniards) in the positions of the colonial government and to make the government more centralized and hierarchical.[45] Second, it reflected the further intertwining of science and empire. Whereas Ruiz and Pavón had been instructed to get involved in the production of quina as one part of a larger scientific expedition, Valdecarzana's "Botanical and Chemical Professors" were to be sent to South America for the sole purpose of acting as the Crown's representatives in the acquisition of high quality quina. Valdecarzana thus further articulated the idea of the empire employing botanists and other scientific practitioners as its agents not just in understanding of American natural resources but also in the exploitation of those resources.

Botanist and Bureaucrat as Agents of Empire

On 7 September 1790, Minister Porlier wrote to the viceroys of New Granada and Peru as well as to the president of Quito to inform them of the reforms to the empire's policies on the cinchona tree. Porlier explained to these officials that the king had decided "not to establish a monopoly of *Quina*." Instead, the Crown had ordered officials in South America to send to Spain "all [quina] from the forests of Loja, Calisaya and other places that produced [bark] of superior quality." [46] Following Valdecarzana's recommendation, Porlier also informed officials in South America that the Crown had appointed a botanist-chemist, Vicente Olmedo, to oversee the implementation of these new policies in Loja. The Crown was also sending another official from Spain, Tómas Ruiz de Quevedo, to serve as the new *corregidor* (royal governor) of Loja. With the appointment of a corregidor and botanist-chemist to work side-by-side in Loja, the royal reserve became a microcosm of the new relationship between science and empire. In addition, a separate set of instructions addressed to the botanist-chemist and corregidor provided insight into how the Crown and its advisers envisioned the dynamics of this relationship. [47]

Although, the reforms to the royal reserve were not as radical as a royal monopoly would have been, they did represent a significant shift in that jurisdiction of the royal reserve was no longer tied to a specific geographical region (Loja). Now the royal reserve had a duty to provide the Crown with all quina "of the highest quality" regardless of its geographical origin. [48] Under this new policy, knowledge and expertise were more integral to the royal reserve than ever before. The new director needed to be able to identify the highest quality quina not only in the "enclosed [forests] of Loja, Cuenca, and Jaen," where the director directly oversaw collection of the bark, but also in the marketplaces where the director now had the authority to purchase on "His Majesty's account" any bark from private traders that was of sufficient quality for royal purposes. The instructions to the royal reserve's new directors also made clear that merchants were to retain "the liberty to trade in all [quina] from Peru on their [own] account[s] and risk." [49] Such measures were most likely concessions to the powerful merchant guild (*consulado*) in Lima, which included many American and Spanish merchants who made their fortune in the quina trade and disliked the Crown meddling in their affairs. [50]

An expert was needed to oversee both the collection and purchasing of bark of highest quality. For this crucial position, the Crown chose not a pharmacist, as it might have done two decades earlier, but a botanist-chemist.

The title was a neologism undoubtedly based on Valdecarzana's original recommendation of appointing a "Botanical and Chemical Professor." Vicente Olmedo proved to be an ideal candidate because he had studied at both the Royal Botanical Garden and the Royal Pharmacy in Madrid. As a result of his time at these two institutions, Olmedo not only developed a good grounding in botanical and chemical techniques but also a good understanding of the emerging nexus of science and empire in Enlightenment Spain. For botanists, Olmedo's appointment represented a new apogee in botany's role in the Spanish Empire, especially because he held a key position in the colonial government. Not surprisingly, Casimiro Gómez Ortega encouraged the Crown and the minister of the Indies to give Olmedo more scientific duties during his time in Loja. In May 1791, he urged Minister Porlier to consider extending Olmedo's "commission" to include "other matters in Botany in general and in Mineralogy, . . . , especially since the natural productions of those Provinces have not yet been studied by a true expert."[51]

As the 1790 instructions made clear, Vicente Olmedo's primary function was to implement the new imperial policy, not necessarily to expand scientific knowledge. Consider those items in the instructions from 1790 that specified how Olmedo was to be trained for his new post. Before leaving Madrid, Olmedo was required to examine quinas from Loja, Cuenca, and Jaen at the Royal Pharmacy. He was also instructed to bring "copies of the instructions communicated by the Royal Pharmacy in 1773 to the Presidency of Quito and to the *corregimiento* of Loja for improved knowledge of Quina [and] the times and method of its cutting, drying, and good packaging." The Crown also ordered Olmedo to bring new instructions from the Royal Pharmacy with information on "all classes of Quina that come from Loja, Cuenca, and Jaen and their effects." In addition to written instructions, Olmedo was ordered to bring "samples of all [the different kinds of quina in the Royal Pharmacy] with notes on their greater or lesser estimation so that [the botanist] may proceed in the selection and development of the best [quina] with all possible security, separating in his shipments, even if the quina is from the same tree, that which is most bathed in the rays of the sun, and always making his learned observations with the new enlightenment which can be acquired from this material."[52] Such evidence makes clear that the Crown wanted Olmedo well trained in the Royal Pharmacy's understanding of quina so that he could implement the Royal Pharmacy's standards at the royal reserve in Loja and displace whatever local standards might exist. Although none of these documents mention Pedro de Valdivieso, these procedures indicate that the Royal Pharmacy was determined to assert the superiority of their knowledge over that of bark collectors in Loja. After being successfully challenged by Pedro de Valdivieso and his bark collectors two decades earlier, this royal pharmacists must have been pleased by the Crown's decision for Olmedo to be trained at the Royal Pharmacy and to implement the Royal Pharmacy's instructions and standards in Loja. In this way, Olmedo functioned as an "agent of empire."[53]

By giving the botanist-chemist jurisdiction over the collection, preparation, and packaging of bark for the annual shipments, the 1790 instructions also indicated how Olmedo was to implement royal standards. He was to be present "especially for the initial cuttings and preparations" of the bark to ensure that all "precautions" were taken to safeguard its "virtue and good quality." He was also to oversee the packaging of every box and to conduct a "detailed and exact examination of all the Quina that is brought to the warehouses."[54]

As to the relationship between Olmedo and the new corregidor, Tómas Ruiz de Quevedo, the instructions explicitly stated that Corregidor Ruiz de Quevedo, "should not get involved with the government, direction, and economy of [this] industry [which is] left to the expertise of the Professor Botanist." The corregidor was directed, however, to seek advice from the botanist-chemist. Olmedo was to conduct "examinations of the forests" of Loja, Cuenca, and Jaen in order to inform the corregidor on the state of the region's cinchona trees for the purposes of developing new regulations. For example, in the case of the "Forests of Cuenca and Jaen," the Crown instructed Ruiz de Quevedo to obtain, presumably from Olmedo, "established and certain information" with the goal of establishing "a system, according to the needs and preferences for acquiring true [and] good quina and paying the locals according to its original and fair value." Using this information, the corregidor was to propose a site for a "factory or warehouse" for "receiving, examining, weighing, boxing and sending Quina with the stamp [of the Crown]." The 1790 instructions explained that Ruiz de Quevedo should consult "the Botanist" in all matters related to quina and the royal reserve that required "expertise and learning."[55] The botanist-chemist and the corregidor were to have the kind of symbiotic relationship that reformers in Madrid envisioned for science and empire at large.

A Botanist-Chemist in the Colonial Andean World

Olmedo's appointment represented an important test of the role of botany and chemistry as sciences of empire. Correspondence between Olmedo and the Corregidor Ruiz de Quevedo as well as a series of reports penned by Olmedo between 1792 and 1794 provide insight into how well the botanist-chemist fared in his endeavors in Loja. In addition to overseeing the collection and preparation of quina for the annual shipments to the Royal Pharmacy, Olmedo's other main task, according to the Crown's instructions, was to lessen the impact of bark extraction on cinchona forests. To this end, the Crown ordered

Olmedo to pursue several strategies. First, he was expected to teach people involved in the quina trade how to extract the bark more effectively. The Crown hoped that, with Olmedo's guidance, merchants, landowners, and bark collectors could reduce the number of cinchona trees destroyed each year by learning how to get more bark from each tree. As a second strategy for making production more efficient, Olmedo was ordered to introduce the production of quina extract to the region. Extraction was a well-known process whereby the medical virtue of a plant product was removed and concentrated into a resinous substance using chemical techniques.[56] With enough processing, even poor quality bark could produce an efficacious extract. Finally, Olmedo was expected to encourage local merchants and landowners to establish plantations of cinchona trees in order to replenish their population. In each of these endeavors, Olmedo met different challenges that provide vital insight into the limits of European botany in the colonial Andean World.

THE INSTRUCTION OF INDIGENOUS LABORERS

In its instructions, the Crown indicated that Olmedo was not being sent to Loja just to represent royal interests. He was to serve also as an adviser to local merchants, landowners, and bark collectors and to promote the quina trade in general. One of his first tasks was to survey the forests of Loja, Cuenca, and Jaen in order to take stock of existing stands of cinchona trees. If he found any new stands of cinchona trees, Olmedo was expected to share this information with local merchants and elites in the region, who had "the resources to develop and conserve [cinchona trees]."[57] He was also instructed to provide landowners and laborers with "formal instruction on the increase, improvement, and conservation of Quina trees."[58] In 1793, Ruiz de Quevedo and Olmedo observed that "wealthy and white people" did not engage in the "work of [collecting] cascarilla" because they abhorred "all kinds of work."[59] Because most harvesting of the bark was done by "Natives or Indians," Olmedo directed his efforts toward them. Olmedo and Ruiz de Quevedo later reported that they were skeptical at first. "Any sudden variation of method or transformation of ideas," they wrote in 1793, "seemed impossible among those natives who lack intelligence, are poorly endowed with little rationality and are unable to accept instruction contrary to their erroneous maxims."[60] Olmedo and Ruiz de Quevedo were not the first to make such observations. Many European observers of the eighteenth-century Americas commented on the ignorance of the indigenous peoples.[61]

Despite their obvious prejudice against indigenous laborers, Olmedo provided the necessary instruction and reported some good results. After a visit to the forests of Cuenca in 1792, Olmedo sent a set of "instructions" to the governor of Cuenca designated for "all those that may have to cut Cascarillas."[62]

Before the 1793 harvesting season, Olmedo gave "bark samples" as well as "verbal and written [instructions] to each and every one of the bark collectors" working for the royal reserve in Loja.[63] "This year, we have reaped the fruit of our labors," he wrote, "in that we have not rejected a single [scrap of] bark [from the bark collectors]."[64] Such results were evidence that bark collectors had learned how to identify and collect bark that met royal standards.[65] Olmedo also reported to the Crown that in regions where bark collectors harvested quina according to his instructions, the cinchona forests were starting to rebound (a characterization that differed from that of bark collectors who emphasized the scarcity of cinchona trees). In 1794, after a visit to Yangana, a small village south of Loja, Olmedo found that the forests were producing "fine Cascarilla in great abundance," a situation that he attributed to the "new regimen for extraction" resulting from his "instruction to the laborers, their assistance and other efforts practice in this interesting matter."[66] Whether Olmedo's instructions were as effective as he had claimed them to be is uncertain; however, these reports show that Olmedo was at least attempting to teach laborers how to improve yields of good quality quina and to introduce more sustainable practices of extraction.[67]

Even as their efforts showed positive results, Olmedo and Ruiz de Quevedo reported that other conditions undermined this apparent success. They quickly discovered that it was difficult to retain laborers from year to year. Because harvesting bark was seasonal work that took place during the drier months of June, July, and August, laborers often migrated to other regions in search of work during the off-season. In some cases, they did not return. In addition, merchants and landowners in the quina-producing regions of the southern sierra of Quito, which often lacked sufficient capital for wages, induced seasonal laborers to collect the bark by advancing them cash or trade goods in exchange for a promise to deliver a specified amount of quina of a specified quality by the end of the harvesting season. If a bark collector could not find sufficient bark to repay the merchant, then he went into debt. It was a local version of the system of debt peonage known as the repartimiento de mercancias, which was common throughout colonial Latin America. Even Olmedo and Ruiz de Quevedo had to resort to this system after reporting that bark collectors "have no fixed residence and there is no one who would voluntarily want to extract cascarilla for His Majesty."[68] Ultimately, the prospect of mounting debt to merchants or to the Crown may have also discouraged seasonal laborers from returing to the sourthern sierra to harvest bark.

Olmedo and Ruiz de Quevedo did little to alleviate the unfavorable working conditions of bark collectors. As a result, they experienced a steady outflow of skilled laborers—the bark collectors that had received training from Olmedo—each year. From year to year, this migration would have only intensified as the local economy in the southern sierra began to contract and the coastal cities, like Guayaquil, offered more economic opportunity. [69] In 1796, Olmedo and Ruiz de Quevedo reported that for every "100 [bark collectors] enlisted in this work" they lost twenty-five of them to death or migration.[70] These lost laborers then had to be replaced by "Boys and Young People" whom Olmedo had "to instruct anew." As a result, Olmedo and Ruiz de Quevedo had "to tolerate [from these inexperienced laborers] some defects in their first attempts" to collect the bark.[71] Ultimately, any improvements wrought by Olmedo's instructions were blunted by the larger social and economic context in which he and the royal reserve operated. Instructions from a botanist-chemist were useless on their own because he depended on laborers to carry them out.

THE INTRODUCTION OF PLANTATIONS

In addition to retraining bark collectors, Olmedo and Ruiz de Quevedo were ordered by the Crown to encourage local merchants and landowners to establish plantations of cinchona trees. The Crown's new interest in cinchona plantations represented a significant shift from past practice and previous understandings of the quina as a finite resource. It is perhaps no coincidence that, in the context of the botanists' increasing role in Spanish imperial governance, officials in Spain finally came to recognize quina as a natural resource amenable to cultivation. According to the Crown's instructions of 1790, Olmedo was to set up a "plantation of Quina trees in order to see and to know by experience if cultivated quina has more or less virtue than that from the forest." The Crown also suggested that the conversion of quina production from extraction to plantation might restore damaged forests.[72] To this end, Olmedo's plantation was meant to serve merely as a prototype to inspire local merchants and landowners by demonstrating that the cultivation of cinchona trees was possible.[73]

In 1794, Olmedo identified the forests of Loja as containing the "best sites" and he soon established a small cinchona plantation in the hopes of demonstrating the "utility of the cultivation of Quina." But the project soon faltered. In their letters to Spain, Olmedo and Ruiz de Quevedo reported that although the mountainous terrain of the region was an significant obstacle on its own, the main challenge to establishing cinchona plantations was the poverty of the region. "Locals of the Villages, which produce *Cascarilla,*" they explained, "find it impossible to care for and develop a plantation." In other instances, they noted that local landowners were discouraged by the uncertainty of such an enterprise, especially because cinchona trees had not been previously grown in plantations. Instead, these landowners were much more likely to grow subsistence crops or established plantation crops such as maize or cacao.[74] Finally, Olmedo and Ruiz de Quevedo cited "poverty" as the main reason why more people in Loja were unable "to develop plantations." "It is evident," they added, "that, lacking the workers and interest to care for the little that they have acquired, [we] have found no good reason to oblige [landowners] to [undertake] the uncertain cultivation [of cinchona trees]."[75] Two years later, in a letter to Spain, Olmedo and Ruiz de Quevedo bitterly described the people of Loja as living "in the grip of laziness" and lacking the "determination" and the "ideas, curiosity and the desire to try or experiment."[76]

The Crown continued to press the issue of cinchona plantations. In a supplementary set of regulations sent to Olmedo and Ruiz de Quevedo in 1796, the Crown described the "reproduction of good Quinas" as an "object of great importance and necessity." To this end, the Crown ordered "landlords of those forests" as well as "Mestizo Indians and other natives" to cultivate cinchona trees "from seeds and from transplants in their own lands and in the royal and unused forests and lands" with instruction from the "Professor Botanist."[77] As incentive, Ruiz de Quevedo and Olmedo were to offer "ample Prizes" to "hacendados and laborers" who have produced the "greatest number of Quina trees" in these plantations.[78] These prizes were likely a pittance in comparison to pervasive poverty in the region as well as a general lack of capital for investment in long-term tree plantations.

Although the results are difficult to assess, existing records suggest that the outcome was mixed. In 1800, Olmedo and Ruiz de Quevedo requested approval to disperse prizes to several individuals who had successfully cultivated cinchona trees.[79] Five years later, in a report to the viceroy, Francisco José de Caldas, a creole naturalist from Bogotá, claimed that Olmedo's efforts had done little to reverse the disappearance of cinchona trees. He also reported that locals in Loja were actively resisting the plantation effort. Caldas recounted the experience of "an honored and curious resident of Loja" who "transported four young [cinchona] plants to the patio of the religious houses of that city where they happily prospered." Upon the death of this "honored" citizen, however, other residents of Loja "introduced horses into the place where the four quina trees grew [and the horses] destroyed [the trees] and made them die."[80]

At first glance, this episode, if true, seems rather puzzling. One would think that the people of Loja would have wanted to promote the cultivation of cinchona trees as quina was one of their main sources of income and an important link to regional trade networks and to the long-distance trade networks of the Atlantic World. At the same time, the transplantation of cinchona trees may have caused anxiety for the residents of Loja. After all, if these trees, which produced superior quina, could be grown in other locations, then Loja might lose its exclusive claim to being the region that produced the best bark. Whether because of poverty or the active resistance of locals, one botanist-chemist was not enough to overcome such obstacles to the establishment of cinchona plantations. On paper in Madrid, the potential of botany as a tool of empire seemed limitless; on the ground in Loja, Olmedo's experiences revealed that the success of European science depended on the interest, support, and labors of the members of colonial Andean society.

THE PRODUCTION OF QUINA EXTRACT

In a supplemental set of instructions produced in 1796, the Crown also directed Olmedo to introduce the chemical processing of the bark as a way to intensify production. It was yet another enterprise in which the botanist-chemist would encounter significant obstacles. Interest in developing an extract of cinchona bark emerged largely from the infusion of chemical ideas and practices into pharmacy in the seventeenth and eighteenth centuries as chemists and pharmacists attempted to isolate and identify the medicinal "virtues" of plants and plant parts.[81] Casimiro Gómez Ortega suggested the idea to the Crown in 1791, citing a recent article on the chemical analysis of cinchona bark by French chemist Antoine Fourcroy.[82] Gómez Ortega suggested that Olmedo had the opportunity to improve upon Fourcroy's work because Olmedo had access to bark fresh from the tree. The idea was that fresher bark would yield a clearer analysis and produce a more potent extract. In 1792, Hipólito Ruiz reinforced this notion in his *Quinología,* where he suggested that extract made in America might have "more virtue" than extract made in Europe because the bark would be fresher.[83] He also noted that many physicians in Europe continued to use pulverized bark therapeutically because they found it to be more potent than the extract made in Europe.[84]

In addition to any potential medical benefits, the extract offered economic benefits as well as another means to make bark harvesting more efficient and effective. According to Ruiz, the quina extract promised "greater exploitation of all [cinchona] barks" because the process provided a means to access even the small amounts of medical "virtue" to be found in the worst barks. "Fabrication of the extract," explained Ruiz, "helps to overcome the loss of more than two thirds of Cascarilla Bark because it is not admitted to commerce" due to its poor quality. In particular, he thought that lesser quality bark, which was not of "equal disposition" as high quality bark, could produce an extract of "equal virtue" to that of the best barks because the process resulted in a concentration of the bark "virtue" in the resinous compound. He also pointed out that the final product was better suited to transportation than dried cinchona bark, which had to be protected from moisture—a difficult task when traveling overland through rain forests and by sea to Europe.[85]

Olmedo had learned of the Crown's interest in the extract as early as 1790. While waiting at Cádiz for his ship to South America, he received a letter from the chamberlain of the Royal Household explaining that "the extract of *Quina* is a prospect that is [most] interesting to His Majesty."[86] Ultimately, implementation of production of the extract proved difficult, especially because Olmedo lacked the proper technology. In 1794, he reported that he was not able to pursue this work for lack of "some important instruments that have not [yet] been sent from the Court" in Spain.[87] He had specifically requested vessels made of glass or tin-plated copper because he knew that many chemists argued that earthenware vessels, which were more commonly available, compromised the purity of an extract.[88] For lack of technology, the attempt to produce quina extract in Loja failed alongside the efforts to introduce cinchona plantations and to instruct bark collectors, in which the botanist-chemist and the corregidor had to contend with the broader challenges of the social and economic dynamics of the colonial Andean World.

Despite the limited impact that Olmedo had in Loja, his appointment as botanist-chemist and codirector of the royal reserve represented an important moment in the emerging relationship between European sciences and empires in the late eighteenth-century Atlantic World. Yet Olmedo's story has received little attention in historical studies. Aside from a tendency in the history of science, until recently, to overlook the Iberian World, the main reason is that his story is told primarily in the Spanish colonial archives.[89] Olmedo did not publish a scientific treatise and left few other traces of his activities even though he met with the famous Prussian scientific traveler Alexander von Humboldt during his time in South America in the early nineteenth century.[90] In this regard, this episode is suggestive of how greater attention to the role of colonial governments as networks of knowledge production could provide new insights into the history of early modern science.

Olmedo's story is indicative of just how intimately intertwined science and empire had become in the Spanish Atlantic World. Because previous studies have focused primarily on the botanical expeditions, the scientific and imperial enterprises in the late eighteenth century often seem distinct endeavors, even if they were connected. Science was not just a tool of empire; nor was empire simply a convenient means to gain access to additional natural phenomena. Instead, the interactions of botanists and colonial officials shaped both enterprises as Ruiz's recommendations influenced the empire's understanding of quina and as Valdecarzana's recommendations influenced the role that the botanist-chemist would play as an agent of empire. In the end, both the Crown and the Royal Botanical Garden claimed Olmedo as their own and envisioned ways in which he could pursue science and empire simultaneously.

The genesis and development of the reforms to the royal reserve of quina in 1790 provide further insight into the ways in which the structure and style of Spanish imperial reform shaped the role that botany would play as a science of empire. Consider the reforms suggested by the Marqués de Valdecarzana in 1789. In many ways, his recommendations against greater government regulation of the quina trade made little sense. After all, the status quo of private trading in the bark, which Valdecarzana sought to preserve, had proven harmful to the cinchona trees as multiple reports of their increasing scarcity suggested. Valdecarzana would have known this since he had a dossier of the empire's accumulated knowledge on quina and the quina trade. At the very least, he would have known that the Crown had recently established a second royal reserve of quina near Bogotá in an attempt to mitigate effects of scarcity. Yet Valdecarzana (and many other officials) remained committed to a policy of laissez-faire in which the colonial government would make minimal intervention into the quina trade aside from the royal reserve of quina in Loja.

One way to make sense of Valdecarzana's recommendations is to put them in the broader context of Spanish imperial reform in the late eighteenth century, known as the Bourbon Reforms. One of the hallmarks of these reforms was a commitment to the liberalization of trade within the empire. Beginning in 1765, imperial reformers began dismantling the old system of imperial trade—a system which, in theory, required that all trade between Spain and Spanish America pass through three ports: Cádiz in Spain, Veracruz in New Spain, and Cartagena in New Granada. In the next few decades, new regulations opened transatlantic imperial trade to additional ports in Spain and Spanish America. Reformers in Spain further committed to this program by placing greater emphasis on taxes on commerce as a source of revenue for the imperial government. In this context, Valdecarzana's recommendations for reform reveal a similar philosophical, even ideological, commitment to the utility of commerce to empire.

As this chapter has shown, this increasingly mercantile vision of empire, prevalent among many Spanish imperial reformers, shaped the role of botanists as imperial agents. For decades, botanists and reformers had envisioned botany primarily as a tool for bringing economic benefit to the empire and its merchants by identifying new and useful plants and plant products.[91] The reforms to the royal reserve in 1790 embraced this vision of empire as well. The instructions from the Crown made clear that the botanist-chemist was to introduce reforms that would not be beneficial only to the royal reserve but would also promote the private trade in quina. Though it appeared that sending a botanist-chemist from Spain to oversee the royal reserve of quina represented an increase in imperial intervention, the Crown's instructions gave him only limited ability to address the problems in the private quina trade that threatened the cinchona trees of South America. When it came to acquiring bark for the shipments to the Royal Pharmacy, the botanist-chemist was expected to act, in a sense, as just another bark buyer with the exception that he represented the king.

At this same time, Olmedo's experience in Loja serves as a reminder that the shortcomings of botany as an imperial science must also be understood within the context of colonial Andean society. Social, environmental, and technological conditions all placed important constraints on his ability to serve as an effective agent of empire in Loja. The local context of the cinchona trees and the royal reserve of quina in the Audiencia of Quito proved too recalcitrant for the Crown and its botanist-chemist to establish effective control. With his experience and training at the Royal Botanical Garden and the Royal Pharmacy, there is no question that Olmedo was up to the task and familiar with the current techniques of European botany, pharmacy, and chemistry. Yet this knowledge was not enough—applying the knowledge effectively required some reorganization of colonial society as well as new understanding of the natural world.

Consider the prospect of establishing a cinchona plantation. Not only did it represent a significant shift in imperial government's conception of quina, but it also required the reshaping of the land (clear cutting a patch of forest to make space for the planting of cinchona trees) and the marshaling of labor (workers to clear the land as well as plant and tend to the trees). How could merchants and landowners in an impoverished region of the southern sierra be expected to gamble precious land, capital, and labor on an enterprise that was largely untested and whose profitability was still in question? The scientific management of the empire's cinchona trees seemed like a good idea on paper, but in practice, the mountainous terrain of the Andes, the persistent poverty in Loja, the migration of skilled laborers, and the lack of technology all represented significant obstacles to the realization this new synergy of science and empire.

The other remarkable feature of the 1790 reform of the royal reserve is that it reflects a transformation in the understanding of the cinchona tree as an imperial natural resource. Given that officials in Madrid were committed to preserving the private trade in quina (by Spanish merchants, of course), their only recourse for addressing the scarcity of cinchona trees was to frame the problem as one of production rather than distribution or consumption. As a result and with the urging of botanists like Hipólito Ruiz, new imperial policies turned their attention to promoting the cultivation of cinchona trees. Yet, in keeping with the mercantile vision of empire, officials in Madrid focused primarily on encouraging merchants and landowners to develop cinchona plantations on their own. Nonetheless, such policies signaled a shift from previous perceptions of quina as a finite resource to an understanding of the bark as a cultivable commodity. This shift was one of the consequences of the entanglement of botany and empire and provides evidence of the influence, even if limited, that botanists were having on imperial policy.

What about the effects of empire on botany? As shown above, imperial ideologies played an important role in defining the ways in which botany was deployed as a tool of empire. In the wake of the failed efforts of the royal pharmacists to assert greater control over the royal reserve from Madrid, the Crown and its advisers recognized the need to send a representative of European science to ensure that the proposed reforms were being implemented. At the same time that this represented a major endorsement by the Crown of the utility of botany to empire, Olmedo's experience in Loja revealed that the apparent usefulness of botany faltered in colonial contexts in which deeper political and economic shifts thwarted the botanist-chemists' efforts to reform land, labor, and technological process involved in the production of quina.

Seven

Regalist and Mercantilist Visions of Empire in the "War of the Quinas"

In September 1800, Francisco Antonio Zea (1766–1822), a botanist from the Viceroyalty of New Granada, published an article in Madrid on the classification of quina in *Annals of Natural History.*[1] Despite its seemingly innocuous title, "Report on quina according to the principles of Mr. Mutis," the article was a polemic. It focused on two recent proposals for the identification and classification of the different kinds of quina. These studies had been published nearly a decade earlier by some of the Spanish Empire's most prominent naturalists and botanists: José Celestino Mutis, director of the Royal Botanical Expedition in New Granada, and Hipólito Ruiz and José Pavón, the former directors of the Royal Botanical Expedition to Peru and Chile. Their publications represented some of the earliest tangible returns from the Spanish Crown's patronage of the botanical expeditions. The only problem was that these botanists proposed divergent systems for identifying and classifying the various kinds of quina. In his article, Zea sought to solve this problem by reducing Ruiz and Pavón's system of classification to that of José Celestino Mutis, who had been Zea's mentor in New Granada. Yet, instead of creating consensus, Zea's article generated controversy.

For the Crown and botanists alike, such an outcome was far from the desired one. After all, for much of the eighteenth century, botanists in Spain, like their contemporaries elsewhere in Europe, had promised that their science would be useful to empire by providing knowledge of plants and their utility. It was based on these promises that the Crown had supported various botanical and natural historical enterprises including expeditions as well as the Royal Botanical Garden and the Royal Cabinet of Natural History in Madrid. Yet when it came to the seemingly simple task of the classification of a single genus of trees (*Cinchona*), botanists could not agree. They had failed the empire when it mattered most, when the Crown and its advisers needed to know which kinds of quina fell under the purview of the royal reserve and were in need of imperial administration. Even a half century after the founding of the royal reserve of quina in Loja, knowing what was the best bark remained essential to imperial management of this valuable natural resource.

This debate over the classification of quina provides further insight into the evolving relationship between science and empire in the late eighteenth-century Spanish Atlantic World. This chapter shows how the integration of botanists into the broader structure of the Spanish Empire undermined efforts to develop a scientific understanding of the cinchona tree and its bark. It argues that a seemingly technical debate among Spanish naturalists was as much about conflicting visions of imperial order as it was about competing claims about the natural world. Ultimately, the disagreement among botanists over the classification of quina was symptomatic of a larger rift between two visions of imperial order—one regalist and another mercantile—each supported by broader social and political networks within the Spanish Empire.[2] The fragility of science became apparent in the face of a seemingly insurmountable divide between these two visions of empire that vexed much of Spanish colonial society in the late eighteenth century as reformers in Spain and elites in Spanish America engaged in debates over how to reconcile the interests of commercial elites in further promoting trade with the interests of reformers and the Crown in further strengthening royal power.

This debate over the classification of quina is significant because it shows how the epistemic culture of the colonial government came to influence the production of knowledge even within the botanical community. Botany and its practitioners had become so intertwined with the Spanish colonial government that they too were subject to the ideological rifts and tensions that permeated the institutions of colonial rule. The epistemic culture of the colonial government reflected the tensions of empire as a result of a commitment on the part of the Crown and colonial officials to seek information and advice from different groups with experience of and an interest in quina. As a result, the colonial bureaucracy provided the institutional context in which various conceptions of nature and empire interacted and competed. Consequently, rather than standing above or outside the divergent perspectives that existed within the empire, Spanish botanists—as they became more closely integrated into the apparatus of colonial governance—assimilated to the agonistic politics of knowledge that characterized the epistemic culture of the Spanish colonial government at large.

Quina in the Botanical and Medical Traditions

At the outset, it is important to recognize that there were real intellectual and scientific issues at stake in the debate over the classification of quina. In the 1770s and 1780s, the Crown had supported the botanical studies of José Celestino Mutis and those of Hipólito Ruiz and José Pavón partly in the hopes that one of them would develop a more reliable system for identifying the different types of quina. The persistent problem of the adulteration of shipments of quina with other kinds of bark made it imperative to find an effective means for distinguishing true bark from false bark as well as good bark from bad bark. At the same time, the Spanish Crown wanted a system of classifying the bark that could be implemented throughout the empire in order to identify the best barks for the royal reserve with consistency.

It was in the 1790s that the Crown's support of botanical research started to bear fruit. In 1792, Hipólito Ruiz published his *Quinología* in Madrid, a comprehensive study of quina that included the botanical description and classification of the seven species of cinchona that Ruiz observed during his time in Peru.[3] It is significant that the first published results from the Royal Botanical Expedition to Peru and Chile focused on quina. Ruiz's study not only signified the continued importance of quina as an object of interest but also provided an opportunity to demonstrate the practical benefit of the Crown's patronage of botany. In 1799, Ruiz and Pavón published the second volume of their *Flora Peruviana et Chilensis* (Flora of Peru and Chile), a work that eventually ran to three volumes and contained the botanical description and classification of plant species that they had studied during their time in South America.[4] The second volume of their *Flora* included a discussion of the genus *Cinchona,* which they had expanded to include an eighth species discovered by Juan Tafalla (1755–1811), the botanist that continued the work of the Royal Botanical Expedition after Ruiz and Pavón had returned to Spain in 1788.

The publication history of Mutis's work on quina is a little more complicated. The first published account of his system for classifying quinas appeared as a summary published anonymously in Cádiz in 1792 under the title *Instrucción formada por un facultativo existente por muchos años en el Perú, relativa de las especies y virtudes de la Quina* (A preliminary investigation of the species and virtues of quina by a physician who has lived in Peru for many years).[5] In 1793 and 1794, Mutis published the full results of his studies of quina in several installments in a periodical based in Bogotá under the title *El Arcano de la Quina* (The secret of quina).[6] In addition to these publications, an unofficial and abbreviated version of Mutis's work on quina appeared in the *Mercurio Peruano,* a journal published in Lima, in 1795 and in the *Seminario de agricultura y artes dirigido a los párrocos,* a Madrid-based journal, in 1798.[7] Because *El Arcano de la Quina* represents the most complete statement of Mutis's system of classification, it is this work that I use in the discussion that follows.

One important difference between the works of Mutis and of Ruiz is that they addressed their texts to different audiences engaged in different ways of studying the cinchona tree and its bark. In short, Mutis wrote his *Instrucción formada* and *El Arcano de la Quina* for physicians and pharmacists, while Ruiz wrote his *Quinología* for botanists. Accordingly, the main traditions in the literature on quina—the medical and botanical—framed different types of questions and problems. Since Europeans had first encountered quina, physicians and pharmacists had grappled with the question of the proper therapeutic administration of the bark and, in the eighteenth century, medical practitioners addressed, debated, and discussed such questions in a wide and extensive literature on fevers and their treatment.[8] Mutis engaged with this medical literature directly, as evidenced by the subtitle of his work, "Discourse that contains the medical part of four species of official quinas, their eminent virtues and legitimate preparation." By contrast, Ruiz's *Quinología* engaged primarily with the literature on the botanical description and classification of the cinchona tree dating back to Charles Marie de la Condamine's 1738 report published by the Royal Academy of Sciences in Paris.

Though related, medical accounts of quina remained distinct from botanical accounts for much of the eighteenth century. Generally, medical works on quina paid little attention to the classification of tree species. Mutis's *El Arcano de la Quina* exemplifies this tradition. While giving a detailed account of the medical utility and therapeutic application of his four official quinas, it provides no botanical description of the tree. By contrast, the botanical literature focused primarily on description and classification of the tree often with little attention to the details of medicinal administration of the bark. *Quinología* is a good example of this botanical literature; Ruiz devotes most of his discussion to the botanical descriptions of different species of cinchona, while only briefly discussing the medical uses of the bark.[9] As a result of this established and implicit division of intellectual labor in the study of the cinchona tree and its bark, contemporaries may not necessarily have read the works of Mutis and Ruiz as presenting contradictory systems of classification—at least not at first. Instead, they may have read these works as complementary in that they approached the study of the cinchona tree and its bark in different ways.

Although they approached quina from different disciplinary perspectives and traditions, Mutis and Ruiz did address similar issues. A central task in any attempt to classify or describe cinchona was to identify which species was the "official" cinchona. The idea that there was an official cinchona came from the Swedish taxonomist Linnaeus, who gave the name *Cinchona officinalis* to the only species of cinchona that he described in his *Species Plantarum* (1753). Linnaeus had based his description and classification of this species on the information and images from Charles Marie de la Condamine's 1738 report to the Royal Academy of Sciences in Paris. The inspiration for the genus name undoubtedly came from the popular myth of the Countess of Chichón, the wife of a seventeenth-century viceroy of Peru, who was allegedly the first European to be treated with the bark and the first to introduce the bark to Europe.[10] According to Mauricio Nieto Olarte, Linnaeus chose the term *officinalis*, which means "of the office" in Latin, as reference to the therapeutic use of the tree's bark by physicians.[11]

To identify a variety as *officinalis* had both scientific and practical implications. For botanists, identification of the official cinchona species was the key to classification of all cinchonas because it would be the standard by which other varieties were judged. Use of this terminology also provided a means for botanists to correlate their classification of cinchona species with that of Linnaeus. Meanwhile, bark collectors and merchants knew that they could fetch a higher price for their bark if it was designated as official—a designation that came to indicate bark of the highest quality. This situation only encouraged confusion and contention in the quina trade as every merchant and bark collector claimed that their bark was from the official species of cinchona. Later, Ruiz and Pavón determined that La Condamine's account of the cinchona tree, on which Linnaeus based his description and classification, was actually a conflation of two distinct species. This error further destabilized the ability of botanists, bark collectors, and merchants to reach consensus on what was official cinchona.[12] "The problem with the word *officinalis*," Nieto Olarte observes, "is that it was simultaneously a classification and an evaluation."[13]

A closer look at their attempts to identify which cinchona was official highlights some of the differences between Mutis's and Ruiz's approaches to the classification of quina. The divergence began with their objects of analysis as reflected in the titles of their publications. For Mutis, the bark was the primary object of classification, and so he called his work *El Arcano de la Quina*. Throughout the eighteenth century, "quina" had referred almost exclusively to the cinchona bark or its powdered form that was used medicinally. Meanwhile, the full title of Ruiz's work, *Quinology or Treatise on the Tree of Quina or Cascarilla*, indicated that the tree was the primary object of classification. Ruiz devoted only one paragraph to the medicinal uses of cinchona bark, while Mutis provided little description of the cinchona tree in his work.

Mutis and Ruiz also had different conceptions of what it meant to designate a species of quina or cinchona as official. For Mutis, such a designation was an indication of medical utility. According to this criterion, Mutis concluded that four of the seven known species of quina were "official." This was the "secret" (*arcano*) alluded to in the title of his work (*El Arcano de la Quina*). Mutis argued that red, orange, yellow, and white quina—all names used by bark collectors and merchants—constituted "four legitimately official species [of bark] in which reside eminent [medicinal] virtues," while the other three species were of "lesser efficacy."[14] Although all species had utility as a febrifuge, orange quina was the most effective febrifuge and red quina was more effective for treating gangrene. In identifying his four official species, Mutis used the methods of both botany and medicine in his studies. Drawing on his skills "as a Botanist," he was able "to distinguish the legitimate species and varieties from other closely-related kinds."[15] Meanwhile, drawing on his skills "as a physician," Mutis was able "to separate the official species from other less virtuous [species]."[16]

For Ruiz, use of the term official was purely a matter of nomenclature. According to the tenets of Linnaean taxonomy, each unique species had to have a unique name to avoid confusion and redundancy.[17] From Ruiz's perspective, there could be only one official cinchona. Different conceptions led to different results. Although Mutis and Ruiz agreed that there were seven identifiable species (whether bark or tree) in need of classification, the former identified four official quinas, and the latter only one official cinchona.

Ultimately, such differences did not necessarily make the two systems inherently incompatible or contradictory. Instead, they show how these texts were part of a broader intellectual division of labor in eighteenth-century literature on quina in which medical texts tended to focus on the bark and its medicinal properties while natural historical texts focused on the tree and its classification. While the conflict between Mutis and Ruiz has been called "the war of the quinas," the use of such language ignores the fact that their two proposed systems of classification existed side by side for nearly a decade without any significant debate.[18] Indeed, in 1792, Ruiz was far from animosity as he awaited Mutis's work with eager anticipation. "What enlightenment," Ruiz wrote in the prologue to his *Quinología*, "would the publication of the Quinology of such a wise Physician and Botanist be able to promise us!"[19] When Mutis did publish his work, Ruiz may have found much to disagree with. But he did not acknowledge that publicly in print until 1802, nearly a decade later.

Zea's "Botanical Synonyms," Ruiz and Pavón's *Supplement*

Public debate over the merits of the two systems of classification erupted with the publication of Francisco Antonio Zea's forty-page article that appeared in the *Annals of Natural History* in 1800.[20] Zea's article was provocative for two reasons. First, he attempted to correlate and reduce Ruiz and Pavón's eight species of cinchona to Mutis's four official quinas. In this, he asserted the superiority of Mutis's system of classification to that of Ruiz and Pavón. Second, he openly questioned Ruiz and Pavón's methods of botanical classification. In 1801, Ruiz and Pavón responded to Zea in their *Suplemento al Quinología* (Supplement to the Quinology).[21] While part of this work provided an update on the botanical classification of cinchona based on the continuing work of Juan Tafalla, the majority (nearly one hundred pages) of the *Suplemento* was devoted to a nearly line-by-line impassioned refutation of Zea's article.

Ruiz and Pavón had good reason to respond so vehemently. After all, Zea had made some serious critiques of their work. The first was his claim that a "botanical synonymy" existed between the four official quinas of Mutis and the various cinchona species identified in the works of several prominent botanists including Ruiz's *Quinología* and the first volume of Ruiz and Pavón's *Flora Peruviana et Chilensis*.[22] For simplicity's sake, table 4 provides an illustration of Zea's botanical synonymy, although it should be noted that he did not present this information in this way. There was nothing especially novel or inflammatory in this. Many eighteenth-century botanical texts provided such explanations to help readers translate between different botanists' proposed classifications of a plant genus. What made Zea's "botanical synonyms" so provocative was his insistence that Mutis's official species were not just medical kinds but also botanical kinds. In addition, he reduced Ruiz and Pavón's cinchona species—in 1800 there were eight—to just three of Mutis's official quinas (table 4). What made this more controversial was that Zea ignored the division between botanical and medical approaches to the study of the cinchona tree and its bark by trying to treat the two systems of classification as comparable if not equivalent.

Zea would have found inspiration for his approach in the work of his mentor. In the first installment of *El Arcano de la Quina*, Mutis included a table that established the correspondence of the names for four different types of quina in botany, in commerce, and in medicine (figure 7). With this table in hand, it would have been a short (but significant) step to Zea's own proposal regarding the equivalence between different systems of classifying quina. Mutis had also expressed his reservations about previous attempts at the botanical classification of cinchona. He noted several "calamities" that had appeared in the work of prominent botanists including the misidentification of noncinchona species as cinchona and the misclassification of bona fide cinchona species.[23] Ultimately, Mutis questioned the reliability of any botanical classification performed before the "true essential character of the genus has been

TABLE 4. Francisco Antonio Zea's "Botanical Synonymy" (1800)

Mutis's Species	Orange Quina (*Cinchona lancifolia*)	Red Quina (*Cinchona oblongifolia*)	Yellow Quina (*Cinchona cordifolia*)	White Quina (*Cinchona ovalifolia*)
			C. purpurea	
	C. officinalis		*C. tenuis*	
	C. glabra		*C. pallescens*	
Ruiz *Quinología**	*C. fusca?*†	*C. lutescens*	*C. micrantha*	—
La Condamine	Quinquina	—	—	—
Linnaeus	*C. officinalis*	—	*C. officinalis?*‡	—
Wahl	*C. officinalis*	—	*C. pubescens*	*C. macrocarpa*

* Zea determined the equivalence of Ruiz and Pavón's cinchona species with those of Mutis through their descriptions in two different publications: Ruiz's *Quinología* and the second volume of their *Flora Peruviana et Chilensis*. To make things more complicated, Ruiz and Pavón changed the species name of some of the cinchona species between the publication of Ruiz's *Quinología* and the second volume of their *Flora Peruviana*. As explained by Arthur R. Steele, Ruiz and Pavón made the following changes (the first name is the name from the *Quinología* and the second is from the *Flora*): 1) *Cinchona officinalis* to *C. nitida*; 2) *C. tenuis* to *C. hirsuta*; 3) *C. glabra* to *C. lanceolata*; 4) *C. lutescens* to *C. magnifolia*; 5) *C. pallescens* to *C. ovata*.

† The question mark appears in Zea's original article indicating some uncertainty as to whether the species is equivalent to the Mutis's quina species.

‡ Linnaeus's *C. officinalis* appears under two of Mutis's species because Zea was uncertain as to which of Mutis's species corresponeded to that of Linnaeus. His use of question mark indicates this uncertainty as well.

fixed."[24] The essence of the genus, in Mutis's view, was the medical efficacy of the bark, a central consideration in his classification system.

Beyond his proposed "botanical synonyms," Zea also challenged the work of Ruiz by questioning his methods. First, Zea accused Ruiz of misunderstanding Linnaeus's classification of cinchona. Whereas Ruiz in his *Quinología* had treated Linnaeus's *Cinchona officinalis* as distinct from La Condamine's *quinquina*,[25] Zea considered the two to be equivalent and argued that Ruiz was trying to divide "synonyms which are inseparable." Zea claimed to have examined the "skeletons" of these cinchonas in the herbarium of Ruiz and Pavón and found that the alleged "difference" between the two samples was "entirely

PROSPECTO

DE LOS NOMBRES Y PROPIEDADES DE LAS QUINAS-OFICINALES.

EN LA BOTANICA.

CINCHONA.

Lancifolia. § Oblongifolia. § Cordifolia. § Ovalifolia.

QUINA.

Hoj. de lanza. § Hoja oblonga. § Hoj. de corazon § Hoja oval.

EN EL COMERCIO.

| Naranjada. § | Roxa. § | Amarilla. § | Blanca. |
| Primitiva. § | Succedánea. § | Substituida. § | Forastera. |

EN LA MEDICINA.

AMARGO.

Aromático. §	Austéro, §	Púro, §	Acérbo.
Balsámica. §	Astringente. §	Acibaráda. §	Xabonosa.
Antipyréctica. §	Antiséptica. §	Cathártica. §	Rhyptica.
Antídoto, §	Polycresta. §	Ecphráctica. §	Prophiláctica.
Nervina, §	Muscular, §	Humoral. §	Visceral.

Febrífuga. §   Indirectamente febrífugas.

Figure 7. "Prospectus" from Mutis's "El Arcano de la Quina" (1793). Biblioteca Nacional de Colombia, Bogotá, Colombia.

accidental," such that "they could be from the same tree." He made a similar observation with four of Ruiz and Pavón's cinchona species (*C. tenuis, C. pallescens, C. purpurea,* and *C. microantha*) that he reduced to two of Mutis's official quinas. As to the difference between these species, Zea observed that it is "so small and accidental that I do not believe that it is worth the attention of the Botanist, according to the principles of Linnaeus." In these instances, Zea supported Mutis's system with the authority of Linnaean taxonomy, while at the same time implying that Ruiz and Pavón did not fully understand the "principles of Linnaeus," whose classification system was the only one officially recognized by the Spanish Crown. Ultimately, he concluded by urging Ruiz and Pavón not to engage in "useless contestation" of his proposal, especially because "not every professor is disposed to reconcile the apparent contradictions that are found on every line of the *Flora Peruviana et Chilensis* and in the *Quinología o tratado del árbol de la Quina*."[26]

THE RESPONSE FROM RUIZ AND PAVÓN

Ruiz and Pavón's response was both prompt and prolific. In 1801, they published their *Suplemento a la Quinología,* which included an extensive refutation of Zea's article.[27] Their first priority was to respond to Zea's critique of their method of classification. At the same time, they elected to leave the "contestation" of Mutis's claims about the medical "virtues" of "his four Quinas orange, yellow, red and white" to the "Professors of Medicine."[28] Ruiz and Pavón focused on Zea's claim that Mutis's orange quina was equivalent to Linnaeus's *Cinchona officinalis.* They argued that Mutis and Zea did not know "botanically" the species described by Linnaeus and La Condamine nor did they possess "exact knowledge" of the species described by Ruiz and Pavón. And what did it mean to know a plant species "botanically"? According to Ruiz and Pavón, it required two methods: direct observation in the field and side-by-side comparison of the live or dried specimens of related species. As for direct observation, they insisted that it must be done in person. To make this point, Ruiz and Pavón contrasted their own research methods with those of Mutis. They wrote:

> We do not doubt that Doctor Mutis had traveled extensively through the forests and fields of Santa Fe [de Bogotá], Mariquita and other [parts] of the New Kingdom of Granada before being named Director of the [Botanical] Expedition [in New Granada]; but we are sure that from that point onward he was almost always content to send Servants and later Students to collect materials [and that] he was not able, on account of his age and ailments, to go out, with Portfolio underarm, collecting plants, observing them, describing them and painting them in their native habitats like we, the Botanists of

Peru, always did; we cannot deny this advantage, along with our eight years of study at the School of the Royal Botanical Garden, in comparison to Dr. Mutis.[29]

They repeatedly emphasized the importance of making observations in the field and stressed that the botanical images from their magnum opus *Flora Peruviana et Chilensis* "were not drawn in the darkness and comfort of a Museum."[30] Here, they employed the association of darkness with ignorance and light with knowledge—a prominent trope during the Enlightenment—to make a pointed attack on Mutis, who no longer traveled out into the field and instead stayed in his study to examine specimens collected by his servants and students.[31]

As to doing botany indoors by comparing dried specimens, Ruiz and Pavón argued that both Mutis and Zea fell short here as well because they did not have the right specimens. Mutis had not performed direct comparisons of "the Drawings, Skeletons [i.e., dried plants], Descriptions, Observations and Barks of his four Quinas with the same materials from our ten [Quinas]."[32] Such a comparison would have been virtually impossible as these two collections were located on opposite sides of the Atlantic.[33] "Why, Mr. Zea, wouldn't Dr. Mutis critique the *Quinología* of Ruiz?" they asked. "Because," they replied, "it is very difficult, if not impossible, to make such a critique, without having the objects present in order to compare them and to see the differences and similarity between them."[34] Zea's problem was that he relied too heavily on his memory. They noted that it had been six years since he had left Bogotá and that was a long time to try to retain the experience of Mutis's specimens in his "imagination."[35] Memory was no match for direct comparison of specimens. Ultimately, Zea (and Mutis) lacked the right kinds of experience in the field and in the natural history museum to make a claim about botanical synonymy.[36]

Although such critiques undermined any possibility of establishing the equivalence of Mutis's species with those of other botanists, Ruiz and Pavón did propose a correlation between Mutis's quinas and the cinchona species identified in other botanical works (table 5). They even conceded that Mutis's official quinas were probably distinct species but still refuted the use of medical utility as the basis for the distinction. They observed, "all Quinas are medicinal" according to "experience" and made note of the "Rule of Linnaeus: *Plantae, quae Genere conveniunt, etiam virtute conveniunt*" (plants of the same genus share the same virtue). To this end, they argued that the medical utility and efficacy of different cinchonas should be determined only after the species had been classified using botanical methods. The medical qualities of a species had nothing to do with its classification. Here, Ruiz revealed that cinchona species in his *Quinología* were listed in order of "greatest efficacy."[37] Yet this ordering had no bearing on the identification and classification of the botanical species. It was simply one manner of presenting the information—no more or less arbitrary than alphabetizing them. Though Ruiz and Pavón could not speak about Mutis's quinas with certainty, they were more than certain that Zea was not only wrong to reduce their cinchona species to those of his mentor but also lacked the knowledge and experience to do so. As for Mutis, they had sharp words for him as well. "The Teacher of Señor Zea," Ruiz and Pavón wrote, "should have shown more modesty and moderation in stating his opinion to the public in such a magisterial tone."[38]

Mutis, Quina de Santa Fé, and the Regalist Vision of Empire

If the debate between Mutis, Ruiz, and their associates had only been about botanical classification, it might have been resolved amicably. After all, Ruiz and Pavón agreed that Mutis's quinas were, in fact, four distinct species. It was only a matter of how to correlate these species with existing classification systems as well as their own. Yet, as it turned out, the intellectual fault lines of this debate overlapped with other divides in the epistemic and political culture of the Spanish Empire. The divergence between Mutis and Ruiz and their supporters involved much more than a question of classification.[39] It also involved questions of political and economic order of the Spanish Empire. As it turned out, the two sides of this debate had conflicting visions of the imperial order and these concerns were as much at play in the question of classifying quina as were questions of scientific method.[40]

Mutis's vision of empire was regalist. In the preceding decades, he had supported the royal reserve and even urged the Crown to establish a full royal monopoly of quina in which all production and distribution of the bark would be controlled by the state. Mutis first expressed his views in letters to Charles III in 1763 and 1764. The primary purpose of these letters was to secure royal patronage for the study of natural history in New Granada. Mutis explained that royal support would benefit both science and empire, especially beccause natural historical research would produce further knowledge of the "natural treasures" in Spain's American territories.[41] He used the case of quina as an example of a natural treasure "bestowed uniquely to the Dominions of Your Majesty," and proposed that natural historical research on quina would restore its reputation among the "Physicians of Europe" by providing "observations" on how to "handle [it] with greater confidence, clarity, and certainty."[42]

Knowledge was important but alone it was not enough to restore the bark's reputation. Mutis also urged direct intervention in the quina trade by the Crown in order to put a stop to "the miseries, which we fundamentally fear, multiply as a result of the ambition of those that trade in this precious

TABLE 5. Ruiz and Pavón's Suggested Botanical Equivalencies for Mutis's *Quinas*

Mutis's Species	Orange Quina (*C. lancifolia*)*	Red Quina (*C. oblongifolia*)	Yellow Quina (*C. cordifolia*)	White Quina (*C. ovalifolia*)
Lopez Ruiz	*C. angustifolia*	—	—	—
Ruiz *Quinología*	—	*C. lutsecens*	*C. pallescens*	—
Wahl	—	—	—	*C. macrocarpa*

Source: These equivalencies are suggested in Hipólito Ruiz and José Pavon, *Suplemento a la Quinología* (Madrid: En La Imprenta de la Viuda e Hijo de Marin, 1801), 44–67.

* Ruiz and Pavón explained that they could not find mention of "orange quina" in any works on cinchona and suggested that Mutis was the first to use the term. They also suggested that Mutis's "orange quina" was equivalent to a species identified by Sebastian José Lopez Ruiz in New Granada as well as a new species identified by Juan Tafalla in Peru.

good."[43] Merchants and bark collectors who harvested and traded in quina for personal gain simply could not be trusted. Moreover, they contributed to the problem of scarcity by cutting down "an entire Quina tree in order to obtain only a small portion [of it]."[44] Of all the problems with the quina trade, Mutis considered the worst to be a bark shortage that resulted from unsustainable harvesting. "There will be a shortage in the third century since its happy discovery," he warned, "if Your Majesty does not apply the most opportune precautions in time."[45]

He never wavered in his support of a total royal monopoly—not just a series of royal reserves. In 1773 and again in 1787, Mutis wrote to the Crown seeking the establishment of a royal monopoly. Part of the impetus for Mutis's support for a royal monopoly came from his connections to the court of the viceroy of New Granada in Santa Fé de Bogotá. The viceroys of New Granada, beginning with José Solís in 1753, consistently supported the idea of a royal monopoly in the interest of bringing economic benefit to the quina-producing regions of New Granada.[46] In 1787, the Audiencia of Santa Fé de Bogotá asked Mutis to write a report on quina and the quina trade for the Crown. Once again, he recommended the establishment of a royal monopoly. The archbishop-viceroy of New Granada endorsed Mutis's plan and, in his letter to Spain, described Mutis as "a man of equal value as political advisor, philosopher, statist and scientist"—a clear indicator of the multiple roles that a naturalist could play in the Spanish Empire. Also in his letter to Spain, the archbishop-viceroy made clear the choice that the Crown faced with regard to quina. He wrote:

> There are only two paths for exploiting quina: free trade and the royal monopoly . . . At almost the same time as Cádiz took charge of the business of quina, the Dutch [East India] company displaced the Portuguese from the business of spices by force. The East India Company gained millions of pounds of cinnamon such that while knocking down the cinnamon trees of Conchinchina and Malabar with one hand, it was planting with the other hand the new [trees] which were necessary to meet demand. The causes, by which the production of cinnamon increases and that of the bark [quina] decreases, are clear: the Dutch concentrate all species and drugs under the state monopoly and determine the profit by official conduct. We must learn from the Dutch![47]

The contrast between Spain and the Netherlands and their respective strategies for managing their most precious natural resources could not be starker. The Dutch profited from cinnamon via a state monopoly, while Spain experienced nothing but the increased scarcity of trees and decreased quina as result of the trade remaining in the hands of merchants in Cádiz. Mutis would have concurred.

With his connections to the viceroy and vocal support for government intervention in the quina trade, it is no wonder that Mutis received the endorsement of Viceroy-Archbishop Antonio Caballero y Góngora when it came time to appoint a new director of the royal reserve of quina near Bogotá in 1785. It was a post that he eagerly accepted as it was another step toward a royal monopoly on quina. For nearly two years, Mutis oversaw shipments of quina from Santa Fé to the Royal Pharmacy in Madrid—a brief realization of his regalist vision of empire.[48] In December 1787, however, the Crown abruptly put an end to the royal reserve of quina near Bogotá after further testing of quina from Santa Fé revealed that it was not the same quality as quina from Loja. It was a move that must have troubled Mutis deeply.

When read in the context of the events of the 1780s, Mutis's claim that there were four official quinas in his *El Arcano de la Quina* takes on new significance. Here, he was openly, if indirectly, rejecting the findings of experts in

Madrid that quina from Santa Fé was of inferior quality to quina from Loja. His text thus supported the established tradition of rejecting European claims and asserting the superiority of the local knowledge of the Andean World—a tradition that traced back to Pedro de Valdivieso's critique of the Royal Pharmacy in 1773 and perhaps even Fernando de la Vega's conversations with Charles Marie de la Condamine in the 1730s and 1740s. More important, the Crown's decision to shut down the royal reserve near Bogotá based on the determination that its quina was inferior had shown that assessments of the bark quality were directly linked to the question of government regulation. Consequently, the revitalization of the royal reserve and the realization of Mutis's regalist vision of empire depended upon his ability to establish that quina from Santa Fé was equivalent to quina from Loja.

Mutis's strategy was both ambitious and ingenious. Rather than publish a tract that was a transparent defense of quina from Santa Fé de Bogotá, he attempted to change the rules for identifying and classifying the bark. By making medical utility the primary characteristic of an official quina, Mutis decreased the importance of classification according to botanical methods or geographical location. Obviously the lack of medical efficacy of quina from Santa Fé had been the main problem for experts in Madrid (even though a few years earlier medical experts in Madrid had approved the bark based on its efficacy). Mutis probably knew that he was not going to convince them directly and, instead, employed a strategy in which he opened the category official to several kinds of quina (rather than just one), and publicized his findings not in scientific journals but in learned periodicals in Santa Fé and Lima, where many quina merchants were to be found. Perhaps by highlighting the prestige and potential profitability of the four official quinas from Santa Fé, Mutis could once again attract the attention of the Crown and achieve a royal monopoly of quina.

Ruiz, Quina from Peru, and the Mercantile Vision of Empire

In contrast to Mutis's regalist or state-centered vision of empire, Ruiz embraced a mercantile or commerce-centered vision of empire in which trade within the empire, even in quina, would be left to merchants. Both visions were consistent with conceptions developed in the broader eighteenth-century discussions about the reform of Spain's economic and political relationship with its American territories, and in the specific discussion of the royal reserve of quina from the 1750s to the 1780s.[49] Indeed, many officials in South America had developed and debated the "two paths for exploiting" quina in the decade before the debate between Mutis and Ruiz.

These discursive precedents and existing imperial policies had led to a fairly clear geographical distribution of the regalist and mercantile visions of the empire. Consider the American responses to the 1751 order that established the royal reserve. In responding to this order, both the residents of Loja and Miguel de Santisteban, special envoy of the viceroy of New Granada, supported the royal reserve. The residents of Loja saw the royal reserve as an opportunity to stop merchants in Piura, a market town, from profiting on Loja's project. At the same time, Miguel de Santisteban argued that the royal reserve would help to protect cinchona bark from the greed of merchants, which only resulted in the increasing scarcity of the trees and the decreasing quality of the product due to mixing of good quality quina from Loja with inferior types of bark.[50]

In his 1753 report to the Crown, the viceroy of Peru offered a different view. Viceroy Manso de Velsaco emphasized possible negative effects on the merchant community in his jurisdiction. He noted that a royal monopoly "eliminates the ability of those, who make living by [the bark], to trade freely."[51] Just as these early replies split between the officials in New Granada advocating royal administration, and officials in Peru advocating for private traders (even if not outright supporting a policy of free trade), so Mutis, closely associated with the viceroy of New Granada, and Ruiz and Pavón, associated with interests in Peru, replicated these views.

The particular situation of Loja highlights the complexity of interactions between policies and practices rooted in the regalist and mercantile visions of imperial order. As we have seen, Loja was a center of bark production. Many considered the region to produce the best bark, especially because it was in Loja that Europeans allegedly first encountered quina. As the quina trade flourished, the province of Loja developed a double identity. On the one hand, Loja was under the political jurisdiction of the Audiencia of Quito, which, in turn, was under the jurisdiction of the Viceroyalty of New Granada. On the other hand, the province of Loja had strong economic ties with the Viceroyalty of Peru, especially because the closest market town (Piura) and Pacific port (Paita) were part of Peru (figure 1). As a result, the viceroys of New Granada and of Peru had different interests and stakes in the quina trade. These historical conditions combined with ideological commitments gave rise to social and political networks of individuals and institutions that supported conflicting visions of the Spanish imperial order.

Mutis and Ruiz's debate was not simply a conflict of visions of the natural and imperial order, but a deeply rooted political conflict between two factions within the colonial government and its epistemic culture. Mutis's commitment to the royal reserve was not only ideological but also the result of his own interests being staked to those of the Viceroyalty of New Granada. But what about Hipólito Ruiz and José Pavón? After all, unlike Mutis, they were not permanent residents in South America, and at the time of the debate, they had not set foot in Peru for over a decade.

Just as Mutis was connected to broader political and social network centered on the vice-regal court in Bogotá, Ruiz was connected to a different political and social network with significant ties to Madrid and Lima. Ruiz and Pavón had close ties to the royal court in Madrid. Royal patronage provided material support for their expedition to Peru and Chile, as well as for their continued work in Madrid on the publication of their multivolume *Flora Peruviana et Chilensis*. As former students of its director, Casimiro Gómez Ortega, both botanists also had strong ties to the Royal Botanical Garden, as reflected in the direct institutional link between the botanical garden and the "Botanical Office," the base of their operations upon their return to Madrid in 1788. When Zea in 1800 cited them as botanists with "special charge from the Government," Ruiz and Pavón did not deny the characterization.[52]

Indeed, such connections to the Crown may have exerted an influence on their views on quina and the proper government policy regarding the extraction of the bark. Like Mutis, Ruiz published his *Quinología* in 1792, just a few years after the Crown had definitively decided to close its royal reserve near Bogotá. As result, the Crown refocused its attention on the royal reserve in Loja. As botanists to the Crown, Ruiz and Pavón may have felt an obligation to show that this decision was justified. Indeed, the botanists remarked, "morally we must believe that [quina] sent to the King from Loja today is the most exquisite."[53] In other words, Ruiz and Pavón started with the assumption that the king was getting the best quina from Loja. Finally, whereas Mutis had often highlighted the "ignorance and malice" of bark collectors and merchants, Ruiz and Pavón valorized the knowledge of the quality of quina from Loja that "natives and Dealers" had passed down for generations.[54]

Such connections provide some explanation as to why Ruiz and Pavón explicitly framed their response to Zea in 1801 as a "Defense of the fine Peruvian Quinas and those from Loja by the Botanists of the Expedition of Perú."[55] This defense involved demonstrating that "orange, red, yellow and white Quinas of Santa Fé are notoriously inferior Species to those from Loja and Peru."[56] They employed several strategies to establish the inferiority of quina from Santa Fé. One strategy was to continually reinforce the distinctions between quina from Santa Fé and quina from Peru through the use of geographical adjectives. Their *Suplemento* is riddled with juxtapositions of "the fine Quinas from Loxa and superior Peruvian [quinas]" with "Quinas from Santa Fé."[57] The equator emerged as an important dividing line between the two locations as they highlighted differences in the terrain and local climatic conditions of Peru and Santa Fé—aspects that many in the quina trade considered essential to assessing the quality of the bark.[58] Ruiz and Pavón also noted that the province of Huánuco, where they studied the cinchona tree and a major quina-producing region in Peru, was closer to Loja than Santa Fé and had similar terrain and climate.[59] Such characteristics of the region, they argued, suggested that Peruvian quinas were of comparable quality to quina from Loja. In other words, Peruvian quinas were better than quina from Santa Fé because of their proximity to Loja.

Ruiz and Pavón cited two additional kinds of evidence of the inferiority of the "quinas from Santa Fe." One was the historical usage of different kinds of bark in commerce and medicine. For example, they wondered how Mutis could claim that "red quina" was official when it "never achieved much esteem in Commerce nor usage in Medicine nor is it likely to ever achieve [esteem or usage] as long as we have more virtuous and efficacious Species."[60] To support their claim of the inferiority of all four of Mutis's official quinas, Ruiz and Pavón cited the results of therapeutic tests conducted in Madrid. "Experiments performed on the order of the Superior Council of the Royal Hospitals of Madrid in 1796," they wrote, "have shown that [the quinas of Santa Fe] do not produce the effects that Mutis claims."[61] They also noted that "other Professors of this Court" had observed a lack of medical efficacy in quina from Santa Fé—undoubtedly a reference to the tests from the late 1780s that convinced the Crown to close down the royal reserve near Bogotá. With such evidence, Ruiz and Pavón defended the Crown's decision to refocus on the royal reserve of Loja.

Yet there had never been a royal reserve in Peru. Why then did Ruiz and Pavón consistently lump Peruvian quinas into their defense of quina from Loja? As it turns out, the Crown was not their only patron. A few years prior to the publication of their *Suplemento*, the two botanists developed direct economic ties to patrons from South America, especially Lima, in the course of soliciting funds for the publication of their *Flora Peruviana et Chilensis*.[62] This work was to be their magnum opus and, in 1792, the Spanish Crown expressed its desire that the *Flora* be a publication of "grandeur and magnificence."[63] Unfortunately, desire alone does not get a work published, and the Crown was unable to provide sufficient funds for the project.

On 17 September 1791, Charles IV asked his American subjects to underwrite the publication of the *Flora*. The royal decree reported, "as the work is so vast, and the expenses required for its execution in typography, engraving, and coloring are so great, the Royal Treasury cannot support them." The Crown urged officials and elites in Spanish America to contribute to this work, which was "principally in honor of [America's] inhabitants."[64] Although the *Flora* focused only on Peru and Chile, the royal solicitation was sent to officials throughout Spanish America and, in turn, funds arrived from all

TABLE 6. Total Contributions from Spanish America for the Publication of *Flora Peruviana et Chilensis* submitted before 1801

Location	Pesos	Reales	Percent of total
Viceroyalty of Peru	17,929	0.5	45.5
Viceroyalty of New Spain	6,377	4	16.2
Viceroyalty of New Granada	4,409	4.5	11.2
Chile	4,160	0	10.6
Cuba	2,893	2	7.3
Viceroyalty of La Plata	2,864	0	7.3
Philippines	586	6	1.5
Venezuela	181	4	0.5
Total sent from Spanish America	39,402	5.0	100.0

Source: Arthur Robert Steele, *Flowers for the King: The Expedition of Ruiz and Pavon and the Flora of Peru* (Durham: Duke University Press, 1964), 218–224 and appendix B.

Note: While Ruiz and Pavón received contributions from Spanish America from 1792 to 1796, the majority of the contributions were submitted in 1792 and 1793.

parts of the American territories, including even the Viceroyalty of New Spain (table 6). The majority of the contributions arrived in the late 1790s, just in time for publication of the first volume of the *Flora,* and well in advance of publication of the *Suplemento a la Quinología* (1801).

The majority of the funds came from the Viceroyalty of Peru. The combined contributions of the collective territories of Peru and Chile accounted for just over half (56.1%) of the total contributions from all of Spanish America (table 6). Clearly, residents of Peru and Chile had an interest in supporting this project. Additional data collected by Arthur Steele provide a more detailed picture of which residents of Peru and Chile in particular supported this project. Table 7 provides a breakdown of contributing groups from the Viceroyalty of Peru using Steele's original categories and titles. Note that the city of Lima provided the majority of the contributions from Peru. Of the total contributions from Spanish America prior to 1801, limeños provided an astounding 38 percent of the funds. Tables 6 and 7 also indicate that the *Consulado* (merchant guild) in Lima alone—an institution representative of

TABLE 7. Contributions from the Viceroyalty of Peru for the Publication of the *Flora Peruviana et Chilensis*, 1792–1793

Location	Donor	Pesos	Reales	Percent of total contribution from Spanish America (prior to 1801)
Lima	Consulado	6,000	0.0	15.2
Lima	Cabildo of Lima	3,000	0.0	7.6
Lima	University of San Marcos	3,000	0.0	7.6
Lima	Townspeople of Lima	2,067	4.5	5.2
Lima	Gabriel de Avilés, inspector general of all military troops in the viceroyalty	100	0.0	0.3
Lima	José Manuel de Tagle Ysoaga, commissar of war and navy, for himself and his uncle, José de Tagle y Bracho, senior *oidor* of the audiencia	100	0.0	0.3
Lima	Archbishop of Lima	200	0.0	0.5
Lima	Viceroy Francisco Gil de Taboada y Lemos	500	0.0	1.3
	Total from Lima	**14,967**	**4.5**	**38.0**
Huamanga	Total from Huamanga	1,787	0.0	4.5
Arequipa	Total from Arequipa	648	0.0	1.6
Cuenca (bishopric)	Total from Cuenca	425	0.0	1.1
Unidentified (probably mostly Huamang)	Total	101	4.0	0.3
	Total from Peru	**17,929**	**0.5**	**45.5**

Source: Arthur Robert Steele, *Flowers for the King: The Expedition of Ruiz and Pavon and the Flora of Peru* (Durham: Duke University Press, 1964), 219–220.

merchant interests—contributed almost as much as the entire Viceroyalty of New Spain and more than any other geographical region in Spanish America (after Peru and New Spain). At the same time, the data do not permit us to determine what proportion of these funds came from merchants trading in quina nor do these data tell us anything about the motivations for contributing to the project.

Overall, the profile of these contributions to Ruiz and Pavón's *Flora Peruviana et Chilensis* reflect a link between Spanish American interests and the Crown's botanists. With significant contributions from Peru, and especially the merchant community in Lima, the two botanists had additional motivation to frame their *Suplemento* as a defense of quinas from Peru, as well as quina from Loja, especially because Mutis's broader definition of the four official quina threatened the de facto monopoly that limeño merchants had in the trade in the highest quality quina from Loja. If official quina could be found elsewhere, such as the port of Cartagena in New Granada on the northern coast of South America, why would European merchants expend the extra time and expense to deal with quina merchants in Lima? In this way, commercial interests were intimately connected to questions of natural knowledge and the limeño merchants may have seen Ruiz and Pavón as potential allies in their attempts to defend their de facto monopoly by defending the quality of Peruvian quinas relative to those from Santa Fé. At the same time that Spanish America "rescue[d] the *Flora*," in the words of Arthur Steele, Ruiz and Pavón came to the rescue of Peruvian quinas. In defending bark from Peru, Ruiz and Pavón returned the favor to their American patrons.[65]

The visions of the natural order espoused by Mutis and Ruiz were intertwined with their visions of the imperial order. Their different perspectives on nature and empire, in turn, were connected to their location in the Spanish Atlantic World. Place was important in terms of what it meant for the social and political network in which Mutis and Ruiz were involved. Mutis's primary network was based in the Viceroyalty of New Granada and its capital city, Santa Fé de Bogotá, where the vice-regal court was located and also where, after 1791, the Royal Botanical Expedition in New Granada was based.[66] Meanwhile, the network centered on Ruiz and Pavón was based, after their return to Spain, at their Botanical Office in Madrid. In terms of patronage, both sets of naturalists had connections to the Crown in Madrid, especially because the king funded their expeditions. However, the strength of these connections differed significantly. In general, Mutis's ties to the Crown were weaker due to his physical distance from Madrid, while Ruiz and Pavón were in close proximity to the court in Madrid. Despite his weaker ties to the Crown, Mutis maintained strong ties to a major center of political power in Spanish America: the vice-regal court in Santa Fé de Bogotá.

At first glance, the debate between Mutis and Ruiz may seem to be a conflict between an imperial center (Madrid) and a colonial periphery (New Granada). Yet such an interpretation overlooks the significant links between Ruiz and Pavón and the commercial elites of South America, especially in the Viceroyalty of Peru. Whereas Mutis had the support of Archbishop-Viceroy Caballero y Góngora, Ruiz and Pavón in the 1790s received direct financial support from officials throughout Spanish America. Such evidence recasts this dispute not as a rift between regional and imperial interests but as a rift between two regions within the empire—New Granada and Peru.

The outline of these social and political networks also highlights the structure of the Spanish colonial government as a network of different institutions, entities, and groups with competing and, at times, contradictory interests. The Crown was just one part of this network, albeit an important one. In this case, it just so happened that by the 1790s the Viceroyalty of Peru had already enlisted the Crown as its ally in pursuit of policies that favored the interests of Lima and Peru over those of Cartagena and New Granada in the quina trade. This imperial alliance was not necessarily conscious or coerced but rather conditioned by a powerful historical precedent in which structures for exploiting quina from Loja had already preceded those for exploiting quina from New Granada. Such historical precedents provided a powerful barrier for Mutis and his associates, even with the support of the viceroy of New Granada, to convincing other botanists and the Crown as well as merchants in the Spanish Atlantic of the efficacy of quina from Santa Fé and its equivalence to quina from Loja.[67] Nonetheless, as it became integrated into the existing structures of imperial governance and knowledge production, science—as embodied by botanists and the botanical expeditions—became a form of statecraft wielded as much by officials in South America, such as the viceroy of New Granada, as by the Crown in Spain.

This debate between Mutis and Ruiz also highlights the difficulties involved in characterizing and, more important, situating the production of natural knowledge in its broader sociocultural contexts. On the one hand, to cast this debate as a purely intellectual dispute disconnected from political concerns, social contexts, and commercial interests would be a gross simplification. On the other hand, to suggest unidirectional lines of causality from the social to the intellectual is just as much a simplification. Mutis and Ruiz articulated their differences in many more registers beyond the language of botany and medicine, but they were not simply shills for different interest groups or the Crown. One solution to this dilemma is to emphasize the mutual influence and interdependence of the content and context of science.[68] Historians of colonial science, as well as science and empire, have benefited and contributed to the development of this approach. For example, in the introduction to their edited volume on colonial botany, Londa Schiebinger and Claudia Swan emphasize the intimate relationship between Europe's scientific and colonial enterprises: "The expanding science of plants depended on access to ever farther-flung regions of the globe; at the same time, colonial profits depended largely on natural historical exploration and the precise identification and effective cultivation of profitable plants."[69] Such a view rightly emphasizes the interdependence of these enterprises.

But, as this chapter has shown, the relationship between science and empire was about more than interdependence because, in this case, both knowledge of nature and visions of empire seem to be the products of the same process. As a result, the relationship between science and empire in the Spanish Atlantic is best described as an instance of "coproduction"—a term that calls attention to the entanglements of knowledge and social order. In a recent volume on cases of coproduction, science and technology studies scholar Sheila Jasanoff explains the concept of coproduction. "Knowledge and its material embodiments," she writes, "are at once products of social work and constitutive of forms of social life; society cannot function without knowledge any more than knowledge can exist without appropriate social supports."[70] As an instance of coproduction, the debate over the classification of quina resembles the seventeenth-century debate between Robert Boyle (1627–1691) and Thomas Hobbes (1588–1679) over the utility of experiment in producing new knowledge about the natural world. Although they do not use the term coproduction, Steven Shapin and Simon Schaffer in their classic study of Boyle and Hobbes observe that "the history of science occupies the same terrain as the history of politics" in part because "solutions to the problem of knowledge are solutions to the problem of social order."[71] This episode in the history of quina shows that Shapin and Schaffer's observations can be extended beyond the social order to the imperial order as well. Ultimately, the idiom of coproduction is useful because it encourages us to recognize that science does not precede the social order, the state, or the empire but is instead a constitutive element of those entities.

One of the distinctive features of the Spanish Empire that makes it difficult, at times, to distinguish science from empire is the way in which practitioners of sciences were so easily integrated into the epistemic culture of the colonial government. With this in mind, one wonders if the notion of "coproduction," while a useful analytical tool to historians, would be surprising to the historical actors. After all, as this book has demonstrated, one of the characteristic features of the imperial knowledge complex in the Spanish Atlantic World was that problems of knowledge were always connected to problems of imperial governance.

The debate between Mutis and Ruiz illustrates the connections between the ordering of the natural and social worlds. Both botanists' proposals were as much about how to order the production of quina (royal vs. private cultivation) as how to produce a proper taxonomic order of the species of cinchona (many vs. one official species). On the one hand, their methods of identifying and classifying official quina or cinchona had consequences for the Crown's policies on the production of quina. If Mutis had convinced officials that he was correct, then the decision to confine the royal reserve to Loja needed reevaluation. After all, the main objective of the royal reserve was to provide the Crown with the best bark to the extent that in the 1790s, Spanish officials redefined the jurisdiction of the royal reserve according to the quality of the bark rather than geographical location. On the other hand, Ruiz's vision of one official quina supported the status quo in which the royal reserve of quina remained centered on Loja—the unique locus of the best bark. Thus, the natural knowledge produced by Mutis and Ruiz had direct implications for the realization of different kinds of imperial order—regalist or mercantile.

Mutis and Ruiz were also embedded in different sociocultural networks that influenced the knowledge they produced. As a physician to the viceroy of New Granada, Mutis focused his work on quina on clarifying its medical application while, at the same time, supporting the extension of the royal monopoly to the region of Santa Fé—a project that was commensurate with Archbishop-Viceroy Caballero y Góngora's plans for state-centered economic development of his viceroyalty. Meanwhile, as botanists in the more direct employ of the Spanish Crown, Ruiz and Pavón coupled their botanical observations of the tree to a defense of quina from Loja being monopolized by the Crown and, in Ruiz's *Quinología*, to an endorsement of private cultivation of the tree. Although there was no necessary connection between their visions of the natural order and their visions of the imperial order, mapping Mutis's and Ruiz's connections to the diverse groups that comprised the imperial state provides a fuller understanding of the rift between the two botanists and their associates. There was much more at stake than the facts of nature.

In the end, botanists and bureaucrats had a similar goal in mind. Both groups thought that knowledge of the proper classification of quina would provide a solid foundation for imperial policies aimed at exploiting this natural resource. Mutis observed, "No [government] order will be effective, as his enlightened Ministry [of the Indies] wishes, while the opinions of distinguished professors, who ought to provide the enlightenment necessary for certainty in resolutions, do not agree."[72] Here, Mutis made explicit his conception of the connection between knowledge and empire. The Crown took a similar view as in its orders to the botanical expeditions in which botanists were instructed to "expel doubts and adulterations."[73] In practice and in defiance of the royal will, the directors of the royal botanical expeditions did little to establish consensus and certainty with regard to quina. How could they? How could these agents of empire act in any consistent or coordinated way when that empire itself was composed of competing interest groups with diverse conceptions of the proper imperial order? Such questions highlight the central irony of the role of botany in the Spanish Empire in the late eighteenth century. As botanists became integrated into the structures and epistemic culture of empire, they became less able to serve the empire—at least in the way that they promised—because integration meant association with some of the interest groups competing to define what "the empire" was. Thus botany failed the empire because it had become, in a sense, too imperial.

Conclusion

The Natures of Empire before the "Drapery" of Modern Science

The story of the royal reserve of quina, like that of the majority of the Spanish Empire, ends in the early decades of the nineteenth century. Consider a prescient report submitted by Carlos Suarez, acting (and self-appointed) "lieutenant of the [quina] commission," to the president of Quito in 1814. Suarez was especially eager to explain to President Torivio Montes why there had been no shipments of quina from Loja for the Royal Pharmacy in the previous three years. He blamed "the circumstances of the times," a reference to the turmoil created by Napoleon's invasion of the Iberian Peninsula in 1808 and the six-year war for independence that ensued.

Smugglers, according to Suarez, took full advantage of the situation, as Spanish colonial governance was crumbling at its foundations. Foreign merchants encouraged bark collectors to trade by claiming that "there is no King [and] there is no Spain," while telling other groups of bark collectors that the Crown had "opened [the harvesting of the bark] to all those that wanted to extract it." Suarez also reported that contraband traders were so convinced of the imminent collapse of Spanish imperial power that they even issued "threats" against those who worked for the royal reserve.[1] These traders guessed correctly. In 1822, the region, formerly known as the Audiencia of Quito, achieved its independence from Spain.

According to early nineteenth-century accounts, the world faced the im-

176

minent and inevitable disappearance of the cinchona tree and with it one of the most valuable botanical remedies ever known. Many of these reports came from British travelers, who visited South America in the wake of the independence movements. In his 1825 *Historical and Descriptive Narrative of Twenty Years' Residence in South America,* William Stevenson noted that bark collection methods used in northern Peru would soon lead to "the destruction of this invaluable vegetable." Stevenson warned his readers that "this highly esteemed production of the new world" would disappear "if the government[s] of America do not attend to the preservation of quina, either by prohibiting the felling of the trees, or obliging the territorial magistrates to enforce cutters to guard them from destruction."[2] Several decades later, British explorer Clements Markham would observe: "The collection of bark in the South American forests was conducted from the first with reckless extravagance; no attempt worthy the name has ever been made either with a view to the conservancy or cultivation of chinchona-trees [*sic*]."[3] According to Lucile Brockway, British observers repeated these claims into the early twentieth century, fostering the notion that intervention "was necessary to save the cinchona trees from extinction because of overcutting and wasteful practices in the Andes."[4]

Subsequent research by historians has demonstrated that claims about the impending disappearance of the cinchona tree were more hyperbole than fact. After all, according to Clements Markham's estimates, the Andean republics exported an estimated 2 million pounds of bark in 1860.[5] In 1881, on the eve of the entrance of Dutch quinine into the world market, exports from South America had risen to 19.84 million pounds according to Margaret Duran-Reynals.[6] Such numbers hardly suggest that extinction of cinchona trees was nigh, though without the purposeful cultivation of the trees, their disappearance was likely. More recently, Kavita Philip has argued that British supporters of cinchona transfer deployed the notion of mismanagement by the Spanish Empire and Latin American republics primarily as a means to justify and garner support for their enterprise.[7] Moreover, because the preservation of the cinchona tree was necessary to ensure continued production of the only effective treatment of malarial fevers, European proponents of efforts to transplant cinchona trees from Latin America to other parts of the globe were able to cast a project that clearly benefited European imperialists as one that was in the interest of all humanity.[8]

In the mid-nineteenth century, the nascent republics in the quina-producing regions of South America would face a new kind of smuggler—agents of the British and Dutch governments looking to transplant the cinchona tree from South America to India and Southeast Asia.[9] In 1852, Dr. John Forbes Royle of the East India Medical Board in Britain recommended the transplantation of cinchona from South America to India. In that same year, Justus Charles

Hasskarl, a botanist and former superintendent of the Dutch botanical gardens in Java, set sail for South America with a charge from the Dutch government to collect cinchona species for transplantation to Southeast Asia. Several British explorers and botanists followed suit in the next decade. Ultimately, their efforts paid off. By century's end, the British and the Dutch had ended their dependence on South American quina, established plantations of cinchona trees, and, in the case of the Dutch, became the world's largest supplier of not only cinchona bark but also quinine—the anti-malarial alkaloid that two French pharmacists isolated from cinchona bark in 1820.

Even since then, the story of quinine as a tool of empire and a miracle of modern medicine has seemed compelling especially as evidence of how a new nexus of scientific knowledge and imperial power provided the foundation of European global hegemony until the mid-twentieht century.[10] While quinine was undeniably one of the most important tools of European imperialism in the nineteenth and twentieth centuries, quina was much more than "quinine's predecessor."[11] Greater understanding of the Spanish Empire's failed efforts to convert science and nature into effective tools of empire in the eighteenth century offer important insights too. In particular, *The Andean Wonder Drug* has shown that even as the Spanish Crown and imperial reformers endorsed the power and utility of science in Madrid, the structure of the Spanish colonial government and the realities of Andean colonial society revealed the fragility of science when it ventured beyond the comfortable confines of royal institutions, such as the Royal Pharmacy and the Royal Botanical Garden, in the imperial center.

By highlighting the fragilities and failures of European science in the Spanish Atlantic World, this book also provides insight into what it took for science to succeed in colonial contexts elsewhere. Here, a key feature of the Spanish Atlantic context seems to be the epistemic culture that pervaded the Spanish colonial government in which officials at all levels actively solicited and, at times, embraced the knowledge of bark collectors, merchants, missionaries, and local officials alongside the knowledge of pharmacists, physicians, and botanists without necessarily taking the superiority of European science for granted. The experiences of botanists, in particular, highlight, the paradox of this situation; at the same time that botanists relied on the Spanish Crown and the colonial government to provide access to information, experiences, and samples of the natural world from the far-flung corners of empire, their integration into Spain's imperial enterprise meant that they were subject to the politics of knowledge that pervaded the epistemic culture of empire. In the eighteenth-century Atlantic World, quina was a junction point between the worlds of science, empire, and commerce and, as such, provides a useful vantage point from which to view the relations between all of these enterprises.[12]

The Politics of Knowledge before the "Drapery of Science"

Nineteenth-century Britain has often served as the primary lens through which science and empire have been examined even in other geopolitical contexts. Britain has long held fascination for historians of science and empire since it achieved global dominance in the nineteenth century, largely through the use of science and technology.[13] In the early decades of historical scholarship on imperial science, Britain and its former colonies proved fertile ground for exploring both science in imperial history and "science as imperial history."[14] Such work is useful for understanding this particular case of imperial science. However, problems arise when the idiosyncrasies of the British case are taken as a model, standard, or yardstick for assessing the complex processes of knowing and governing in other geopolitical contexts, especially as sciences and empires varied greatly across time and space. By highlighting the unique assemblages of imperial governance and knowledge production in the late eighteenth-century Spanish Atlantic World, *The Andean Wonder Drug* reveals not a derivative instance of science and empire, but a distinctive one. From the perspective of the Southern Atlantic, the British case appears as just one instance of the various ways in which science and empire interacted.

Nonetheless, some comparison with nineteenth-century Britain helps to illuminate the importance of the case of the royal reserve of quina. Let us consider natural history in Victorian Britain as described in a recent biography of Joseph Hooker, naturalist and director of Kew Gardens.[15] Hooker's career as a botanist can be understood in the context of the larger story of an expanding empire and British botany's path to becoming one of England's "great imperial sciences" in the nineteenth century.[16] This shift was a significant change of status as many members of the British scientific establishment in the 1820s and 1830s considered botany to have a relatively low place in the hierarchy of sciences.[17] In an address to the British Association for the Advancement of Science, philosopher William Whewell illuminated this hierarchy of the sciences by making a distinction between "fact-gathering" and "science." "The mere gathering of raw facts," Whewell said, "may be compared to the gatherings of the cotton from the tree. The separate filaments must be drawn into a connected thread, and the threads woven into an ample web, before it can form the drapery of science."[18] For Whewell and many others, botany only became scientific and philosophical when its practitioners began to study questions of "systematics (the principles and laws of classification), plant anatomy and physiology (structure), and plant distribution (particularly as a way of discovering the laws that governed vegetation)."[19] In this nineteenth-century view, botany prior to 1800 was neither philosophical nor scientific because its practitioners focused too much on cataloging natural phenomena, and not enough on understanding their underlying causes.[20]

The notion that to be "scientific" a discipline must engage in more than just identifying and classifying natural phenomena is still common today. From this perspective, many of the activities connected to the royal reserve of quina would not count as science—even the activities of Spanish botanists. What *The Andean Wonder Drug* has shown is that such definitions underestimate the philosophical and political dimensions of the processes of identification and classification and, as a result, may lead us to overlook important episodes in the history of the natural sciences. After all, much of the debate and disagreement among and between botanists, bureaucrats, and bark collectors in the late eighteenth-century Spanish Atlantic was over the identity of cinchona bark and, more important, the proper techniques for identifying and assessing samples of quina. Although these debates did not take place in the context of a European academy of sciences, or on the pages of a scientific journal, they were no less philosophical. Moreover, the case of quina shows how establishing the identity of a natural entity in a context of practice could be just as fraught with ontological and epistemological questions as any philosophical debate that occurred in Europe's formal institutions of learning and knowledge.

It is also important to recognize that in the various debates about quina that took place in the epistemic culture of the Spanish colonial government, there was no outside "truth" to adjudicate the divergent views. Knowledge was political, and its veracity and authority were made through social processes, and not simply discovered in the natural world. Context mattered, and in the different sociocultural contexts of the Spanish Atlantic, various experts saw quina in different ways—even if they were supposedly looking at the exact same thing. Once stripped from the tree, quina became a malleable object subject to the influence of a bewildering variety of physical, social, and cultural forces. Because the colonial government engaged a wide variety of local and learned experts in its efforts to know and to regulate quina, there was no single fixed point from which to examine and identify the bark. As a result, there were as many quinas, in a sense, as there were groups of advisers to the Crown and its colonial officials. This situation was in part a product of the distinctive epistemic culture of the Spanish colonial government. At the same time, as this book has argued, it was a product of a deeper history of quina in which the bark had become embedded in Andean and Atlantic networks of knowledge production. As a result, a pharmacist's quina was not the same as a botanist's quina, which, in turn, differed from that of a merchant, bark collector, or even a government official in New Granada or Peru. As observed by Löic Charles and Paul Cheney, the problem was one of coordination of different understandings of the natural world, not necessarily the accumulation or production of knowledge.[21]

The Entanglement of Science and Empire in the Spanish Atlantic World

Knowledge of nature was political in other ways, too. Imperial projects like the royal reserve of quina reveal just how closely intertwined science and empire could be in the eighteenth-century Atlantic World. After all, colonial governments served as important networks facilitating the production, accumulation, and circulation of knowledge.[22] The Spanish colonial government fostered forms of social organization, networks of circulation, and techniques of collection for the production of knowledge of various kinds, not just knowledge of the natural world. Moreover, the institutions of colonial governance fostered an epistemic culture that valued direct observation of natural phenomena and observations from many different groups throughout the empire. The royal reserve of quina in its daily operations exemplified this approach with the added twist that questions about the identity and quality of cinchona bark often had direct implications for imperial policies regarding the production and distribution of this precious natural resource. Such practical engagements of botanists and other scientific practitioners were just as important and, in some cases, more important than any of their activities that we might be tempted to construe as more traditionally scientific.

The colonial government and its epistemic culture also fostered debate and discord among the Crown's various expert advisers as botanists, pharmacists, physicians, bark collectors, and officials jockeyed to assert their authority and influence relative to each other. From this perspective, lack of consensus among experts was not necessarily an instance of failure but rather a sign that this epistemic culture was operating as intended. After all, royal advisers and colonial officials actively sought multiple perspectives and opinions in the course of making decisions about which policies to endorse. Governance in this view was the weighing and balancing of the variable effects that any policy—such as a royal reserve or royal monopoly—might have on the interested parties. Moreover, as Bourbon reformers worked to reconstitute royal power after 1750, they increasingly made the Crown and its ministries the locus for the adjudication of competing claims of all kinds.[23] The power to adjudicate was the essence of royal power in the Spanish Atlantic.[24] Because the production of knowledge about New World nature relied on the long-distance networks established and maintained by the Spanish colonial government, it should be no surprise that botanists and other experts were also subject to these processes of negotiation and adjudication that characterized governance in the Spanish Empire more broadly. In the end, bureaucrats, as much as botanists, played a crucial role in the making of knowledge about quina and other American botanical products.

The royal reserve of quina and other projects like it provided a context in which botanists and pharmacists had to contend with local experts like Pedro de Valdivieso and the bark collectors of Loja. Here a comparison with pharmacists and botanists in places like London, Paris, or Rome is revealing. These groups did not have the same kind of direct link to bark collectors in South America that Spanish botanists and pharmacists did. In theory at least, a Spanish botanist, like Hipólito Ruiz, could exert influence over the actual production of the bark. For a pharmacist in England, however, the feedback loop was mediated through the complex and convoluted networks of trade that crisscrossed the Atlantic Ocean. In the context of the royal reserve of quina, the practical application and policy implications of knowledge were much more immediate for Spanish botanists, pharmacists, and physicians. This was knowledge produced in a "context of application," and use of this phrase suggests that a mode of knowledge production thought to be distinctive to late twentieth century might have a deeper geneaology in the imperial and colonial enterprises of the early modern world.[25] As a result, knowledge and politics were virtually indistinguishable. Questions of knowledge were questions of politics when it came to imperial control of quina.[26] In many cases, officials in Spain often requested information and policy recommendations from learned and local experts in the same questionnaire or royal order

Spanish botanists offered no objection to this arrangement. Instead, they embraced it. Note that at no point did any botanist claim that they just did botany. Yes, their expertise on the natural world provided the basis for their initial involvement with imperial governance. However, as a part of the epistemic culture of the colonial government, botanists like Casimiro Gómez Ortega, Hipólito Ruiz, José Celestino Mutis, and Vicente Olmedo moved easily from making claims about the natural world to making recommendations on imperial and economic policies. For example, Hipólito Ruiz and José Pavón's books were chocked full of suggestions on how to reorganize and improve the production and circulation of cinchona bark. These recommendations were, in turn, supported by their visions of the proper imperial order—a vision in which commerce was central. Other physicians, pharmacists, and local experts also shifted easily from knowledge claims to policy recommendations and back again. Most of the knowledge about quina—as well as other American botanical products—was produced in conjunction with debates over new imperial policies, making it difficult, if not impossible, to impose a strict divide between knowledge or science, on the one hand, and the politics or empire, on the other.[27]

In the case of quina, the politics of knowledge also had an important spatial dimension as well as a social one. Officials in the colonial government relied on experts that were dispersed throughout the empire, from botanists in Madrid to bark collectors in Loja. Consequently, knowledge was political in the sense that it was the product of negotiations between and within different localities of the empire. The forests of Loja, the "axis" of health in the Andean World, the merchant guild in Lima, the vice-regal court in Santa Fé de Bogotá, or the Court in Madrid—all these places constituted different sociocultural contexts that supported distinctive bodies of knowledge and ways of knowing quina in particular and the natural world in general. The broader social, political, and commercial networks of empire provided the larger context through which all of these places were connected. The question was: Which place, with its associated methods and standards for identifying and assessing quina, would achieve prominence throughout the network?

In 1751, when the Crown established the royal reserve of quina, the answer to this question was far from clear, and at first officials in Spain seemed content to let local officials decide what counted as the best quina. As the decades progressed, botanists, and to a lesser extent pharmacists, in Madrid challenged the authority of local experts as they emphasized their own. Such developments and tensions effectively intertwined the politics of knowledge and the politics of quina (the social, economic, and political structures that fostered certain modes of production of the bark). As a result, the royal reserve of quina represented a "regime of hybrid productivity" in which it is difficult to make a rigid distinction between the production of knowledge about quina, the production of the bark as a botanical commodity, and the production of imperial power over this natural resource.[28] More important, when one kind of production failed, the others were likely to fail, too. To make a rigid distinction between any of these elements obscures the interesting and important ways in which these processes were connected. Ultimately, our understanding of colonial practices like the *repartimento de mercancias*—the system of debt peonage through which merchants, landowners, and government officials exploited indigenous labor for the royal reserve of quina and other enterprises—is enriched by attention to ways in which the production of natural knowledge about quina supported or undermined such enterprises. Knowledge about quina—its distribution, abundance, and quality of different species—had implications for its role as an imperial natural resource and its circulation as a botanical commodity in the Atlantic World.

Each of these dimensions of the politics of knowledge in Spain's royal reserve of quina help to address one final question: Why did relations between science and empire persist even when practical results were not forthcoming? In practice, the two enterprises were almost indistinguishable. The structures of the colonial government often served simultaneously to produce knowledge and enact governance. In addition, when botanists, pharmacists, and other learned experts on the natural world got involved with the colonial government, they became integrated into an extant epistemic culture with its accompanying methods and structures of knowledge production. Discourses of empire and commerce that circulated in Enlightenment Spain *also* fostered this entanglement of science and empire. Much of the economic discourse at the time offered a new vision of empire that privileged commerce over conquest—a vision of the reinvigoration of Spain through reaping the economic benefits of its American territories. Looking to the model of the Dutch, the great economic success story of seventeenth-century Europe, Spanish political and economic writers in the eighteenth century suggested that Spain's imperial salvation lay in the development and exploitation of American commodities, especially botanical ones. For short, we might call this the discourse of commercial imperialism.[29]

This discourse fostered the intertwining of science and empire because of a remarkable convergence between economic and scientific visions of empire and nature. Both imperial bureaucrats and Spanish naturalists emphasized utility and construed American nature generally as a vast collection of useful things to be exploited commercially and studied scientifically.[30] Certainly, it was not uncommon in the early modern period for the enterprises of science, commerce, and empire to go hand in hand.[31] Yet this convergence of imperial visions around the concept of utility also rendered science and empire at times indistinguishable. As a result, Spanish botany became a hybrid enterprise serving scientific, economic, and political ends simultaneously.[32] One unique feature of the Spanish Atlantic is that the relations between science and empire achieved a level of interconnection not seen in other imperial contexts prior to 1800—and the discourse of commercial imperialism provided key ideological support for the forging of this alloy of knowledge and power.

Unpacking Imperial Epistemologies

Quina remained a contested object well into the nineteenth century. One of the last quina shipments from the royal reserve arrived in Madrid in 1806 just a few years before the dissolution of the royal reserve and Spain's empire in America. Following the usual protocol, the Crown asked a group of experts to review the bark and determine its quality and utility. This group included two botanists, Hipólito Ruiz and José Pavón, and a pharmacist of the Royal Chamber, Castor Ruiz del Cerro.

Upon reviewing three of the boxes of quina, the examiners produced divergent assessments of the shipment. Although the pharmacist, Ruiz del Cerro, found the bark to be "sufficient" for royal purposes, Ruiz and Pavón withheld their endorsement of the bark's utility.[33] These differences derived from their views on what was in the boxes. Cerro asserted that this quina "was the same species that had always come to the Royal Pharmacy without any other [species] mixed in," whereas Ruiz and Pavón asserted that the quina was actually a mixture of two kinds of bark known as "*cascarilla colorada* and *amarilla*" (colored and yellow cascarilla).[34] Once again, lack of expert consensus left the Crown with no clear course of action. If Ruiz del Cerro was right, then the Royal Pharmacy could use the bark. If Ruiz and Pavón were right, then the bark must be burned or, at best, sold as a dyestuff. Despite certain measures that had been put in place to protect the bark from physical degradation and fraud, the empire's experts still disagreed on fundamental issues, such as the number of species of cinchona tree, which kind of bark was the best, and how to distinguish different kinds of bark. Such questions about the identity of this scientific object proved just as intractable as the ones that nineteenth-century botanists would tackle regarding plant systematics, physiology, and distribution.

When reflecting on the disagreement over this 1806 shipment of bark, one contemporary observer, Gregorio Bañares, a respected pharmacist with ties to the Crown, the Army, the Royal Academy of Medicine in Madrid, and Spain's Royal College of Physicians, considered the botanists to be the root of the problem.[35] In his 1808 report to the Royal Pharmacy, Bañares castigated the botanists for failing to provide the Crown with a definitive assessment of whether the shipment should be used or destroyed. Instead, Ruiz and Pavón merely observed that the shipment "seem[ed]" to contain a mixture of two different kinds of bark.[36] Bañares took the botanists to task for their choice of words. He declared that Ruiz and Pavón had put "everything in doubt" by using the verb "to seem" (*parecer*), which, in Bañares's view, was akin "to saying nothing [at all]."[37] Yet Ruiz and Pavón had good reason for their equivocation. After all, as dutiful followers of Linnaeus, they would have considered it impossible to make a definitive identification of a tree species by its bark alone.[38] Certain classification for botanists required at the very least a sample of the tree's flowers.[39] Bañares saw things differently. Ruiz and Pavón's strict adherence to botanical method was an obstacle to the production of useful knowledge and, by extension, an obstacle to effective governance. In trying to be good botanists, they ended up being bad advisers to the Crown.

Opening boxes was a central, if mundane, material practice for those associated with the royal reserve of quina. Opening boxes has also been a central conceptual practice in this book. Following boxes of bark and their openings at various locations throughout the Spanish Empire has suggested that some opening up of concepts like science and empire is warranted. At various levels from the geopolitical to the institutional, *The Andean Wonder Drug* has explored the difficulties of making rigid distinctions between nature, knowledge, and empire, especially prior to 1800.

In the royal reserve of quina, the main function of boxes was to mitigate the forces of impurity that threatened their contents. After all, bark collectors and imperial officials worked hard to produce these little pieces of the natural world, and so wanted to keep them separate from the rest of nature. In this sense, quina was an artifact, not simply a natural object.[40] Europeans also used boxes, literally and conceptually, to impose divisions on American nature to make it knowable, governable, and useful. In this way, botanists, bureaucrats, and bark collectors were engaged in analogous and, at times, overlapping enterprises. Boxes, thus, were an important imperial technology and are a useful metaphor for some imperial epistemologies.

Historians have also made use of boxes to produce and preserve the purity of things like science and empire or knowledge and politics. All attempts to divide and contain social or natural worlds rely on some principle of selection. Consequently, not everything ends up in boxes, but only those things that historians (and imperialists) deem useful or valuable. *The Andean Wonder Drug* has offered some exploration of some of the things that do not quite fit into our conceptual boxes through an analysis of the boxes of the royal reserve of quina and disputes over their contents. Upon opening the boxes of science and empire, we find much that supposedly does not belong there; upon opening the shipments of the royal reserve, the royal pharmacist in eighteenth-century Madrid found that boxes alone were no guarantee of the stability and durability of quina, an object that was simultaneously powerful and fragile, much like European science in the eighteenth-century Atlantic World.

Notes

Introduction: The Power and Fragility of European Science in the Spanish Atlantic World

1. Archbishop-Viceroy Antonio Caballero y Góngora to José García de Leon y Pizarro, Santa Fe de Bogotá, 2 October 1783, Archivo Nacional de Ecuador-Quito (ANE/Q), Fondo Especial, Box 80, vol. 203, no. 5294, fol. 36r–v. Correspondence between Caballero y Góngora and Valdivieso was often mediated by José García de Leon y Pizarro, the president of the *Audiencia* of Quito. For the purposes of clarity in this vignette, I have omitted references to Leon y Pizarro as the mediator.

2. José Ramón Marciada and Juan Pimentel, "Green Treasures and Paper Floras: The Business of Mutis in New Granada (1783–1808)," *History of Science* 52 (2014): 277–296; Daniela Bleichmar, *Visible Empire: Botanical Expeditions and Visual Culture in the Hispanic Enlightenment* (Chicago: University of Chicago Press, 2012), 17–42; José Antonio Amaya, *Mutis, apóstol de Linneo. Historia de la botánica en el virreinato de la Nueva Granada (1760–1783)*, 2 vols. (Bogotá: Instituto columbiano de antropologiá e historia, 2005).

3. James L. A. Webb Jr., *Humanity's Burden: A Global History of Malaria* (Cambridge: Cambridge University Press, 2008), 1–17; Christopher Hamlin, *More Than Hot: A Short History of Fever* (Baltimore: Johns Hopkins University

Press, 2014); Randall Packard, *The Making of a Tropical Disease: A Short History of Malaria* (Baltimore: Johns Hopkins University Press, 2011).

4. The Royal Pharmacy served as the clearinghouse for the quina that was used and distributed by the king. Only a small portion of the Royal Pharmacy's bark was used to treat members of the royal family (~0.7%). The majority of it was sent to royal pharmacies throughout Spain (~44.5%), where it was made available to the public, especially the poor. Military pharmacies in Spain received the next largest portion of the Crown's bark (~14.9%) and used the bark to treat soldiers afflicted with the malarial fevers that regularly visited the Iberian Peninsula. These estimates are from María Andrés del Turrión, "Quina del Perú para la Real Hacienda Española (1768–1807): Notas sobre su estanco," in Antonio González Bueno, ed., *La expedición botánica al Virreinato del Perú (1777–1788),* vol. 1 (Barcelona: Lunwerg, 1988), 71–84; see also M. E. Alegre Pérez, "La asistencia social en la Real Botica durante el último cuarto del siglo XVIII," *Boletín de la Sociedad Española de Historia de la Farmacia* 139 (1984): 199–211.

5. Caballero y Góngora to García de Leon y Pizarro, Santa Fe de Bogotá, 2 October 1783, fol. 36r.

6. On the economic importance of quina to Loja, see Luz del Alba Moya Torres, *El Arbol de la Vida: Auge y Crisis de la Cascarilla en la Audiencia de Quito. Siglo XVIII.* (Quito: Facultad Latinoamericana de Ciencias Sociales Sede Ecuador, 1994); M. Petitjean and Y. Saint-Geours, "La economía de la cascarilla en el corregimiento de Loja," *Cultura: Revista del Banco Central del Ecuador* V, no. 15 (1983): 171–207.

7. Sources indicate that Pedro José Javier de Valdivieso y Torres Hinojosa was born in Loja in 1723 to Ignacio Valdivieso and Francisca Torres, both of whom were probably born in Loja as well. All four of his grandparents were born in South America. While his maternal grandfather was born in Loja, his paternal grandfather was born in Piura, an important entrepôt in the regional trade network of northern Peru and southern New Granada that linked Loja to the coast. Two of his great-grandparents immigrated to South America from different regions of Spain (Biscay and Castile) probably in the early seventeenth century. See Marcia Stacey de Valdivieso, *La polemica sangre de los Riofrío: la Casa de Riofrío en Segovia, Ecuador, Perú, Chile* (Quito: M. Stacey Ch., 1997); Alfonso Anda Aguirre, "Don Pedro Javier Valdivieso y Torres, Corregidor de Loja," in *Corregidores y Servidores Públicos de Loja* (Quito: Banco Central de Ecuador, 1987), 134–136.

8. Martin Minchom, "The Making of a White Province: Demographic Movement and Ethnic Transformation in the South of the *Audiencia* de Quito (1670–1830)," *Bulletin de l'Institut Française d'Estudes Andines* 12 (1983): 23–39.

9. Pedro Xavier de Valdivieso y Torres to José García de Leon y Pizarro,

Loja, 22 November 1783, Archivo Nacional del Ecuador, Quito (ANE/Q), Fondo Especial, Box 80, vol. 203, no. 5294, fol. 47r.

10. Valdivieso y Torres to García de Leon y Pizarro, Loja, 22 November 1783, fol. 47r.

11. Caballero y Góngora to García de Leon y Pizarro, Santa Fe de Bogotá, 2 October 1783, fol. 36r.

12. Valdivieso y Torres to García de Leon y Pizarro, Loja, 22 November 1783, fol. 47r.

13. Valdivieso y Torres to García de Leon y Pizarro, Loja, 22 November 1783, fol. 47r.

14. "Royal Order" [Draft] Madrid, 27 August 1751, Archivo General de Indias (AGI), Indiferente General 1552, fol. 343r-347v.

15. Marc Bloch, *The Royal Touch,* translated by J. E. Anderson (New York: Dorset Press, [1924] 1989).

16. On the Bourbon Reforms in the Spanish Empire, see Allan J. Kuethe and Kenneth J. Andrien, *The Spanish Atlantic World in the Eighteenth Century: War and the Bourbon Reforms, 1713–1796* (Cambridge: Cambridge University Press, 2014); Jorge Cañizares-Esguerra, "'Enlightened Reform' in the Spanish Empire: An Overview," in *Enlightened Reform in Southern Europe and Its Atlantic Colonies, c. 1750–1830,* ed. Gabriel Paquette (Burlington: Ashgate, 2009), 33–37; Gabriel Paquette, *Enlightenment, Governance, and Reform in Spain and Its Empire, 1759–1808* (New York: Palgrave Macmillan, 2008); David J. Weber, *Bárbaros: Spaniards and Their Savages in the Age of Enlightenment* (New Haven: Yale University Press, 2005); John Fisher, *Bourbon Peru, 1750–1824* (Liverpool: Liverpool University Press, 2003); Stanley J. Stein and Barbara H. Stein, *Apogee of Empire: Spain and New Spain in the Age of Charles III, 1759–1789* (Baltimore: Johns Hopkins University Press, 2003); Stanley J. Stein and Barbara H. Stein, *Silver, Trade, and War: Spain, America, and the Making of Early Modern Europe* (Baltimore: Johns Hopkins University Press, 2000); Allan J. Kuethe, "The Early Reforms of Charles III in the Viceroyalty of New Granada, 1759–1776," in *Reform and Insurrection in Bourbon New Granada and Peru* (Baton Rouge: Louisiana State University Press, 1990), 19–40; David A Brading, "Bourbon Spain and Its American Empire," in *The Cambridge History of Latin America,* vol. 1: *Colonial Latin America,* ed. Leslie Bethell (Cambridge: Cambridge University Press, 1984), 389–439.

17. Francisco Javier Puerto Sarmiento, *La Ilusión Quebrada: Botánica, Sanidad y Política Científica en la España Ilustrada* (Madrid: CSIC, 1988).

18. Emily Berquist Soule, *The Bishop's Utopia: Envisioning Improvement in Colonial Peru* (Philadelphia, University of Pennsylvania Press, 2014); Daniela Bleichmar, *Visible Empire: Botanical Expeditions and Visual Culture in the Hispanic Enlightenment* (Chicago: University of Chicago Press, 2012); Juan

Pimentel, "The Iberian Vision: Science and Empire in the Framework of a Universal Monarchy, 1500–1800," *Osiris* 15 (2000):17–30; Londa Schiebinger, *Plants and Empire: Colonial Bioprospecting in the Atlantic World* (Cambridge: Harvard University Press, 2004); Richard Drayton, *Nature's Government: Science, Imperial Britain, and the "Improvement" of the World* (New Haven: Yale University Press, 2000). It is important to remember that not all of Spanish science was necessarily imperial science or connected to the activities of empire (see John Slater and Andrés Prieto, "Introduction: Was Spanish Science Imperial?" *Colorado Review of Hispanic Studies* 7 (2009): 3–10.

19. This observation draws on the recent interest in the importance of place to the production of knowledge, see Peter Burke, "Locating Knowledge: Centres and Peripheries," in *A Social History of Knowledge: From Gutenberg to Diderot* (Oxford: Polity Press, 2000), 53–80; Steven Shapin, "Placing the View from Nowhere: Historical and Sociological Problems in the Location of Science," *Transactions of the Institute of British Geographers* 23 (1998): 5–12.

20. On the concept of the "geography of knowledge" especially as it relates to history of science, see David N. Livingstone, *Putting Science in Its Place: Geographies of Scientific Knowledge* (Chicago: University of Chicago Press, 2003); see also Steven J. Harris, "Long-Distance Corporations, Big Sciences, and the Geography of Knowledge," *Configurations* 6 (1998): 269–304.

21. Harold J. Cook, *Matters of Exchange: Commerce, Medicine, and Science in the Dutch Golden Age* (New Haven: Yale University Press, 2007).

22. Loïc Charles and Paul Cheney have recently made a similar argument regarding science and empire in the French Atlantic. See Loïc Charles and Paul Cheney, "The Colonial Machine Dismantled: Knowledge and Empire in the French Atlantic," *Past & Present* 219 (2013): 127–163.

23. Maria M. Portuondo, *Secret Science: Spanish Cosmography and the New World* (Chicago: University of Chicago Press, 2009); Antonio Barrera-Osorio, *Experiencing Nature: The Spanish American Empire and the Early Scientific Revolution* (Austin: University of Texas Press, 2006); Simon Varey, Rafael Chabrán, and Dora Weiner, eds., *Searching for the Secrets of Nature: The Life and Works of Dr. Francisco Hernández* (Stanford: Stanford University Press, 2000); Barbara Mundy, *The Mapping of New Spain: Indigenous Cartography and the Maps of the Relaciones Geográficas* (Chicago: University of Chicago Press, 1996); Raquel Alvarez Peláez, *La conquista de la naturaleza americana* (Madrid: Consejo Superior de Investigaciones Científicas, 1993).

24. This phrase was coined by Karin Knorr-Cetina in *Epistemic Cultures: How the Sciences Make Knowledge* (Chicago: University of Chicago Press, 1999).

25. The Royal Pharmacy in Madrid had developed ties to the Spanish imperial enterprise in the late seventeenth and early eighteenth centuries.

26. Several recent studies have focused on the institutions and techniques

of knowledge production employed by the Spanish colonial government or in conjunction with Spanish imperial rule. See Arndt Brendecke, *Imperio e información. Funciones del saber en el dominio colonial español* (Madrid: Iberoamericana, 2012); Kathryn Burns, *Into the Archives: Writing and Power in Colonial Peru* (Durham: Duke University Press, 2010); Antonio Barrera-Osorio, "Empire and Knowledge: Reporting from the New World," *Colonial Latin American Review* 15 (2006): 39–54; Irene Silverblatt, *Modern Inquisitions: Peru and the Colonial Origins of the Civilized World* (Durham: Duke University Press, 2004); Mundy, *The Mapping of New Spain*; Alvarez Peláez, *La conquista de la naturaleza americana.*

27. Paquette, *Enlightenment, Governance, and Reform*; Weber, *Bárbaros;* Anthony Pagden, *Spanish Imperialism and the Political Imagination: Studies in European and Spanish-American Social and Political Theory, 1513–1830* (New Haven: Yale University Press, 1990).

28. *The Miraculous Fever-Tree* by Fiammetta Rocco devotes a chapter to the eighteenth century that focuses primarily on botanists' efforts to identify and classify the different species of cinchona. Meanwhile, Andreas-Holger Maehle uses cinchona bark as one of the case studies in his excellent book on the development of eighteenth-century pharmacology; see Fiammetta Rocco, *The Miraculous Fever-Tree: Malaria and the Quest for a Cure That Changed the World* (New York: HarperCollins, 2003); Andreas-Holger Maehle, *Drugs on Trial: Experimental Pharmacology and Therapeutic Innovation in the Eighteenth Century* (Amsterdam: Rodopi, 1999). See also Luis Alfredo Baratas Díaz, *Conocimiento botánico de las especies de Cinchona entre 1750 y 1850: relevencia de la obra botánica española en América* (Salamanca: Consejería de Educación y Cultura de la Junta de Castilla y León, 1998).

29. Harold J. Cook, "Markets and Cultures: Medical Specifics and the Reconfiguration of the Body in Early Modern Europe," *Transactions of the Royal Historical* Society 21 (2011): 123–145; Fernando Crespo Ortiz, "Fragoso, Monardes, and pre-Cinchona Knowledge of Cinchona," *Archives of Natural History* 22 (1995): 169–181; Saul Jarcho, *Quinine's Predecessor: Francesco Torti and the History of Cinchona* (Baltimore: Johns Hopkins University Press, 1993); A. W. Haggis, "Fundamental Errors in the Early History of Cinchona," *Bulletin of the History of Medicine* 10 (1941): 417–459, 568–592; Missouri Botanical Garden, *Proceedings of the Celebration of the Three Hundredth Anniversary of the First Recognized Use of Cinchona* (St. Louis: Missouri Botanical Garden, 1931).

30. Daniel R. Headrick, *The Tools of Empire: Technology and European Imperialism in the Nineteenth Century* (New York: Oxford University Press, 1981); Lucile Brockway, *Science and Colonial Expansion: The Role of the British Royal Botanic Garden* (New York: Academic Press, 1979).

31. Robert S. Desowitz, *The Malaria Capers: Tales of Parasites and People* (New York: W. W. Norton 1993).

32. Mark Honigsbaum, *The Fever Trail: In Search of the Cure for Malaria* (New York: Farrar, Straus and Giroux, 2001); Rocco, *The Miraculous Fever-Tree*.

33. Brockway, *Science and Colonial Expansion: The Role of the British Royal Botanic Garden* (New York: Academic Press, 1979); Headrick, *The Tools of Empire: Technology and European Imperialism in the Nineteenth Century* (New York: Oxford University Press, 1981); Daniel R. Headrick, *Power over Peoples: Technology, Environments, and Western Imperialism 1400 to the Present* (Princeton: Princeton University Press, 2012).

34. For example, see Teodoro S. Kaufman and Edmundo A. Rúveda, "The Quest for Quinine: Those Who Won the Battles and Those Who Won the War," *Angewandte Chemie International Edition* 44 (2005): 854–885; Mark Honigsbaum, *The Fever Trail: In Search of the Cure for Malaria* (New York: Farrar, Straus and Giroux, 2002).

35. Moya Torres, *El Arbol de la Vida: Auge y Crisis de la Cascarilla en la Audiencia de Quito, Siglo XVIII* (Quito: Facultad Latinoamericana de Ciencias Sociales Sede Ecuador, 1994); Carlos Contreras, *El Sector Exportador de una Economia Colonial: La Costa del Ecuador entre 1760 y 1820* (Quito: Facultad Latinoamericana de Ciencias Sociales/ABYA-YALA, 1990); M. Petitjean and Y. Saint-Geours, "La economía de la cascarilla en el corregimiento de Loja," *Cultura: Revista del Banco Central del Ecuador* V (1983): 171–207.

36. María Luisa de Andrés del Turrión and Maria Rosario Terreros Gómez, "Organización administrativa del Ramo de la Quina para la Real Hacienda española en el virreinato de Nueva Granada," in *Medicina y Quina en la España del Siglo XVIII*, ed. Juan Riera Palmero (Salamanca: EUROPA Artes Gráficas, 1997), 37–43; M. E. Alegre Pérez, "La asistencia social en la Real Botica durante el último cuarto del siglo XVIII," *Boletín de la Sociedad Española de Historia de la Farmacia* 139 (1984): 199–211; M. E. Alegre Pérez, "La Real Botica y las especies americanas (siglo XVIII)," *Boletín de la Sociedad Española de Historica de la Farmacia* 140 (1984): 225–244.

37. Daniela Bleichmar and Mauricio Neito Olarte have situated the case of cinchona bark in the context of the Spanish Empire and the Spanish Atlantic World. Yet these accounts do not offer a comprehensive history of the bark in the eighteenth century; instead, they use the bark as one case study among many to highlight broader themes in the history of science in the eighteenth-century Spanish Atlantic. See Daniela Bleichmar, *Visible Empire: Botanical Expeditions and Visual Culture in the Hispanic Enlightenment* (Chicago: University of Chicago Press, 2012); Mauricio Nieto Olarte, *Remedios para el imperio: Historia natural y la apropiación del nuevo mundo* (Bogotá: La Imprenta Nacional de Colombia, 2000).

38. The literature on science and empire is extensive. Some good starting points include James E. McClellan, *The Colonial Machine: French Science and Overseas Expansion in the Old Regime* (Turnout: Brepolis, 2011); Sujit Sivasundaram, ed., "Focus: Global Histories of Science," *Isis* 101 (2010): 95–158; James Delbourgo and Nicholas Dew, eds., *Science and Empire in the Atlantic World* (New York: Routledge, 2008); Londa Schiebinger, *Plants and Empire: Colonial Bioprospecting in the Atlantic World* (Cambridge: Harvard University Press, 2007); Londa Schiebinger, ed., "Focus: Colonial Science," *Isis* 96 (2005): 52–63; Londa Schiebinger and Claudia Swan, eds., *Colonial Botany: Science, Commerce and Politics in the Early Modern World* (Philadelphia: University of Pennsylvania Press, 2005); Richard Drayton, *Nature's Government: Science, Imperial Britain, and the "Improvement" of the World* (New Haven: Yale University Press, 2000); Roy MacLeod, ed., "Nature and Empire: Science and the Colonial Enterprise," *Osiris* 15 (2000); Bernard S. Cohn, *Colonialism and Its Forms of Knowledge* (Princeton: Princeton University Press, 1996); Patrick Petitjean, Catherine Jami, and Anne Marie Moulin, eds., *Science and Empires: Historical Studies about Scientific Development and European Expansion* (Dordrecht: Kluwer Academic, 1992); Roy MacLeod, "On Visiting the 'Moving Metropolis': Reflections on the Architecture of Imperial Science," in *Scientific Colonialism: A Cross-Cultural Comparison*, ed. Nathan Reingold and Marc Rothenberg (Washington, DC: Smithsonian Institution Press, 1987) 217–249.

39. Suman Seth, "Putting Knowledge in Its Place: Science, Colonialism and the Postcolonial," *Postcolonial Studies* 12 (2009): 373–388.

40. Londa Schiebinger, *Nature's Body: Gender in the Making of Modern Science* (Boston: Beacon Press, 1993).

41. Charles and Cheney, "The Colonial Machine Dismantled."

42. Jorge Cañizares-Esguerra, *Nature, Empire, and Nation: Explorations of the History of Science in the Iberian World* (Stanford: Stanford University Press, 2006); Kenneth Maxwell, "The Atlantic in the Eighteenth Century: A Southern Perspective on the Need to Return to the 'Big Picture,'" *Transactions of the Royal Historical Society* 3 (1993): 209–236.

43. Cañizares-Esguerra, *Nature, Empire, Nation*, 23–25; Víctor Navarro Brotóns and William Eamon, eds., *Mas allá de la leyenda Negra: España y la revolución científica/Beyond the Black Legend: Spain and the Scientific Revolution* (Valencia: Instituto de Historia de la Ciencia y Documentación López Piñero, 2007).

44. J. H. Elliott, *Empires of the Atlantic: Britain and Spain in America, 1492–1830* (New Haven: Yale University Press, 2007).

45. Jorge Cañizares-Esguerra, *How to Write the History of the New World: Historiographies, Epistemologies, and Identities in the Eighteenth-Century Atlantic World* (Stanford: Stanford University Press, 2001).

46. Gabriel Paquette, ed., *Enlightened Reform in Southern Europe and Its Atlantic Colonies, 1750–1830* (Burlington: Ashgate, 2009).

47. Bleichmar, *Visible Empire*, 28; Margaret Ewalt, *Peripheral Wonders: Nature, Knowledge and Enlightenment in the Eighteenth-Century Orinoco* (Lewisburg: Bucknell University Press, 2008); Jose de la Sota Ruis, "Spanish Science and Enlightenment Expeditions," in *Spain in the Age of Exploration, 1492–1819*, ed. Chiyo Ishikawa (Lincoln: University of Nebraska Press, 2004), 159–188; Iris Engstrand, *Spanish Scientists in the New World: The Eighteenth-Century Expeditions* (Seattle: University of Washington Press, 1981); Arthur Robert Steele, *Flowers for the King: The Expedition of Ruiz and Pavón and the Flora of Peru* (Durham: Duke University Press, 1964).

48. Alexander von Humboldt, *Ensayo político sobre el Reino de la Nueva España* (México: Porum, 1966), 120.

49. Bleichmar, *Visible Empire*.

50. On the new vision of science and empire among Spanish reformers, see Helen Cowie, *Conquering Nature in Spain and Its Empire, 1750–1850* (New York: Manchester University Press, 2011); Daniela Bleichmar, "Atlantic Competitions: Botany in the Eighteenth-Century Spanish Empire," in *Science and Empire in the Atlantic World*, ed. James Delbourgo and Nicholas Dew (New York: Routledge, 2008), 225–252; Paula De Vos, "Research, Development, and Empire: State Support of Science in the Later Spanish Empire," *Colonial Latin American Review* 15 (2006): 55–79; Antonio Lafuente and Nuria Valverde, "Linnaean Botany and Spanish Imperial Biopolitics," in *Colonial Botany: Science, Commerce, and Politics in the Early Modern World* (Philadelphia: University of Pennsylvania Press, 2005), 134–147; Francisco Javier Puerto Sarmiento, *La Ilusión Quebrada: Botánica, Sanidad y Política Científica en la España Ilustrada* (Madrid: CSIC, 1988).

51. The rubric of expertise is useful here for two reasons. First, this terminology reinforces the methodological commitment of avoiding an a priori assumption of the preeminence of science. As far as the Crown and its representatives were concerned, the status of "expert" could be conferred as much to an individual with firsthand knowledge of quina but no formal credentials as to a pharmacist with formal knowledge of medicaments but no direct experience of the cinchona tree in its American habitat. A second reason for employing the rubric of expertise is that its cognates were actors' categories. According to a prominent contemporary dictionary, eighteenth-century Castilian included both the noun "expertise" (*pericia*) and the adjective "expert" (*perito*). Contemporary definitions of these terms associate them with formal training or education in a specific field of knowledge or trade. For example, the *Diccionario de Autoridades* defines *perito* as "knowledgeable, experienced, able and skilled (*acertado*) in some science or art." Similarly, *pericia* is defined as "knowledge, experience,

practice, and ability in any science or art." Use of these terms was much more flexible. For example, when the Crown or its Ministry of the Indies ordered officials in Latin America to consult with an hombre perito, this phrase referred in most, if not all, cases to a person, usually a man, with experiential knowledge of the cinchona tree and its bark. At least, this is how imperial officials understood these terms as evidenced by their employment of missionaries, merchants, other local officials, and bark collectors to acquire information about the cinchona tree. See Real Academia Española, *Diccionario de Autoridades*, facsimile edition, vol. 3 (Madrid: Editorial Gredos, [1726–1739] 1984), 223 and 225.

52. Shawn William Miller, *An Environmental History of Latin America* (Cambridge: Cambridge University Press, 2007); Alfred Crosby, *The Colombian Exchange: Biological and Cultural Consequences of 1492* (Westport, CT: Greenwood Press, 1973).

53. Mark Carey, "Commodities, Colonial Science and Environmental Change in Latin America," *Radical History Review* 107 (2010): 185–194; Mark Carey, "Latin American Environmental History: Current Trends, Interdisciplinary Insights and Future Directions," *Environmental History* 14 (2009): 221–252; Mark Carey, "The Nature of Place: Recent Research on Environment and Society in Latin America," *Latin American Research Review* 42 (2007): 251–264; J. R. McNeill, "Observations on the Nature and Culture of Environmental History," *History and Theory* 42 (2003): 5–43.

54. Richard Grove has observed a similar development of an ethic of resource conservation in other European imperial contexts in the early modern world. See Richard Grove, *Green Green Imperialism: Colonial Expansion, Tropical Island Edens and the Origins of Environmentalism, 1600–1860* (Cambridge: Cambridge University Press, 1995).

55. Although there has been an increasing interest in object studies in the history of science and commodities studies in the history of the Atlantic World, object-centered approaches are less common in the histories of colonial science, imperial science, or science and empire. For object studies in history of science, see D. Graham Burnett, *Trying Leviathan: The Nineteenth-Century New York Case That Put the Whale on Trial and Challenged the Order of Nature* (Chicago: University of Chicago Press, 2010); Juan Pimentel, *El rinoceronte y el megaterio: un ensayo de morfologia historica* (Madrid: Abada, 2010); Lorraine Daston, *Things That Talk: Object Lessons from Art and Science* (New York: Zone Books, 2004); Lorraine Daston, ed., *Biographies of Scientific Objects*. Chicago (University of Chicago Press, 2000). Object-centered studies can also be found in the history of medicine, the history of drugs, and the history of commodities. See Gabriela Soto-Lavega, *Jungle Laboratories: Mexican Peasants, National Projects and the Making of the Pill* (Durham: Duke University Press, 2009); Steven Topik, Carlos Marichal, and Zephyr Frank, eds., *From Silver to Cocaine: Latin*

American Commodity Chains and the Building of the World Economy, 1500–2000 (Durham: Duke University Press, 2006); Jordan Goodman and Vivian Walsh, *The Story of Taxol: Nature and Politics in the Pursuit of an Anti-Cancer Drug* (Cambridge: Cambridge University Press, 2001); Sidney Mintz, *Sweetness and Power: The Place of Sugar in Modern History* (New York: Viking, 1985).

56. Marwa Elshakry, "When Science Became Western: Historiographical Reflections," *Isis* 101 (2010): 98–109.

57. Brian Keith Axtel, ed., *From the Margins: Historical Anthropology and Its Futures* (Durham: Duke University Press, 2002).

58. Ann Laura Stoler, *Along the Archival Grain: Epistemic Anxieties and Colonial Common Sense* (Princeton: Princeton University Press, 2010); Frederick Cooper, *Colonialism in Question: Theory, Knowledge, History* (Berkeley: University of California Press, 2006).

59. For this definition of science, see Sandra Harding, *Is Science Multicultural?* (Bloomington: Indiana University Press, 1998), 10.

Chapter 1. Quina as a Medicament from the Andean World

1. Charles Marie de la Condamine, "Sur l'Arbre du Quinquina," in *Histoire de l'Académie Royale des Sciences. Année M.DCCXXXVIII* (Paris: l'Imprimerie Royale, 1740), 226–243. La Condamine mentioned Fernando de la Vega by name in a description of his second visit to Loja in 1743, see Charles Marie de la Condamine, *Journal du voyage fait par ordre du Roi, a l'équateur servant d'introduction historique a la mesure des trois premiers degres du méridien* (Paris: De L'Imprimerie Royale, 1751), 185–186. For a discussion of this encounter from the Andean perspective, see Eduardo, Estrella, "Ciencia ilustrada y saber popular en el conocimiento de la quina en el siglo XVIII," in *Saberes Andinos: Ciencia y tecnología en Bolivia, Ecuador, y Perú*, ed. Marcos Cueto, 37–57 (Lima: Instituto de Estudios Peruanos, 1995).

2. Larrie D. Ferreiro, *Measure of the Earth: The Enlightenment Expedition That Reshaped Our World* (New York: Basic Books, 2011); Neil Safier, *Measuring the New World: Enlightenment Science and South America* (Chicago: University of Chicago Press, 2008).

3. La Condamine, "Sur l'Arbre du Quinquina," 227.

4. Before the publication of La Condamine's article on the cinchona tree in 1738, a handful of images of the tree appeared in European publications. These early printed images were not based on firsthand observation but on written descriptions that compared the cinchona tree to European tree species. In his essay, "Sur l'Arbre du Quinquina," La Condamine cited several printed accounts of the tree that he may have read before embarking on his journey to South America. These include Sebastian Badus, *Anastasis cortices Peruviae* (Genoa: Calenzani, 1663); Diego de Mendoza, *Chronica de la Provincia de S. Antonio de*

los Charcas del Orden de n[uest]ro seraphico P.S. Francisco en la Indias Occidentales reyno del Peru (Madrid, 1665); Antonio de la Calancha, *Coronica moralizada del orden de San Augustin en el Peru* (Barcelona: Pedro Lacavalleria, 1639). None of these works contained an image of the tree.

5. European ignorance of the geography of North America is a prominent feature in Paul Mapp's *The Elusive West and the Contest for Empire, 1713–1763* (Chapel Hill: University of North Carolina Press, 2011). Mary Louise Pratt has also noted how eighteenth-century Europeans knew little about the interiors of the Americas and other regions around the globe; see Mary Louise Pratt, *Imperial Eyes: Travel Writing and Transculturation* (New York: Routledge, 1992), 15–36.

6. Mary Terrall, "Heroic Narratives of Quest and Discovery," *Configurations* 6 (1998): 223–242.

7. Teodoro S. Kaufman and Edmundo A. Rúveda, "The Quest for Quinine: Those Who Won the Battles and Those Who Won the War," *Angewandte Chemie International Edition* 44 (2005): 854–885.

8. Carolus Linnaeus, *Species Plantarum,* 2 vols. (Holmiae: Imprensis Laurentii Salvii, 1753).

9. Recent attempts to write the history of the cinchona tree give short shrift to indigenous Andeans and their knowledge of the bark; see Fiammetta Rocco, *The Miraculous Fever-Tree: Malaria, Medicine and the Cure that Changed the World* (New York: HarperCollins, 2003), and Mark Honigsbaum, *The Fever Trail: In Search of the Cure for Malaria* (New York: Farrar, Straus and Giroux, 2001).

10. Saul Jarcho, *Quinine's Predecessor: Francesco Torti and the Early History of Cinchona* (Baltimore: Johns Hopkins University Press, 1993), 1–12.

11. David N. Livingstone, *Putting Science in Its Place: Geographies of Scientific Knowledge* (Chicago: University of Chicago Press, 2003); Steven Shapin, "Placing the View from Nowhere: Historical and Sociological Problems in the Location of Science," *Transactions of the Institute of British Geographers* 23 (1998): 5–12; Steven Shapin, "The House of Experiment in Seventeenth-Century England," *Isis* 79 (1988): 373–404.

12. Timothy D. Walker, "The Medicines Trade in the Portuguese Atlantic World: Acquisition and Dissemination of Healing Knowledge from Brazil (c. 1580–1800)," *Social History of Medicine* 26 (2013): 403–431.

13. There is still no published research on the influence of local environmental conditions on the cinchona trees and the medical efficacy of their bark. Most scientific studies focus on the classification of different varieties and species of *Cinchona*; see Lennart Anderson, "Tribes and Genera of the Cinchoneae Complex (Rubiaceae)," *Annals of the Missouri Botanical Garden* 82 (1995): 409–427.

14. Unfortunately, for lack of historical evidence, we do not know exactly

how these healers would have used the bark before the arrival of Europeans in South America.

15. The sources of Andean medicine before contact with the Spanish are few and consist primarily of archeological evidence. Sources from the Spanish colonial period provide some insight into the healing practices and medical theory of precolonial Andean World; see David Sowell, *The Tale of Healer Miguel Pedromo Neira: Medicine, Ideologies, and Power in Nineteenth-Century Andes* (Lanham, MD: Rowman & Littlefield, 2001). Much of what is known about precontact Andean medicine relates to the Inca medical tradition; see Constance Classen, *Inca Cosmology and the Human Body* (Salt Lake City: University of Utah Press, 1993).

16. Rainer Bussman and Douglas Sharon, "Traditional Medicinal Plant Use in Northern Peru: Tracking Two Thousand Years of Healing Culture," *Journal of Ethnobiology and Ethnomedicine* 2 (2006): 47–65.

17. Bussman and Sharon, "Traditional Medicinal Plant Use in Northern Peru," 47. See also Charles C. Mann, *1491: New Revelations of the Americas Before Columbus* (New York: Vintage, 2005), 272–273.

18. Bussman and Sharon, "Traditional Medicinal Plant Use in Northern Peru," 47.

19. Eduardo Estrella, *Medicina Aborigen: La Practica Medica Aborigen De La Sierra Ecuatoriana* (Quito: Editorial Epoca, 1977), 179–181. On the various terms for Andean religious specialists, including healers, see Claudia Brosseder, *The Power of Huancas: Change and Resistance in the Andean World of Colonial Peru* (Austin: University of Texas Press, 2014), 1–3.

20. Lupe Camino, *Cerros, plantas, y lagunas poderosas: La medicina la norte de Piura* (Piura: Cipca, 1992), 153.

21. Michael Taussig, "Folk Healing and the Structure of Conquest in Southwest Colombia," *Journal of Latin American Lore* 6 (1980): 217. For a more complete statement on Taussig's views of the significance of folk healers in modern Latin America, see Michael Taussing, *Shamanism, Colonialism and the Wild Man: A Study in Terror and Healing* (Chicago: University of Chicago Press, 1987).

22. Nicholas Griffiths, "Andean *curanderos* and Their Repression: The Persecution of Native Healing in Late Seventeenth- and Early Eighteenth-Century Peru," in *Spiritual Encounters: Interactions between Christianity and Native Religions in Colonial America,* ed. Nicholas Griffiths and Fernando Cervantes (Birmingham: University of Birmingham Press, 1999), 185–197; Frank Salomon, "Shamanism and Politics in Late-Colonial Ecuador," *American Ethnologist* 10 (1983): 413–428.

23. Camino, *Cerros, plantas y lagunas poderosas,* 39. The research of Rainer Bussman of the Missouri Botanical Garden and his collaborators have supported Camino's claim for the existence of an "axis of health." See Bussman

and Sharon, "Traditional Medicinal Plant Use in Northern Peru"; R. W. Bussman and A. Glenn, "Cooling the Heat: Traditional Remedies for Malaria and Fever in Northern Peru," *Ethnobotany Research and Application* 8 (2010): 125–134; Rainer Bussman and Douglas Sharon, "Shadows of the Colonial Past—Diverging Plant Use in Northern Peru and Southern Ecuador," *Journal of Ethnobiology and Ethnomedicine* 5 (2009): 1–17; Rainer W. Bussman and Douglas Sharon, "Traditional Medicinal Plant Use in Loja Province, Southern Ecuador," *Journal of Ethnobiology and Ethnomedicine* 2 (2006): 44–55.

24. Camino, *Cerros, plantas y lagunas poderosas*, 39.

25. Although the consensus seems to be that these elements are unique to Andean culture, anthropologists have debated the extent to which this humoral theory is derivative of the humoral theory introduced by Europeans in the colonial period. See Joseph W. Bastien, "Exchange between Andean and Western Medicine," *Social Science and Medicine* 16 (1982): 795–803; Joseph W. Bastien, "Qollahuaya-Andean Body Concepts: A Topographical-Hydraulic Model of Physiology," *American Anthropologist* 87 (1985): 595–611; George M. Foster, "On the Origin of Humoral Medicine in Latin America," *Medical Anthropology Quarterly* 1 (1987): 355–393; George M. Foster, "The Validating Role of Humoral Theory in Traditional Spanish-American Therapeutics," *American Ethnologist* 15 (1988): 120–135; Joseph W. Bastien, "Differences between Kallawaya-Andean and Greek-European Humoral Theory," *Social Science and Medicine* 28 (1989): 45–51; E. N. Anderson, "Why Is Humoral Medicine so Popular?" *Social Science and Medicine* 25 (1987): 331–337.

26. Nancy Siraisi, *Medieval and Early Renaissance Medicine* (Chicago: University of Chicago Press, 1990); P. M. Teigen, "Taste and Quality in Fifteenth- and Sixteenth-Century Galenic Pharmacology," *Pharmacy in History* 29 (1987): 60–68; John Riddle, *Dioscorides on Pharmacy and Medicine* (Austin: University of Texas Press, 1985).

27. Bussman and Glenn, "Cooling the Heat."

28. Irene Silverblatt, "The Evolution of Witchcraft and the Meaning of Healing in Colonial Andean Society," *Culture, Medicine, and Psychiatry* 7 (1983): 417–418.

29. Silverblatt, "The Evolution of Witchcraft," 417–423.

30. Silverblatt draws her description of health and healing practices primarily from colonial accounts of Inca religion and history ("The Evolution of Witchcraft," 419).

31. Joseph Bastien, *Healers of the Andes: Kallawaya Herbalists and the Plants* (Salt Lake City: University of Utah Press, 1987), 46.

32. Bastien, *Healers of the Andes*, 46.

33. J. Van Kessel, "Ayllu y ritual terapéutico en la medicine andina," *Chungara: Revista de Antropología Chilena* 10 (1983): 165–176.

34. Bastien, "Exchange between Andean and Western Medicine," 45.

35. Camino, *Cerros, plantas y lagunas poderosas,* 65–77.

36. Estrella, *Medicina Aborigen,* 77–97.

37. Susan Scott Parrish, "Diasporic African Sources of Enlightenment Knowledge," in *Science and Empire in the Atlantic World,* ed. James Delbourgo and Nicholas Dew (New York: Routledge, 2007), 281–310; Antonio Barrera-Osorio, "Empiricism in the Spanish Atlantic World," in *Science and Empire in the Atlantic World,* ed. James Delbourgo and Nicholas Dew (New York: Routledge, 2007), 177–202; Antonio Barrera-Osorio, "Empire and Knowledge: Reporting from the New World," *Colonial Latin America Review* 15 (2006): 39–54.

38. Nicolas Monardes, *Segunda parte del libro de las cosas que se traen nuestras Indias Occidentales, que sirven al uso de medicina* (Seville, 1571); Fernando Crespo Ortiz, "Fragoso, Monardes and pre-Cinchona Knowledge of Cinchona," *Archives of Natural History* 22 (1995): 169–181.

39. La Calancha, *Corónica Moralizada,* bk. 1, 59.

40. Matthew James Crawford, "An Empire's Extract: Chemical Manipulations of Cinchona Bark in the Eighteenth-Century Spanish Atlantic World," *Osiris* 29 (2014): 218–220.

41. Camino, *Cerros, plantas y lagunas poderosas*, 37–48. Camino writes, "se establecen los posibles puntos geográficos que comprehenderían el eje, siendo éstos: Quito, Zamora, y Loja en el Ecuador, Las Huarinjas, Huancabamba, Sondor, Sondorillo, Huarmaca, Salas, Penachí, el desierto de Olmos, Mochumí, Monsefú, y quizás terminaría en Puerto Eten en el Perú," 40.

42. Bussman and Sharon, "Shadows of the Colonial Past."

43. Bussman and Sharon, "Traditional Medicinal Plant Use in Loja Province," 44–45.

44. In the eighteenth century Creole elites in South America were well aware of this biological diversity and imagined their regions as natural warehouse supplying all kinds of products to the world. See Jorge Cañizares-Esguerra, "Eighteenth-Century Spanish Political Economy: Epistemology and Decline," in *Nature, Empire, and Nation: Explorations of the History of Science in the Iberian World,* 96–111 (Stanford: Stanford University Press, 2006).

45. Linda Newson, *Life and Death in Early Colonial Ecuador* (Norman: University of Oklahoma Press, 1995), 55–58.

46. Bussman and Sharon, "Traditional Medicinal Plant Use in Loja Province," 44–45.

47. On modern-day networks of exchange that support the movement of medicinal plants in the region, see Z. Revene, R.W. Bussman, and D. Sharon, "From Sierra to Coast: Tracing the Supply of Medicinal Plants in Northern Peru—A Plant Collector's Tale," *Ethnobotany Research and Application* 6 (2008): 15–22.

48. Kenneth Andrien, *Andean Worlds: Indigenous History, Culture, and Consciousness under Spanish Colonial Rule, 1532–1825* (Albuquerque: University of New Mexico Press, 2001), 11–40; María Rostworowski, *A History of the Inca Realm* (Cambridge: Cambridge University Press, 1998).

49. Newson, *Life and Death in Early Colonial Ecuador*; Noble David Cook, *Born to Die: Disease and New World Conquest, 1492–1650* (Cambridge: Cambridge University Press, 1998); Suzanne Austin Alchon, *Native Society and Disease in Colonial Ecuador* (Cambridge: Cambridge University Press, 1991).

50. Randall Packard, *The Making of a Tropical Disease: A Short History of Malaria* (Baltimore: Johns Hopkins University Press, 2011); Herbert M. Gilles and David A. Warrell, *Bruce-Chwatt's Essential Malariology* (London: Edward Arnold, 1993).

51. Newson, *Life and Death in Early Colonial Ecuador*, 144–153.

52. Camino, *Cerros, plantas y lagunas poderosas*, 43.

53. Camino, *Cerros, plantas y lagunas poderosas*, 179–181.

54. Unfortunately, Camino, one anthropologist who has studied the medical and botanical cosmology of this group of healers, does not explicitly discuss how these Andean healers would have made sense of cinchona trees found in other regions. After all, cinchona trees could be found in Andean forests from Santa Fé de Bogotá, Colombia, to La Paz, Bolivia (map 1).

55. There was much debate about the use of the bark in late seventeenth-century Europe. See Jarcho, *Quinine's Predecessor*, chaps. 3 and 4.

56. Because many healers also performed spiritual rituals or had a religious function in their communities, European missionaries recognized them as competitors for indigenous believers. In addition, Andean healers were often persecuted for heterodox beliefs especially during the campaigns to extirpate idolatry in seventeenth-century Peru, see Irene Silverblatt, *Modern Inquisitions: Peru and the Colonial Origins of the Civilized World* (Durham: Duke University Press, 2004), 141–160; Kenneth Mills, *Idolatry and Its Enemies: Colonial Andean Religion and Extirpation, 1640–1750* (Princeton: Princeton University Press, 1997).

57. See "De los Ministros de Idolatria," in Pablo José de Arriaga, *La Extirpación de la Idolatria del Piru* (Lima: Geronymo de Contreras, 1621), ch. 3.

58. La Calancha, *Corónica Moralizada*, bk. 1, 59.

59. Arriaga, *La Extirpación de la Idolatria*, 74.

60. Arriaga, *La Extirpación de la Idolatria*, 74.

61. La Calancha, *Corónica Moralizada*, bk. 1, 59

62. La Calancha, *Corónica Moralizada*, 59.

63. A recent article by Leo Garofalo that focuses on spiritual or magical healing examines the integration of coca usage into the practices of Afro-Peruvian healers who were popular in colonial Peru. See Leo J. Garofalo, "Conjuring with

Coca and the Inca: The Andeanization of Lima's Afro-Peruvian Ritual Specialists, 1580–1690," *Americas* 63, no. 1 (July 1, 2006): 53–80.

64. Andrés I. Prieto, *Missionary Scientists: Jesuit Science in Spanish South America, 1570–1810* (Nashville: Vanderbilt University Press, 2011); Sabine Anagnostou, "Jesuits in Spanish America: Contributions to the Exploration of the American Materia Medica," *Pharmacy in History* 47 (2005): 3–17; Sabine Anagnostou, "Jesuit Missionaries in Spanish America and the Transfer of Medical-Pharmaceutical Knowledge," *Archives Internationales d'Histoire des Sciences* 52 (2002): 176–197.

65. One exception would be Andrés Prieto's *Missionary Scientists* on Jesuit missionaries in South America.

66. Garcilaso de la Vega, *Primera Parte de los Comentarios reales* (Lisbon: En la oficina de Pedro Crasbeeck, 1609). The quotations are from an early eighteenth-century Spanish edition. See Garcilaso de la Vega, *Primera Parte de los Comentarios Reales, que tratan, de el origen de los Incas, reies, que fueron del Perú, de su idolatria, leies, y govierno, en pas, y en guerra, de sus vidas, y conquistas, y de todo lo que fue aquel imperio,* 2nd ed. (Madrid: Oficina Real y á Costa de Nicolas Rodriguez Franco, Impresor de Libros, 1723), 64.

67. Garcilaso de la Vega, *Primera Parte,* 64.

68. Other writers in the seventeenth and eighteenth centuries copied Garcilaso de la Vega's account of healers under the Inca including the Jesuit Juan de Velasco, who, in his *Historia del Reino de Quito,* copied Vega's account almost verbatim. See Juan de Velasco, *Historia del reino de Quito en la América meridional* (Quito: Casa de la Cultura Ecuatoriana "Benjamin Carrión," 1994).

69. Rolena Adorno, *Guaman Poma: Writing and Resistance in Colonial Peru* (Austin: University of Texas Press, 1988).

70. Felipe Guaman Poma de Ayala, *El primer corónica* [sic] *i buen gobierno,* 1615, fol. 331 [333], http://www.kb.dk/permalink/2006/poma/333/en/text/.

71. This reading is supported by another source: the *Huarochirí Manuscript* written by a Quechua scribe probably at the request of Father Francisco de Avila as part of the broader effort to extirpate idolatry in Peru by documenting non-Christian practices among the indigenous communities in the region of Huarochirí near Lima, the vice-regal capital. In chapter 23 of the manuscript, the author describes how the Inca "Tupay Inga Yupanqui" gave a litter to Maca Uisa, a representative of Huarochirí that had been summoned by the Inca. To serve in the litter and carry Maca Uisa, the author reports: "The people called Calla Uaya were chosen by the Inca because they were all very strong." "These people," the manuscript continues, "could carry him, in a few days, a journey of many days." Frank Salomon and George Urioste, eds., *The Huarochirí Manuscript: A Testament of Ancient and Colonial Andean Religion* (Austin: University of Texas Press, 1991), 115.

72. Saolomon and Urioste, *The Huarochirí Manuscript*, 115, n. 581.

73. [Hipólito Unanue], "Introduccion á la Descripcion Científica de las Plantas del Peru," *Mercurio Peruano* vol. 2, no. 43 (29 May 1791): 68–76.

74. Unanue, "Introduccion," 71, n. 3.

75. Unanue does not indicate his source for this information.

76. Bastien, *Healers of the Andes*; Bussman and Sharon, "Traditional Medicinal Plant Use in Northern Peru."

77. La Condamine, "Sur l'Arbre du Quinquina," 227.

78. La Condamine, *Journal du voyage fait par ordre du Roi*, 186.

79. Miguel de Santisteban, "Relación informativa práctica de la quina de la ciudad de Loxa," Santa Fé de Bogotá, 4 June 1753, Biblioteca del Palacio Real de Madrid, II/2823, fols. 83r–87v. Santisteban had previously visited Loja during a tour of Spanish South America. See Miguel de Santisteban, *Mil leguas por América: De Lima a Caraca, 1740–1741,* Diario de don Miguel de Santisteban, ed. David J. Robinson (Bogotá: Banco de la República, 1992).

80. Eduardo Estrella is one of the few historians of quina to recognize the significance of Fernando de la Vega and Miguel de Santisteban in the communication of "popular knowledge" about the cinchona tree to Europeans; see Eduardo Aguirre Estrella, "Expedición Geodesica: Mito y Realidad de la Quina," in *Anales de las II Jornadas de Historia de la Medicina Hispanoamericana* (Cádiz: Servicio de Publicaciones de la Universidad de Cádiz, 1989), 25–32. A transcription of the information collected from Vega to Santisteban is available as an appendix to Eduardo Estrella's "Ciencia Ilustrada y saber popular." In contrast to Estrella, who portrays the encounter between Vega and La Condamine as an interaction between "popular knowledge" and "Enlightenment science," it is better to characterize this episode as the meeting of two specialists or experts from divergent traditions of knowledge and expertise. To cast Vega's healing knowledge as "popular knowledge" is a mischaracterization because curanderos, like indigenous healers elsewhere, often closely guarded their healing knowledge and only transmitted this knowledge to apprentices or other healers.

81. For example, see James Sweet, *Domingos Álvares, African Healing and the Intellectual History of the Atlantic World* (Chapel Hill: University of North Carolina Press, 2011).

82. The literature on science and popular culture has shown that a similar cleavage between science and superstition emerged in early modern Europe. See Stephen Pumfrey, Paolo Rossi, and Maurice Slawinski, eds., *Science, Culture and Popular Belief in Renaissance Europe* (Manchester: Manchester University Press, 1991).

83. Marwa Elshakry, "When Science Became Western: Historiographical Reflections," *Isis* 101 (2010): 98–109; Helen Tilley, "Global Histories, Vernacular Science and African Genealogies; Or, Is the History of Science Ready for the

World?" *Isis* 101 (2010): 110–119; Sujit Sivasundaram, "Sciences and the Global: On Questions, Methods, and Theory," *Isis* 101 (2010): 146–158.

84. Recent examples include Kelly Wisecup, *Medical Encounters: Knowledge and Identity in Early American Literature* (Amherst: University of Massachusetts Press, 2013); James Delbourgo and Nicholas Dew, eds., *Science and Empire in the Atlantic World* (New York: Routledge, 2007); Susan Scott Parrish, *American Curiosity: Cultures of Natural History in the Colonial British Atlantic World* (Chapel: University of North Carolina Press, 2006).

85. Among the vast literature on this subject, recent works include Harold Cook, *Matters of Exchange: Commerce, Medicine and Science in the Dutch Golden Age* (New Haven: Yale University Press, 2008), and Kapil Raj, *Relocating Modern Science: Circulation and the Construction of Knowledge in South Asia and Europe, 1650–1800* (New York: Palgrave Macmillan, 2010).

86. Sowell, *The Tale of Healer Miguel Pedromo Neira*; Classen, *Inca Cosmology*.

Chapter 2. Quina as a Product of the Atlantic World

1. Anthony van Leeuwenhoek, "Microscopical Observations on the Cortex Peruvianus," *Philosophical Transactions* 312 (1707): 2446–2455.

2. The classification of cinchona bark as hot or cold according to Galenic theory was explained in several seventeenth-century works on the bark. See: Gaspar de la Heredia, *De pulvere febrifugio occidentalis Indiae (1663),* ed. and trans. by José María López Piñero and Francisco Calero Calero (Valencia: Instituto de Estudios Documentales e Históricas sobre la Ciencia, 1992). See also John M. Riddle, *Discorides on Pharmacy and Medicine* (Austin: University of Texas Press, 1985).

3. For an overview of the early debates on the therapeutic use of the bark, see Saul Jarcho, *Quinine's Predecessor* (Baltimore: Johns Hopkins University Press, 1993), chap. 3; Andres-Holger Maehle, *Drugs on Trial* (Amsterdam: Rodopi, 1999), 225–233. On early modern European fever theory, see William F. Bynum, and Vivian Nutton, eds., *Theories of Fever from Antiquity to the Enlightenment* (London: Wellcome Institute for the History of Medicine, 1981), 19–120. Several works printed in early modern Europe explained the pharmacological classification of plants. Many of these draw on the work of the ancient Roman physician, Pedanius Dioscorides (40–90 CE), whose work appeared in modern printed editions in the sixteenth century. In sixteenth-century Spain, Andrés de Laguna produced an edition of Dioscorides' work; see Andrés de Laguna, *Pedacion Dioscórides Anazerbeo (1555)* (Madrid: Instituto de España, 1968).

4. In *Quinine's Predecessor,* Saul Jarcho describes several early cases in which cinchona bark proved ineffective. As news of these failures circulated, some physicians became skeptical of the utility of the new medicament from Peru.

5. Daniel Margoscy, *Commercial Visions: Science, Trade and Visual Culture in the Dutch Golden Age* (Chicago: University of Chicago Press, 2014).

6. Harold Cook, *Matters of Exchange: Commerce, Medicine, and Science in the Dutch Golden Age* (New Haven: Yale University Press, 2007), chap. 4; Jordan Goodman, "Excitantia: Or, How Enlightenment Europe Took to Soft Drugs," in *Consuming Habits: Drugs in History and Anthropology,* ed. Jordan Goodman, Paul E. Lovejoy, and Andrew Sherratt (London: Routledge, 1995), 126–148; Anthony Grafton, April Shelford, and Nancy Siraisi, *New Worlds, Ancient Texts: The Power of Tradition and the Shock of Discovery* (Cambridge: Belknap Press, 1992), chap. 4; Charles Talbot, "America and the European Drug Trade," in *First Images of America: The Impact of the New World on the Old,* ed. Fredi Chiappelli et al. (Berkeley: University of California Press, 1976), 833–845.

7. "Royal Order" [Draft], Madrid, 27 August 1751, Archivo General de India (AGI), Indiferente General 1552, fol. 344r.

8. Kenneth Andrien, *Andean Worlds: Indigenous History, Culture, and Consciousness under Spanish Colonial Rule, 1532–1825* (Albuquerque: University of New Mexico Press, 2001), 11–40; María Rostworowski, *A History of the Inca Realm* (Cambridge: Cambridge University Press, 1998).

9. Linda A. Newson, *Life and Death in Early Colonial Ecuador* (Norman: University of Oklahoma Press, 1995); Noble David Cook, *Born to Die: Disease and New World Conquest, 1492–1650* (Cambridge: Cambridge University Press, 1998); Suzanne Austin Alchon, *Native Society and Disease in Colonial Ecuador* (Cambridge: Cambridge University Press, 1991).

10. Newson, *Life and Death in Early Colonial Ecuador,* 144–153.

11. In the twentieth century, scientific research showed that quinine, the primary antimalarial alkaloid in the bark, interrupts the lifecycle of the malaria-causing parasite that takes place in the human body.

12. James L. Webb Jr., *Humanity's Burden: A Global History of Malaria* (Cambridge: Cambridge University Press, 2008), 3. Webb writes, "the term *malaria* rather confusingly bundles together the disease consequences of four different parasites that have broad biological similarities." A fifth species of malaria parasite, *Plasmodium knowlesi,* has recently been identified among some human populations in Southeast Asia, see http://www.ncbi.nlm.nih.gov/pubmedhealth/PMH0001646/#adam_000621.disease.symptoms.

13. World Health Organization, *World Malaria Report 2014* (Geneva: World Health Organization, 2014).

14. Manuel Lima, *The Book of Trees: Visualizing Branches of Knowledge* (New York: Princeton Architectural Press, 2014), 15–43.

15. Francesco Torti, *Therapeutice specialis* (Mutinae [Modena]: Typis B. Soliani, 1712), 666; Jarcho, *Quinine's Predecessor,* 125–153.

16. Webb, *Humanity's Burden,* 4.

17. The possibility that malaria sufferers can be infected with more than one species of plasmodium at a time may account for some of the complexity in early modern European classification of intermittent fevers.

18. J. R. McNeill, *Mosquito Empires: Ecology and War in the Greater Caribbean, 1620–1914* (Cambridge: Cambridge University Press, 2010), 5. According to the World Health Organization, there were an estimated 198,000,000 cases of malaria worldwide in 2013 with an estimated 584,000 of these cases resulting in death. See World Health Organization, *World Malaria Report 2014*.

19. Randall Packard, *The Making of a Tropical Disease: A Short History of Malaria* (Baltimore: Johns Hopkins University Press), 19; Webb, *Humanity's Burden*, 1.

20. Webb, *Humanity's Burden*, 66.

21. Webb, *Humanity's Burden*, 12–14.

22. Alfred Crosby, *The Colombian Exchange: Biological and Cultural Consequences of 1492* (Westport, CT: Greenwood Press, 1973), 35–63.

23. Webb, *Humanity's Burden*, 72–73.

24. Webb, *Humanity's Burden*, 68.

25. Webb, *Humanity's Burden*, 21.

26. McNeill, *Mosquito Empires*, 22–31, 52–59.

27. For an overview of the eighteenth-century trade in medicinal plants from Spanish America, in which the case of cinchona bark features prominently, see Stefanie Gänger, "World Trade in Medicinal Plants from Spanish America, 1717–1815," *Medical History* 59 (2015): 44–62.

28. Patrick Wallis, "Exotic Drugs and English Medicine: England's Drug Trade, c. 1550–c. 1800," *Social History of Medicine* 25 (2011): 25–46.

29. Harold J. Cook and Timothy Walker, "Circulation of Medicine in the Early Modern Atlantic World," *Social History of Medicine* 26 (2013): 337–351.

30. Wallis, "Exotic Drugs and English Medicine," 23–25.

31. Wallis, "Exotic Drugs and English Medicine," 33.

32. Wallis, "Exotic Drugs and English Medicine," 26.

33. Wallis, "Exotic Drugs and English Medicine," 36.

34. On Spanish colonial trade from 1717 to 1778, see Antonio García-Baquero González, *Cádiz y el Atlántico: El comercio colonial español bajo el monopolio gaditano,* 2 vols. (Cádiz: Diputación Provincial de Cádiz, [1976] 1988). For Spanish colonial trade after 1778, see John R. Fisher, *Commercial Relations between Spain and Spanish America in the Era of Free Trade, 1778–1795* (Liverpool: Centre for Latin-American Studies, University of Liverpool, 1985).

35. Stanley J. Stein and Barbara Stein, *Silver, Trade and War: Spain and America in the Making of Early Modern Europe* (Baltimore: Johns Hopkins University Press, 2003), 180–199.

36. The various wars of the eighteenth century were the biggest obstacle to

the successful operation of the annual convoy. Records of the flota exist for the following years: 1718, 1721, 1724, 1727, 1730, 1734, 1737, 1758, 1761, 1767, 1770, 1774, and 1778.

37. An arroba was a unit of measure equal to approximately twenty-five pounds.

38. García-Baquero González, *Cadíz y el Atlántico,* vol. 2, tables 19–21, 131–152.

39. García-Baquero González, *Cadíz y el Atlántico,* vol. 2, tables 19–21, 131–152.

40. Stein and Stein, *Silver, Trade and War,* 3–39.

41. The cinchona tree was not fully domesticated until the mid-nineteenth century when the British and the Dutch transplanted cinchona trees from Latin America to Asia and established cinchona plantations in India and Indonesia respectively.

42. See chapter 6 in this volume.

43. On quina in the regional economies of Loja and the system of *repartimiento de mercancias,* see Carlos Contreras, *El Sector Exportador de una Economía Colonial: La Costa del Ecuador entre 1760 y 1820* (Quito: Facultad Latinoamericana de Ciencias Sociales/ABYA-YALA, 1990); John Fisher, *Bourbon Peru, 1750–1824* (Liverpool: Liverpool University Press, 2003), 33–34, 43–45, 58–59; Luz del Alba Moya Torres, *El Arbol de la Vida: Auge y Crisis de la Cascarilla en la Audiencia de Quito, Siglo XVIII* (Quito: Facultad Latinoamericana de Ciencias Sociales Sede Ecuador, 1994), 94–104.

44. Kenneth Andrien, *The Kingdom of Quito, 1690–1830: The State and Regional Development* (Cambridge: Cambridge University Press, 2002), 15–32.

45. Andrien, *The Kingdom of Quito,* 18.

46. Andrien notes that during the height of the textile economy in the northern highlands of Quito, "approximately 10,000 workers produced an average of over 200,000 *varas* [of paña azul]" in the late sixteenth century; see Andrien, *The Kingdom of Quito,* 18.

47. Andrien, *The Kingdom of Quito,* 20; Andrien's estimates are based on Robson Tyrer, "The Demographic and Economic History of the Audiencia of Quito: Indian Population and the Textile Industry, 1600–1800," PhD diss., University of California, Berkeley, 1978, 35–58.

48. Andrien, *The Kingdom of Quito,* 15.

49. Because collection of the bark was not often the central activity of landowners and merchants, records of labor and production are sparse.

50. Andrien, *The Kingdom of Quito,* 27.

51. Andrien, *The Kingdom of Quito,* 28.

52. Andrien, *The Kingdom of Quito,* 28.

53. On the French trade, see Andrien, *The Kingdom of Quito,* 28. See also

Stein and Stein, *Silver, Trade and War,* 106–146; Paul Mapp, *The Elusive West and the Contest for Empire, 1713–1763* (Chapel Hill: University of North Carolina Press, 2013), chap. 4.

54. Andrien, *The Kingdom of Quito,* 29.

55. Andrien, *The Kingdom of Quito,* 29.

56. Andrien, *The Kingdom of Quito,* 80.

57. Andrien, *The Kingdom of Quito,* 94–95.

58. John Gray, William Arrot, and Philip Miller. "An Account of the Peruvian or Jesuits Bark, by Mr. John Gray, F.R.S. now at Cartagena in the Spanish West-Indies; extracted from the Papers given him by Mr. William Arrot, a Scotch Surgeon, who had gather'd it at the Place where it grows in Peru. Communicated by Mr. Phil. Miller, F.R.S. &c.," *Philosophical Transactions* 446 (July–December 1737): 81–86. According to Raymond Stearns, Gray was associated with the Navy Office in London and went to Cartagena in 1733 as a factor for the South Sea Company. Stearns suggests that "for five or six years [Gray] made astronomical observations, weather observations, and collected other scientific data in his 'out of office' hours"; see Raymond Phineas Stearns, "Colonial Fellows of the Royal Society of London, 1661–1788," *Notes and Records of the Royal Society of London* 8, no. 2 (1951): 178–246. I thank Kathleen Murphy and Claire Gherini for sharing information on John Gray and other aspects of eighteenth-century British colonial science.

59. It is likely that this information circulated orally among bark collectors, merchants, missionaries, physicians, pharmacists, and consumers of the bark. However, written records of such oral exchanges are often difficult to find if they existed in the first place.

60. Graciela Márquez, "Commercial Monopolies and External Trade," in *The Cambridge Economic History of Latin America* (Cambridge: Cambridge University Press, 2006), 928–1004.

61. Gray et al., "An Account of the Peruvian or Jesuits Bark," 83.

62. Francisco Suarez de Ribera, *Clave Botanica, o Medicina Botanica Nueva y Novissima* (Madrid: Manuel de Moya, 1738).

63. Suarez de Ribera, *Clave Botanica,* 241–251.

64. Suarez de Ribera, *Clave Botanica,* 241.

65. Christopher Parsons and Kathleen S. Murphy, "Ecosystems under Sail: Specimen Transport in the Eighteenth-Century French and British Atlantics," *Early American Studies* 10 (2012): 503–539.

66. La Condamine, "Sur l'arbre de quinquina," *Historie de l'Académie Royale des Sciences* (1738): 230–231.

67. For other examples of the exchange of knowledge in the Spanish Atlantic and critique of the center-periphery and diffusion model, see Daniela Bleichmar, "Atlantic Competitions: Botany in the Eighteenth-Century Spanish Empire," in

Science and Empire in the Atlantic World, ed. James Delbourgo and Nicholas Dew (New York: Routledge, 2008), 225–252; Paula De Vos, "From Herbs to Alchemy: The Introduction of Chemical Medicine to Mexican Pharmacies in the Seventeenth and Eighteenth Centuries," *Journal of Spanish Cultural Studies* 8 (2007): 135–168.

68. Several examples can be found in Delbourgo and Dew's *Science and Empire in the Atlantic World.*

69. On the transimperial circulations of knowledge, see Neil Safier, *Measuring the New World: Enlightenment Science and South America* (Chicago: University of Chicago Press, 2008); Neil Safier, "Spies, Dyes and Leaves: Agro-Intermediaries, Luso-Brazilian Couriers, and the Worlds They Sowed," in *The Brokered World: Go-betweens and Global Intelligence, 1770–1820,* ed. S. Schaffer, L. Roberts, K. Raj, and J. Delbourgo (Sagamore Beach: Science History Publications, 2009).

70. Royal Decree, Madrid, 27 August 1751, AGI, Indiferente General 1552, fols. 343r–347v.

Chapter 3. Quina as a Natural Resource for the Spanish Empire

1. Daniela Bleichmar and Peter Mancall, eds., *Collecting Across Cultures: Material Exchanges in the Early Modern Atlantic World* (Philadelphia: University of Pennsylvania Press, 2013); Paula Findlen, *Possessing Nature: Museums, Collecting, and Scientific Culture in Early Modern Italy* (Berkeley: University of California Press, 1994).

2. Juan Francisco Toro, "Copia del informe que hizo el P[ad]re Juan Fran[cis]-co Toro," Lima, 4 June 1748, Archivo General de Indias (AGI), Indiferente General 1552, fols. 74r–75r; "Memoria de Generos para la Botica de el Rey n[uest]-ro S[eñ]or que se crian en el Reyno de el Peru y Provincias immediates," Buen Retiro, 24 November 1746, AGI, Indiferente General 1552, fols. 22r–23r; Joseph Manso to Zeñon de Somodevilla y Bengoechea, Marqués de la Ensenada, Lima, 8 August 1748, AGI, Indiferente General 1552, fols. 70r–71v.

3. Toro, "Copia del informe," fols. 74r–75r.

4. A protomedicato was an official appointed by the Spanish Crown to regulate the medical professions. While Avendaño was the protomedicato for Peru, there were also protomedicatos for all the major kingdoms in America and Europe that comprised the Spanish Empire. See John Tate Lanning, *The Royal Protomedicato: The Regulation of the Medical Professions in the Spanish Empire,* ed. John TePaske (Durham: Duke University Press, 1985).

5. Paula De Vos, "Research, Development and Empire: State Support of Science in the Later Spanish Empire," *Colonial Latin American Review* 15 (2006): 55–79.

6. Marcy Norton, *Sacred Gifts, Profane Pleasures: A History of Tobacco and Chocolate in the Atlantic World* (Ithaca: Cornell University Press, 2010).

7. Daniela Bleichmar, *Visible Empire: Botanical Expeditions and Visual Culture in the Hispanic Enlightenment* (Chicago: University of Chicago Press, 2012); James Delbourgo and Nicholas Dew, *Science and Empire in the Atlantic World* (New York: Routledge, 2007); Londa Schiebinger, *Plants and Empire: Colonial Bioprospecting in the Atlantic World* (Cambridge: Harvard University Press, 2004); Richard Drayton, *Nature's Government: Science, Imperial Britain, and the "Improvement" of the World* (New Haven: Yale University Press, 2000).

8. Lisbet Koerner, *Linnaeus: Nature and Nation* (Cambridge: Harvard University Press, 1999)

9. Bleichmar, *Visible Empire*; Helen Louise Cowie, *Conquering Nature in Spain and Its Empire, 1750–1850* (Manchester: Manchester University Press, 2011); Emily Berquist Soule, *The Bishop's Utopia: Envisioning Improvement in Colonial Peru* (Philadelphia: University of Pennsylvania Press, 2014).

10. Cowie, *Conquering Nature*; Jorge Cañziares-Esguerra, "How Derivative Was Humboldt? Microcosmic Narratives in the Early Spanish America and the (Other) Origins of Humboldt's Ecological Sensibility," in *Nature, Empire and Nature: Explorations in the History of Science in the Iberian World* (Stanford: Stanford University Press, 2006), 112–128; Juan Pimentel, "The Iberian Vision: Science and Empire in the Framework of a Universal Monarchy, 1500–1800," *Osiris* 15 (2000): 17–30; Antonio Lafuente, "Enlightenment in an Imperial Context: Local Science in the Late Eighteenth-Century Hispanic World," *Osiris* 15 (2000): 155–173.

11. Barrera-Osorio, *Experiencing Nature*; María M. Portuondo, *Secret Science: Spanish Cosmography and the New World* (Chicago: University of Chicago Press, 2009).

12. Paula De Vos, "Research, Development, and Empire: State Support of Science in the Later Spanish Empire," *Colonial Latin American Review* 15 (2006): 55–79; Paula De Vos, "The Science of Spices: Empiricism and Economic Botany in the Early Spanish Empire," *Journal of World History* 17 (2006): 399–427.

13. For a new interpretation of the purpose of the knowledge and information collected by the colonial government, see Arndt Brendecke, *Imperio e información: Funciones del saber en el dominio colonial español* (Madrid: Vervuet/Iberoamericana, 2015).

14. Portoundo, *Secrete Science*; Barrera-Osorio, *Experiencing Nature: The Spanish American Empire and the Early Scientific Revolution* (Austin: University of Texas Press, 2006).

15. Barbara Mundy, *The Mapping of New Spain: Indigenous Cartography and the Maps of the Relaciones Geográficas* (Chicago: University of Chicago

Press, 1996); Raquel Alvarez Peláez, *La conquista de la naturaleza americana* (Madrid: Consejo Superior de Investigaciones Científicas, 1993).

16. Jacob Stoll, *The Information Master: Jean-Baptiste Colbert's Secret State Intelligence System* (Ann Arbor: University of Michigan Press, 2011); Eric Ash, ed., "Expertise: Practical Knowledge and the Early Modern State," *Osiris* 25 (2010); Eric Ash, *Power, Knowledge and Expertise in Elizabethan England* (Baltimore: Johns Hopkins University Press, 2004); Peter Dear, "Mysteries of State, Mysteries of Nature: Authority, Knowledge and Expertise in the 17th Century," in *States of Knowledge: The Co-Production of Science and the Social Order*, ed. Sheila Jasanoff (New York: Routledge, 2004), 206–224; Chandra Mukerji, "Cartography, Entreprenuerialism and Power in the Reign of Louis XIV: The Case of the Canal du Midi," in *Merchants & Marvels: Commerce, Science and Art in Early Modern Europe*, ed. Pamela Smith and Paula Findlen (New York: Routledge, 2002), 248–276; Richard Drayton, *Nature's Government: Science, Imperial Britain, and the "Improvement" of the World* (New Haven: Yale University Press, 2000).

17. See Karin Knorr-Cetina, *Epistemic Cultures: How the Sciences Make Knowledge* (Chicago: University of Chicago Press, 1999).

18. "Royal Order" [Draft] Madrid, 27 August 1751, Archivo General de India (AGI), Indiferente General 1552, fol. 344r.

19. Philip Curtin, *The Rise and Fall of the Plantation Complex: Essays in Atlantic History* (Cambridge: Cambridge University Press, 1990).

20. In 1826, a copy of the report originally submitted by Ulloa and Juan, was printed in London; see: Jorge Juan and Antonio Ulloa, *Noticias Secretas de America* (London: En La Imprenta de R. Taylor, 1826). Ulloa and Juan also published an official travel account that was much less critical of Spanish America; see Jorge Juan and Antonio Ulloa, *Relacion Historica del Viage a la America Meridional*, 4 vols. (Madrid: Antonio Marin, 1748). See also Kenneth Andrien, "The *Noticias secretas de America* and the Construction of a Governing Ideology for the Spanish American Empire," *Colonial Latin American Review* 7 (1998): 175–192.

21. Juan and Ulloa, *Noticias Secretas*, 572–573. Ulloa and Juan expressed a similar sentiment, though with less urgency, in *Relacion Historica*, vol. 2, 440–441.

22. Neil Safier, *Measuring the New World: Enlightenment Science and South America* (Chicago: University of Chicago Press, 2008); Antonio Lafuente and Antonio Mazuecos, *Los caballeros del punto fijo: ciencia, política y aventura en la expedición geodésica hispanofrancesa al virreinato del Perú en el siglo XVIII* (Barcelona: Serbal-CSIC, 1987); Antonio Lafuente and Eduardo Estrella, "Scientific Enterprise, Academic Adventure and Drawing-Room Culture in the Geo-

desic Mission to Quito," in *Cross Cultural Diffusion of Science: Latin America*, ed. Juan José Saldaña (Mexico City: Cuadernos de Quipu, 1987), 13–31.

23. Heidi Scott, *Contested Territory: Mapping Peru in the Sixteenth and Seventeenth Centuries* (Notre Dame: University of Notre Dame Press, 2009).

24. Miguel de la Piedra et. al., "Capitulos que ynterponer ante los poderosos S[eño]res de la R[ea]l Audiencia de Quito los vecinos de la ciudad de Loxa," Loja, 28 April 1752, Archivo Nacional del Ecuador, Quito (ANE/Q), Fondo Especial, Caja 106, vol. 253, no. 6338.

25. Stanley J. Stein and Barbara H. Stein, *Apogee of Empire: Spain and New Spain in the Age of Charles III, 1759–1789* (Baltimore: Johns Hopkins University Press, 2003); Stanley J. Stein and Barbara H. Stein, *Silver, Trade and War: Spain and America in the Making of Early Modern Europe* (Baltimore: Johns Hopkins University Press, 2000).

26. "Ordén del Rey" [Draft], Madrid, 27 August 1751, AGI, Indiferente General 1552, 343r.

27. Matthew James Crawford, "'Para Desterrar las Dudas y Adulteraciones': Scientific Expertise and the Attempts to Make a Better Bark for the Royal Monopoly of *Quina* (1751–1790)," *Journal of Spanish Cultural Studies* 8, no. 2 (2007): 193–212.

28. "Ordén del Rey" [Draft], Madrid, 27 August 1751, AGI, Indiferente 1552, fol. 344r–v.

29. Negotiation of competing interests was a structural feature of the Spanish colonial government before and during the Bourbon reforms of the late eighteenth century. Officials in the colonies were given quite a bit of latitude to handle matters as they saw fit as evidenced by the fact that officials, upon receipt of a order from the Crown, could reply: "I obey but do not execute." See Colin MacLachlan, *Spain's Empire in the New World: The Role of Ideas in Institutional and Social Change* (Berkeley: University of California Press, 1988); J. H. Fernández-Santamaria, "Reason of State and Statecraft in Spain (1595–1640)," *Journal of the History of Ideas* 41 (1980): 353–379.

30. "Ordén del Rey" [Draft], Madrid, 27 August 1751, AGI, Indiferente 1552, fol. 346r–v.

31. "Real Cedula," Madrid, 27 August 1751, AGI, Indiferente 1552, fol. 346v.

32. "Real Cedula," Madrid, 27 August 1751, AGI, Indiferente 1552, fol. 347v–348r.

33. The best source of biographical information on Santisteban is David Robinson's introduction to a modern Spanish edition of Santisteban's diary of his travels from Lima to Caracas in 1740–1741. See David J. Robinson, ed., *Mil leguas por América: De Lima a Caracas 1740–1741: Diario de don Miguel de Santisteban* (Bogotá: Banco de la República, 1992).

34. Miguel de Santisteban, "Relación informativa práctica de la quina de la

ciudad de Loxa," Santa Fe, 4 June 1753, Biblioteca del Palacio Real de Madrid (BPRM), II/2823, fol. 83r.

35. On empiricism in Spanish colonial institutions, see Antonio Barrera-Osorio, "Empiricism in the Spanish Atlantic World," in *Science and Empire in the Atlantic World*, ed. James Delbourgo and Nicholas Dew (New York: Routledge, 2008), 177–202; Antonio Barrera-Osorio, "Empire and Knowledge: Reporting from the New World," *Colonial Latin American Review* 15 (2006): 39–54.

36. In 1776, when the Crown was considering yet another round of reforms to the *estanco de quina*, the minister of the Indies sent copies of Santisteban's 1753 report back to officials in New Granada and Quito and asked for opinions on the feasibility of its proposals. In the early 1790s, when there was another spate of discussion and debate over the estanco de quina in connection with the ascendancy of Charles IV, Santisteban's report was, again, copied and circulated throughout the imperial bureaucracy. See Miguel de Santisteban, "Informe" [copy], Santa Fe, 4 June 1753, AGI, Indiferente General 1555, fols. 511r–517v.

37. Santisteban, "Relación informativa," fol. 83r.

38. Santisteban, "Relación informativa," fol. 85r; Charles Marie de La Condamine, "Sur l'arbre du quinquina," *Mémoires de L'Académie des Sciences* (1738): 226–243.

39. Much scholarship on the production and circulation of commodities has made the assumption that producers had empirical knowledge while consumers have primarily evaluative or qualitative knowledge. As noted by Arjun Appadurai, producers and consumers of commodities often develop both empirical and evaluative knowledge of things. See Arjun Appadurai, "Introduction," in *The Social Life of Things*, ed. Arjun Appadurai (Cambridge: Cambridge University Press, 1988), 3–63.

40. Santisteban, "Relación Informativa," fol. 88r.

41. The first edition of the dictionary was published in 1704. Although it is impossible to know which edition Santisteban had in hand, he may have had one of the later editions: *Dictionnaire universel françois et latin, contenant la signification et la définition tant des mots de l'une & de l'autre langue, avec leurs différens usages*. Nouvelle édition corrigée et considerablement augmentée (Paris: Chez la Vueve Delaune, rue S. Jacques, 1743).

42. The classic statement on the diffusion of scientific knowledge from the centers of Europe to its peripheries comes from an article by George Basalla, "The Spread of Western Science," *Science* 156 (May 1967): 611–622. For one of the early critiques of Basalla's model, see Roy MacLeod, "'On Visiting the 'Moving Metropolis': Reflections on the Architecture of Imperial Science," in *Scientific Colonialism: A Cross-Cultural Comparison*, ed. Nathan Reingold and Marc Rothenberg (Washington, DC: Smithsonian Institution Press, 1987), 217–249.

43. Santisteban, "Relación informativa," fol. 85v. By "divine providence,"

Santisteban means that God put the cinchona trees in South America specifically as a reward or gift to the Spanish Empire.

44. On the Bourbon reforms in general, see Gabriel B. Paquette, *Enlightenment, Governance, and Reform in Spain and Its Empire, 1759–1808* (New York: Palgrave Macmillan, 2008); John Fisher, *Bourbon Peru, 1750–1824* (Liverpool: Liverpool University Press, 2003); Allan J. Kuethe, "The Early Reforms of Charles III in the Viceroyalty of New Granada, 1759–1776," in *Reform and Insurrection in Bourbon New Granada and Peru*, ed. John R. Fisher, Allan J. Kuethe, and Anthony McFarlane (Baton Rouge: Louisiana State University Press, 1990), 19–40; John Lynch, *Bourbon Spain, 1700–1808* (Oxford: Basil Blackwell, 1989); D. A. Brading, "Bourbon Spain and Its American Empire," in *The Cambridge History of Latin America*, ed. Leslie Bethell (Cambridge: Cambridge University Press, 1984); Geoffrey J. Walker, *Spanish Politics and Imperial Trade, 1700–1789* (London: Macmillan, 1979).

45. José Manso de Velasco to the Marqués de la Ensenada, Lima, 4 November 1753, AGI, Indiferente General 1552, fols. 329r–351r.

46. Manso de Velaso to the Marques de Ensenada, 4 November 1753, fol. 331r. Manso de Velasco had probably read La Condamine's report on the cinchona tree, which identified the Mountain of Caxanuma as the source of the best bark. He attributed the high quality of Loja's bark to the local climate. He noted that cinchona trees in Loja enjoyed "a simultaneous abundance of rain, warm air, and terrain in which to enjoy the mild rays of the sun."

47. Manso de Velaso to the Marques de Ensenada, 4 November 1753, 332v, 333r–v.

48. Manso de Velaso to the Marques de Ensenada, 4 November 1753, 333v.

49. Manso de Velaso to the Marques de Ensenada, 4 November 1753, fol. 334r.

50. Manso de Velaso to the Marques de Ensenada, 4 November 1753, 337v, 334v.

51. Manso de Velaso to the Marques de Ensenada, 4 November 1753, fol. 334v.

52. All copies of this royal order seem to have been lost. Its existence is inferred from references in the official correspondence of the viceroys of Peru and New Granada.

53. Santisteban, "Relación informativa," fol. 83r.

54. José Alonso Pizarro to Marqués de Ensenada, Santa Fe, 8 June 1753, AGI, Indiferente General 1552, fol. 302v. Later in the letter, Pizarro notes that he awaits "more *tercios* [of quina] from different places" which would include samples from the trunk and roots (303r).

55. José António Manso de Velasco to the Marqués de la Ensenada, Lima, 23 December 1753, AGI, Indiferente general 1552, 366r. One concern was that the Portuguese might gain access to this quina, develop their own trade in the bark, and break the Spanish monopoly.

56. José Manso de Velasco [to the Marqués de Ensenada], Lima, 3 December 1754, AGI, Indiferente general 1552, fol. 417v.

57. Manso de Velasco [to the Marqués de Ensenada], 3 December 1754, fol. 419r. The bark was tested by José Martínez Toledano, the head pharmacist at the Royal Pharmacy, Dr. Joseph Ximenez, a physician to the Royal Family, and Dr. Campillo, a "physician of the Chamber." In November 1754, Martínez Toledano informed Julian de Arriaga, minister of the Indies, that "it works well against many intermittent Fevers." See Joseph Martínez Toledano to Julian de Arriaga, Buen Retiro, 24 November 1754, AGI, Indiferente general 1552, 368r.

58. José Manso de Velasco to Julian de Arriaga, Lima, 20 July 1755, AGI, Indiferente general 1552, fol. 373r.

59. José Ortega (d. 1761), according to Arthur Steele, was the founder and perpetual secretary of the Academia de Medicina in Madrid and served as "first pharmacist of the army." Ortega also served as director of the botanical garden at Migas Calientes founded by Ferdinand VI on 21 October 1755. In 1771, Ortega's nephew, Casimiro Gómez Ortega (1740–1818) would come to occupy the directorship of the Royal Botanical Garden. See Arthur Robert Steele, *Flowers for the King: The Expedition of Ruiz and Pavon and the Flora of Peru* (Durham: Duke University Press, 1964), 31–32, 36–37.

60. José Ortega to Julian de Arriaga, Madrid, 17 May 1757, AGI, Indiferente general 1552, fol. 472r.

61. As part of the provisions of the Treaty of Utrecht (1713) that brought the War of Spanish Succession to an end, the South Sea Company of Britain, which had the monopoly on the trade with Middle and South America, was granted the right to send one ship to Spanish America annually and maintain a "factory" or trading post in the port of Portobelo on the Isthmus of Panama.

62. On Jussieu's botanical misadventures in South America, see Neil Safier, "Fruitless Botany: Joseph de Jussieu's South American Odyssey," in *Science and Empire in the Atlantic World*, ed. James Delbourgo and Nicholas Dew (New York: Routledge, 2005) 203–224.

63. José Ortega to Julian Arriaga, 17 May 1757, fol. 472r–v.

Chapter 4. Loja's Bark Collectors, the King's Pharmacists, and the Search for the Best Bark

1. Miguel de Muzquiz to Julián de Arriaga, Aranjuez, 26 April 1773, Archivo General de Indias (AGI), Indiferente General 1554, fols. 12r-15v; Manuel Gonzalez Garrido et al., "Copia de la Representación que han hecho los Ayudantes, y demas Dependientes de la Botica de S[u] M[ajestad] al Boticario Mayor D[o]n Josef M[a]r[ti]n[e]z Toledano," Madrid, 5 February 1773, AGI, Indiferente General 1554, fols. 16r–19r.

2. Muzquiz Arriaga, 26 April 1773, fol. 14r–15r.

3. Manuel Gonzalez Garrido, Diego Lopez Manzera, Juan Daiz, Antonio Sanchez, Luis Blet, and Leandro Martin Sandoval to José Martinez Toledano, Head Pharmacist, Madrid, 5 February, 1773, AGI, Indiferente General 1554, fol. 19r.

4. Muzquiz Arriaga, 26 April 1773, fol. 15r.

5. Although recent scholarship has contested the idea that in the early modern geography of knowledge Europe was the center and the Americas were merely a periphery, this was how many people in Enlightenment Europe understood the geography of knowledge. See Charles W. J. Withers, *Placing the Enlightenment: Thinking Geographically about the Age of* Reason (Chicago: University of Chicago Press, 2007), 87–11; David J. Weber, *Bárbaros: Spaniards and Their Savages in the Age of Enlightenment* (New Haven: Yale University Press, 2006), 19–51.

6. Lee Alan Dugatkin, *Mr. Jefferson and the Giant Moose: Natural History in Early America* (Chicago: University of Chicago Press, 2009); Susan Scott Parrish, *American Curiosity: Cultures of Natural History in the Colonial British Atlantic World* (Chapel Hill: University of North Carolina Press, 2006), 307; Antonio Lafuente, "Enlightenment in an Imperial Context: Local Science in the Late Eighteenth-Century Hispanic World," *Osiris* 15 (2000): 155–173. For a classic overview of the Enlightenment debate over the Americas, see Antonello Gerbi, *The Dispute over the New World: The History of a Polemic, 1750–1900,* trans. Jeremy Moyle (Pittsburgh: University of Pittsburgh Press, 2010).

7. In a 1967 article on the history of colonial science, George Basalla also employed a center-periphery model to characterize the spread of European science; see George Basalla, "The Spread of Western Science," *Science* (1967): 611–622. In his classic *Science in Action,* Bruno Latour also characterized networks of knowledge production in terms of centers and peripheries in an attempt to explain how and why modern science had become so powerful. To this end, he described the ways in which modern science relies upon practices that lead to the accumulation of artifacts and inscriptions at central locations, "centres of calculation"; see Bruno Latour, *Science in Action: How to Follow Scientists and Engineers through Society* (Cambridge: Harvard University Press, 1987), 232–257. John Law emphasizes similar themes in "On the Social Explanation of Technical Change: The Case of the Portuguese Maritime Expansion," *Technology and Culture* 28 (1987): 227–252; and in his "On the Methods of Long-Distance Control: Vessels, Navigation and the Portuguese Route to India," in *Power, Action and Belief: A New Sociology of Knowledge?* (London: Routledge & Kegan Paul, 1986), 234–263. More recently, historians of science and empire have rejected the center-periphery model in favor of focusing on the interactions between different sites of knowledge production. See, for example, Marcelo Aranda et al., "The History of Atlantic Science: Collective Reflections from the 2009

Harvard Seminar on Atlantic History," *Atlantic Studies* 7 (2010): 493–509; see also contributions in James Delbourgo and Nicholas Dew, eds., *Science and Empire in the Atlantic World* (New York: Routledge, 2008); Miruna Achim, ed., "Science in Translation: The Commerce of Facts and Artifacts in the Transatlantic Spanish World," Special Issue of *Journal of Spanish Cultural Studies* 8, no. 2 (2007).

8. Philip II (r. 1559–1598) established the Royal Pharmacy in 1594—one of the many consequences of his efforts to support the sciences and to centralize Spanish imperial governance by making Madrid the capital city of Spain and its empire; see David C. Goodman, *Power and Penury: Government, Technology, and Science in Philip II's Spain* (Cambridge: Cambridge University Press, 1988). On the establishment of the Royal Pharmacy, see Maria del Pilar García de Yébenes Torres, *La Real Botica durante el reinado de Felipe V (1700–1746),* PhD diss., Universidad Complutense de Madrid, 1994, 34. García de Yébenes cites a document from 6 December 1594, titled "Instrucción para que la Botica nueva de S.M. pueda comenzar a servir y dar recaudo." Originally, the Royal Pharmacy was supposed to have two separate locations—one to serve the Royal Family and the other to serve everyone else. Only one location was established at the Royal Palace.

9. Carmen Sánchez Tellez, "Estudio histórico de la botica del Palacio como institución real," PhD diss., Universidad de Granada, 1979.

10. Carmen Añon Feliu, *Real Jardína Botánico de Madrid: Sus Origines, 1755–1781* (Madrid: Real Jardín Botánico, CSIC, 1987).

11. "Remisión de espcies medicinales para la Real Botica," Indiferente General 1552, AGI.

12. Daniela Bleichmar, "Atlantic Competitions: Botany in the Eighteenth-Century Spanish Empire," in *Science and Empire in the Atlantic World,* ed. James Delbourgo and Nicholas Dew (New York: Routledge, 2008), 225–252.

13. José del Campillo y Cosío, *Nuevo sistema de gobierno económico para América* (Madrid: En la imprenta de Benito Cano, 1789), 151–152. Although Campillo y Cosío's work was not published until 1789, it circulated in manuscript among elites and government officials in the early 1740s. In addition, Bernardo Ward included a substantial part of Campillo y Cosío's work in his *Proyecto económico: en que se proponen varias providencias dirigidas a promover los intereses de España* (Madrid: Joachin Ibarra, 1779).

14. Latour, *Science in Action,* 215–257.

15. David Livingstone, *Putting Science in Its Place: Geographies of Scientific Knowledge* (Chicago: University of Chicago Press, 2003) 17–86; E. C. Spary, *Utopia's Garden: French Natural History from Old Regime to Revolution* (Chicago: University of Chicago Press, 2000).

16. José Alonso Pizarro, Marqués de Villar, to the Marqués de Ensenada,

Santa Fe, 8 June 1753, AGI, Indiferente General 1552, fol. 302v–303r. Pizarro noted that he was waiting for "more [quina] from different places."

17. In the late eighteenth century, the Crown supported several botanical expeditions under the direction of the Royal Botanical Garden. See Bleichmar, *Visible Empire*; Alejandro R. Díez Torre et al., eds., *La ciencia española en ultramar. Actas de las I Jornadas sobre "España y las expediciones científicas en América y Filipinas"* (Madrid: Doce Calles, 1991); Iris Engstrand, *Spanish Scientists in the New World: The Eighteenth-Century Expeditions* (Seattle: University of Washington Press, 1981).

18. José Diguja to Pedro Valdivieso, Quito, 5 October 1768, Archivo Nacional del Ecuador Quito (ANE/Q), Fondo Especial, Box 24, vol. 68, no. 2897, fols. 151r–154r.

19. Local officials in the Audiencia of Quito had complained repeatedly to President Diguja that Daza y Fominaya was unfit to oversee the royal monopoly because, as an appointee from Spain, he lacked sufficient knowledge of or experience with the cinchona tree and the quina trade. See Matthew James Crawford, "A Cure for Empire? An Andean Wonder Drug and the Politics of Knowledge in the Eighteenth-Century Spanish Empire," in *Eighteenth Century: Theory and Interpretation* (forthcoming).

20. Pedro Valdivieso to José Diguja, Loja, 30 November 1768, ANE/Q, Fondo Especial, box 24, vol. 67, no. 2858, fols. 89r–90r; Pedro Valdivieso to José Diguja, Loja, 20 January 1769, ANE/Q, Fondo Especial, box 26, vol. 72, no. 2970-4, fols. 77r–78v.

21. Shortly after their return, six of the explorers gave testimony before Valdivieso and Casimiro Castilla, a royal scribe. These included Carlos Xaramillo, "Testimonio," Loja, 3 December 1768, ANE/Q, Fondo Especial, vol. 67, no. 2858, fols. 109v–110r; Pedro de Abarca, "Testimonio," Loja, [3 December 1768], ANE/Q, Fondo Especial, vol. 67, no. 2858, fols. 110r–v; Manuel de Mora, "Testimonio," Loja, [3 December, 1768], ANE/Q, Fondo Especial, vol. 67, no. 2858, fols. 110v–111r; Pedro Calderon, "Testimonio," Loja, 5 December 1768, ANE/Q, Fondo Especial, vol. 67, no. 2858, fols. 111r–v; Fernando Calderon, "Testimonio," Loja, [5 December, 1768], ANE/Q, Fondo Especial, vol. 67, no. 2858, fols. 111v–112r; Alberto de Leon, "Testimonio," Loja, 6 December 1768, ANE/Q, Fondo Especial, vol. 67, no. 2858, 112r–113r.

22. Pedro de Valdivieso, "Decreto," Loja, 3 December 1768, ANE/Q, Fondo Especial, vol. 67, no. 2858, fols. 105r–106r. Valdivieso implemented differential punishments based on the race and ethnicity of the offender. "Whites," who were caught harvesting bark, were to receive "a fine of twenty five pesos," "two months in prison," and "the loss of all cascarilla," which would be given to the Royal Pharmacy. "Indians or Mulattos," by contrast, would face the "loss of [their] carscarilla," "100 lashes," and "one month in prison."

23. Luz del Alba Moya Torres identifies three main epochs of preferences for different types of cinchona bark. In the first epoch, from the 1630s to c. 1700, there was little knowledge of the different species of trees and different types of bark from the different parts of the tree so the only preference was for cinchona bark from Loja. In the second epoch (c. 1700–1740), *quina roja* (red quina) was considered the best while merchants and consumers had a preference for "fat bark" which probably meant bark from the trunk. After 1740, tastes shifted to *quina amarilla* (yellow quina) and the bark of tender and thin branches. See Luz del Alba Moya Torres, *El arbol de la vida: Auge y crisis de la cascarilla in la Audiencia de Quito, siglo XVIII* (Quito: Facultad Latinoamericana de Ciencias Sociales Sede Ecuador, 1994), 50–51.

24. José Diguja to Pedro de Valdivieso, Quito, 8 February 1769, ANE/Q, Fondo Especial, vol. 72, no. 2972-3, fol. 195r.

25. Pedro de Valdivieso to José Diguja, Loja, 8 March 1769, ANE/Q, Fondo Especial, vol. 72, no. 2970-13, fol. 89r.

26. Pedro de Valdivieso to José Diguja, Loja, 8 March 1769, ANE/Q, Fondo Especial, vol. 72, no. 2970-13, fols. 89r–v. The "Illustrious Feijoo" is the Spanish Benedictine monk Benito Jerónimo Feijoo who was one of the important members of the Spanish Enlightenment in the early eighteenth century. Feijoo mentions the use of quina in "Carta 13" of *Tomo Primero* (1742) and "Carta 21" of *Tomo Quinto* (1760) of his *Cartas Eruditas y Curiosas*. See Benito Jerónimo Feijoo, *Cartas erudítas y curiosas, en que por la mayor parte se continua designio de el Theatro critico universal, impugnando, o reduciendo a dudosas varias opinions comunes*, 5 vols. (Madrid: F. del Hierro, 1742–1760).

27. Pedro de Valdivieso to José Diguja, Loja, 8 March 1769, ANE/Q, Fondo Especial, vol. 72, no. 2970-13, fol. 88v.

28. Valdivieso to Diguja, 8 March 1769, fol. 88v.

29. Valdivieso to Diguja, 8 March 1769, fols. 89v–90r.

30. Valdivieso to Diguja, 8 March 1769, fols. fol. 90v. Here, he referenced the work of José Alsinet, a physician at the court in Spain, who had published a work on the therapeutic use of quina in 1763. See José Alsinet, *Nuevas utilidades de quina* (Madrid: A. Muñoz del Valle, 1763).

31. Valdivieso to Diguja, 8 March 1769, fol. 90r.

32. Valdivieso to Diguja, 8 March 1769, fol. 91r.

33. Pedro de Valdivieso to José Diguja, Loja, 26 April 1769, ANE/Q, Fondo Especial, vol. 72, no. 2970-24, fol. 104v.

34. Valdivieso to Diguja, 26 April 1769, fol. 105r.

35. Pedro de Valdivieso, "Factura instructiva de la Cascarilla q[u]e se [h]a acopiado p[ar]a la Real Botica este año de 1769," Loja, 26 October 1769, ANE/Q, Fondo Especial, vol. 72, no. 2970-67, fols. 155r–157v.

36. Pedro de Valdivieso, "Factura instructiva," fols. 155r–157v.

37. One tube contained *cascarilla amarilla* and the other contained *cascarilla colorada*. Valdivieso attested to their purity explaining that the samples were "separated from all useless material" (*limpia de toda brosa*). Yet, unlike the rest of the cascarilla, which was shipped as dried whole bark, these samples had been "ground and pulverized." With these samples, Valdivieso hoped that the pharmacy would make a comparison of the quality of cascarilla amarilla versus cascarilla colorada. He also hoped that the pharmacists would compare these samples "with that [bark] which arrives intact and investigate if this bark has equal activity to that which is pulverized and accommodated with [these] new safeguards from Humidity [i.e., the tubes of bamboo.]." If the bark was found to be better quality, he suggested that this could be designated "for the Royal Family and in subsequent [shipments] one or more *quintales* of this cascarilla could be sent." See Pedro de Valdivieso, "Factura instructiva," fols. 155r–157v.

38. I am grateful to the archivists at the Archivo del Palacio Real de Madrid, which houses the records of the Royal Pharmacy, for searching for documents related to the testing of the contents of this box. No evidence was found.

39. In March, as his assistants were composing the instructions for bark collectors in Loja, José Martínez Toledano sent samples from the 1773 shipment for testing at Royal Hospitals in Madrid. By allowing the bark to dry out, Martínez Toledano was able to recover 37 arrobas (925 pounds) of quina. He then sent one pound of this bark as a powder to Patricio Bustos, administrator of the Royal Hospitals, for testing. On 20 March 1773, Dr. Josef Salomon y Morales and Dr. Vicente Velinchon reported that they had "experimented" with the quina on "various patients" and found that it was of good quality. Encouraged by these results, Martínez Toledano identified an additional 240 arrobas (6,000 pounds) of quina that seemed useful. Another round of testing at the Royal Hospitals confirmed that this quina was medically efficacious but not as good as the first sample. In the end, Martínez Toledano found that only 70 arrobas (1,750 pounds) were absolutely useless and needed to be destroyed. This was quite an improvement over the original assumption that the entire shipment of over 15,000 pounds of quina was useless. For the reports on these additional tests, see Josef Salomon y Morales and Vicente Velinchon to José Martínez Toledano, [Madrid], 20 March 1773, Archivo del Palacio Real de Madrid (APRM), Reinados, Carlos III, legajo 197-1; Eugenio Escolano, Bartolome de Siles, and Josef Salomon [y] Morales to José Martínez Toledano, Madrid, 14 May 1773, APRM, Reinados, Carlos III, legajo 197-1; José Martínez Toledano to Duque de Losada, Aranjuez, 15 May 1773, APRM, Reinados, Carlos III, legajo 197-1.

40. José Martínez Toledano to Duque de Losada, Madrid, 15 May 1773, APRM, Reinados, Carlos III, legajo 197-1.

41. María Luisa de Andrés Turrión, "Quina del Peru para la Real Hacienda Española (1768–1807): Notas sobre su "Estanco," in *La Expedición Botánica al*

Virreinto del Perú (1777–1788), ed. Antonio González Bueno (Barcelona: Lunwerg Editores, 1988), 71–84; María Luisa de Andrés Turrión and Maria Rosario Terreros Gómez, "Organización administrativa del Ramo de la Quina para la Real Hacienda española en el virreinato de Nueva Granada," in *Medicina y Quina en la España del Siglo XVIII*, ed. Juan Riera Palmero (Salamanca: EUROPA Artes Gráficas, 1997), 37–43.

42. Merchants, bark collectors, and botanists often commented on the fickle and seemingly arbitrary tastes of European consumers especially as the desired traits of the bark—mainly color and thickness—fluctuated over time.

43. On connoisseurship, see Ursula Klein, "Technoscience avant la lettre," *Perspectives on Science* 13 (2005), 226–266; J. V. Pickstone, "Thinking over Wine and Blood: Craft-Products, Foucault, and the Reconstruction of Enlightenment Knowledges," *Social Analysis* 41 (1997): 99–108.

44. Out of the eight propositions of the pharmacists' instructions, all except the first and fifth propositions specifically addressed the problem of humidity. See Manuel Gonzalez Garrido and Diego Lopez Manzera, "Copia de la Ynstruccion," [Madrid], 16 March 1773, AGI, Indiferente General 1554, fols. 20r–v.

45. Manuel Gonzalez Garrido and Diego Lopez Manzera, "Copia de la Ynstruccion . . . ," [Madrid], 16 March 1773, AGI, Indiferente General 1554, fols. 20r–v.

46. José Diguja to Pedro de Valdivieso, Quito, 4 September 1773, AGI, Quito 239, fol. 123r. Diguja at this point revealed to Valdivieso that he had an alternate theory of the cause of the corruption. He suggested that substitution of the bark occurred at the warehouses in the port of Callao, near Lima. Officials in South America conducted an investigation into the royal warehouse in Callao but the results were ultimately inconclusive.

47. Pedro de Valdivieso, *Auto*, [Malacatos], 16 September 1773, AGI, Quito 239, fols. 123v–124r.

48. Because moisture compromised the bark's medical virtue, bark collection occurred during the dry season from April to August; see M. Petitjean and Y. Saint-Geours, "La economía de la cascarilla en el corregimiento de Loja," *Cultura: Revista del Banco Central del Ecuador* V, no. 15 (1983), 171–207. Valdivieso related the conditions of his receipt of the "little box" (*cajonsillo*) from the president of Quito at the beginning of an *auto* that he issued on 16 September 1773; see Pedro de Valdivieso, "Auto," [Malacatos], 16 September 1773, AGI, Quito 239, fols. 123v–124r.

49. Valdivieso testified that he opened the box containing the sample "in the presence of various white men and many natives, all with experience [harvesting the bark], who were gathered [at Malacatos] to deliver their respective quantities of [quina]"; see Pedro de Valdivieso, "Decree," [Malacatos], 16 September 1773, AGI, Quito 239, fols. 123v–124r. Testimonies from other

witnesses provided estimates that twenty to thirty people were present at the examination.

50. Valdivieso, "Auto," 16 September 1773, fols. 123v–124r.

51. Examiners included Antonio Blanco de Alvardo (a forty-eight-year-old resident of the City of Loja), Alexandro Toledo (thirty-six years old), Juan de Aguirre de Dicastillo (a forty-four-year-old resident of the Jaen), Pedro Cevallos (a "native" of Vilcabamba, over thirty years old), Mathias de Salazar (a resident of Loja and owner of a hacienda in Vilcabamba), Nicolas Carpio (a resident of Loja and owner of a hacienda in Malacatos), and Matheo Benites (the Alcalde Mayor de Naturales in Malacatos). Their testimonies were taken on 17 September 1773. See AGI, Quito 239, fols. 124r-130r.

52. Pedro Cevallos, "Testimonio," Malacatos, 17 September 1773, AGI, Quito 239, fols. 126r–127r.

53. Cevallos, "Testimonio," 17 September 1773, fols. 126r–127r; Mathias de Salazar, "Certficación," Malacatos, 17 September 1773, AGI, Quito 239, fols. 127v–129r.

54. Nicolas Carpio, "Testimonio," Malacatos, 17 September 1773, AGI, Quito 239, fols. 127r–v; Matheo Benites, "Certificación," Malacatos, 17 September 1773, AGI, Quito 239, fols. 129r–130r.

55. Antonio Blanco de Alvarado came to the conclusion that the bark was from Jaen based on the physical features of the bark—especially its hardness—which Alvarado and other witnesses claimed were a product of the way in which cascarilla was dried in Jaen. The procedure in Jaen, according to Alvarado, was to chop down the whole tree and let the bark "air out" for "six or seven days" with the tree lying on the forest floor; see Antonio Blanco de Alvarado, "Testimonio," Malacatos, 16 September 1773, AGI, Quito 239, fol. 124v. Nicolas Carpio, who did not have direct experience harvesting cascarilla in Jaen, built on Alvarado's testimony by explaining that such a production process could not be used in Loja. Since cascarilla trees in Loja were so scarce, explained Carpio, bark collectors did not have the luxury of indiscriminately chopping down the tree lest the supplies become scarcer. Carpio further explained that it took bark collectors one to two months to produce one arroba (twenty-five pounds) of quina whereas Carpio recalled collecting one arroba in two to three days fifteen years earlier; see Carpio, "Testimonio," 17 September 1773, fols. 127r–v.

56. Antonio Blanco de Alvarado, Alexandro Toledo, and Juan de Aguirre de Dicastillo testified to having experience harvesting bark in Jaen.

57. Antonio Blanco de Alvarado, "Testimonio," Malacatos, 16 September 1773, AGI, Quito 239, fols. 124r–125r.

58. Alexandro Toledo, "Testimonio," Malacatos, 16 September 1773, AGI, Quito 239, fols. 125r–v; Juan de Aguirre de Dicastillo, "Testimonio," Malacatos, 16 September 1773, fols. 125v–126r. All these witnesses seem to assume that

the Royal Pharmacy's bark sample came from the bark that Valdivieso sent in 1770. There is no evidence to indicate from where the Royal Pharmacy received its sample.

59. Along with the Royal Pharmacy's sample, Valdivieso sent "five bundles [of bark] from the Province of Jaen four of which were sent by my request from the Governor [of Jaen] Don Patricio de Vega and one from Don Bernardo de Andrade Lieutenant of Zumba," see Pedro de Valdivieso, "Exorto del comisionado," Loja, 20 September 1773, AGI, Quito 239, fols. 130r–v.

60. Manuel de Riofrio et al., "Reconocimiento," Loja, 20 September 1773, AGI, Quito 239, fols. 130v–131r.

61. Francisco Palacios Vallejo et al., "Reconocimiento," Loja, 21 September 1773, AGI, Quito 239, fols. 131v–132v.

62. Pedro de Valdivieso to José Diguja, Loja, 24 September 1773, AGI, Quito 239, fol. 138v.

63. This passage echoes the phrase, "Obedezco pero no cumplo" (I obey but do not comply), which was used by colonial officials to acknowledge but not enact orders from the Crown that would be detrimental to local conditions. See Colin M. MacLachlan, *Spain's Empire in the New World: The Role of Ideas in Institutional and Social Change* (Berkeley: University of California Press, 1988).

64. Meanwhile, they hailed Valdivieso's quina as "the best of that which is known to be harvested in Loja." Ignacio Checa et al., "Reconocimiento," Quito, 9 November 1773, AGI, Quito 239, fol. 140v.

65. Checa et al., "Reconocimiento," fol. 140v.

66. José Diguja to Julián de Arriaga, Quito, 20 December 1773, AGI, Quito 239, fol. 102r. Diguja's letter to the minister of the Indies stretched to twelve manuscript pages and focused on several themes. Convinced that Valdivieso's techniques in harvesting and packaging the bark were impeccable and that the corruption was the result of theft and substitution of the bark while in transit, Diguja directed Arriaga's attention to the port of Callao in the Viceroyalty of Peru and suggested that the shipment had been compromised while sitting in a warehouse in Callao awaiting shipment to Spain.

67. Diguja to Arriaga, 20 December 1773, fol. 104r.

68. Later in his letter, Diguja made a similar comment stating that this bark of the Royal Pharmacy's sample "does not have the least value nor use among those who know how to identify it and only Merchants and those ignorant of this specific send it, for the purpose of doing business, and sell it in Europe to those who have no knowledge of its poor quality." See Diguja to Arriaga, 20 December 1773, fols. 108r–109v.

69. Diguja to Arriaga, 20 December 1773, fols. 102v–103r.

70. On the devaluation of the knowledge possessed or produced by these groups in the eighteenth-century Iberian Atlantic, see Jorge Cañizares-Esguerra,

How to Write the History of the New World: Histories, Epistemologies and Identities in the Eighteenth-Century Spanish Atlantic World (Stanford: Stanford University Press, 2001), 11–59

71. Löic Charles and Paul Cheney, "The Colonial Machine Dismantled: Knowledge and Empire in the French Atlantic," *Past & Present* 219 (2013): 127–163.

72. Latour, *Science in Action*; Londa Schiebinger and Claudia Swan, eds., *Colonial Botany: Science, Commerce, and Politics in the Early Modern World* (Philadelphia: University of Pennsylvania Press, 2005); Pamela Smith and Paula Findlen, eds., *Merchants & Marvels: Commerce, Science, and Art in Early Modern Europe* (New York: Routledge, 2002).

73. Schiebinger, *Plants and Empire*, 105–193.

74. Neil Safier, "Fruitless Botany: Joseph de Jussieu's South American Odyssey," in *Science and Empire in the Atlantic World*, ed. James Delbourgo and Nicholas Dew (New York: Routledge, 2007), 203–224; Safier, *Measuring the New World*, 57–92.

75. For invoices of shipments from 1774 to 1779, see Pedro Valdivieso, "Factura," Loja, 6 December 1774, AGI, Quito 239, fols. 192r–193r; Pedro Valdiviso, "Factura," Loja, 8 November 1776, AGI, Quito 239, fols. 242r–243r; Pedro Valdivieso, "Factura," Loja, 7 Novembber 1777, AGI, Quito 239, fols. 321r–322v; Pedro Valdivieso, "Factura," Loja, 17 October 1778, AGI, Quito 239, fols. 354r–355v; Pedro Valdivieso, "Factura," Loja, 7 November 1779, ANE/Q, Cascarilla, box 2, expediente 4, fols. 7r–8v.

76. The pharmacy's conception of superior quina persisted. In 1790, during another period of reform to the monopoly, the Crown sent the pharmacy's 1773 instructions to Loja again.

Chapter 5. Botanists as the Empire's New Experts in Madrid

1. Miguel García de Cáceres, "Informe," Guayaquil, 16 March 1779, Archivo General de Indias (AGI), Quito 240, N. 36a, fols. 181r–192v. In 1786, the Crown produced a printed copy of Cáceres's report and sent copies to officials in South America. For more on this, see Matthew James Crawford, "A 'Reasoned Proposal' against 'Vain Science': Creole Negotiations of an Atlantic Medicament in the *Audiencia* of Quito (1776–1792)," *Atlantic Studies* 7, no. 4 (2010): 397–419.

2. Manuel Sellés, José Luis Peset, and Antonio Lafuente, eds., *Carlos III y la ciencia de la Ilustración* (Madrid: Alianza Editorial, 1988); Antonio Domínguez Ortiz, *Carlos III y la España de la Ilustración* (Madrid: Alianza Editorial, 1988); Puerto Sarmiento, *La Ilusión Quebrada: Botánica, Sanidad y Política Científica en la España Ilustrada* (Madrid: CSIC, 1988).

3. On eighteenth-century notions of the utility of the natural sciences, see

Margaret Schabas and Neil De Marchi, eds., *Oeconomies in the Age of Newton* (Durham: Duke University Press, 2003); Staffan Müller-Wille, "Nature as Marketplace: The Political Economy of Linnaean Botany," *History of Political Economy* 35 (2003): 154–172; Richard Drayton, *Nature's Government: Science, Imperial Britain, and the "Improvement" of the World* (New Haven: Yale University Press, 2000).

4. Daniela Bleichmar, *Visible Empire: Botanical Expeditions and Visual Culture in the Hispanic Enlightenment* (Chicago: University of Chicago Press, 2012); Helen Cowie, *Conquering Nature in Spain and Its Empire, 1750–1850* (Manchester: University of Manchester Press, 2011); Daniela Bleichmar, "Painting as Exploration: Visualizing Nature in Eighteenth-Century Colonial Science," *Colonial Latin American Review* 15 (2006): 81–104; Londa Schiebinger and Claudia Swan, eds., *Colonial Botany: Science, Commerce, and Politics in the Early Modern World* (Philadelphia: University of Pennsylvania Press, 2005); Schiebinger, *Plants and Empire: Colonial Bioprospecting in the Atlantic World* (Cambridge: Harvard University Press, 2004); Drayton, *Nature's Government*; Mauricio Nieto Olarte, *Remedios para el imperio: Historia natural y la apropiación del nuevo mundo* (Bogotá: La Imprenta Nacional de Colombia, 2000); N. J. Jardine, A. Secord, and E. C. Spary, eds., *Cultures of Natural History* (Cambridge: Cambridge University Press, 1996); D. P. Miller and P. H. Reill, eds., *Visions of Empire: Voyages, Botany, and Representations of Nature* (Cambridge: Cambridge University Press, 1996).

5. Michael Adas, *Machines as the Measure of Men: Science, Technology, and Ideologies of Western Dominance* (Ithaca: Cornell University Press, 1989); Daniel R. Headrick, *The Tools of Empire: Technology and European Imperialism in the Nineteenth Century* (New York: Oxford University Press, 1981); Lucile Brockway, *Science and Colonial Expansion: The Role of the British Royal Botanic Gardens* (New York: Academic Press, 1979).

6. Cowie, *Conquering Nature*; Schiebinger, *Plants and Empire*.

7. Antonio González Bueno, *Tres botánicos de la Ilustración: Gómez Ortega, Zea, Cavanilles: la ciencia al servicio del poder* (Nivola: Tres Cantos, 2002); Francisco Javier Puerto Sarmiento, *Ciencia de Cámara: Casimiro Gómez Ortega (1741–1818): El Científico Cortesano* (Madrid: CSIC, 1992).

8. By "therapeutic testing," I mean that physicians used the quina in question on a patient as a means to assess its medical efficacy. I thank Dr. Judith Farquhar for suggesting this terminology.

9. In March, as his assistants were composing the instructions for bark collectors in Loja, José Martínez Toledano sent samples from the 1773 shipment for testing at Royal Hospitals in Madrid. He received the results of these tests in March and May. See Josef Salomon y Morales and Vicente Velinchon to José Martínez Toledano, [Madrid], 20 March 1773, Archivo del Palacio Real

de Madrid (APRM), Reinados, Carlos III, legajo 197-1, nn.; Eugenio Escolano, Bartolome de Siles, and Josef Salomon [y] Morales to José Martínez Toledano, Madrid, 14 May 1773, APRM, Reinados, Carlos III, legajo 197-1, nn.; José Martínez Toledano to Duque de Losada, Aranjuez, 15 May 1773, APRM, Reinados, Carlos III, legajo 197-1.

10. Duque de Losada to José Martínez Toledano, Aranjuez, 30 April 1776, APRM, Reinados, Carlos III, 197-3. Losada's communiqué provides a summary of these events but provides little detail on how much quina was tested and which classes were tested at the General Hospital. Additional archival research is needed to find supporting documentation for Losada's account.

11. APRM, Reinados, Carlos III, 198-3.

12. APRM, Reinados, Carlos III, 198-3. This document also describes how the Crown instructed the head pharmacist to give a portion of the quina from the 1774 shipment that the pharmacy had dubbed "useless" to the General Hospital for testing.

13. On early modern medical practices of human and animal experimentation, see Schiebinger, *Plants and Empire,* 156–166; Andreas-Holger Maehle, *Drugs on Trial: Experimental Pharmacology and Therapeutic Innovation in the Eighteenth Century* (Amsterdam: Rodopi, 1999), 2, 311–317.

14. This response by the physicians at the General Hospital derived in part from their perception that the quina donated by the king had declined in quality in recent years. However, because the quality of different kinds of quina was under such intense debate and scrutiny at this time, it is difficult to establish definitively whether a particular shipment or kind of quina was poor quality or not. For a contemporary account of the complaints from the General Hospital, see Duque de Losada to Duqe de Híjar, San Ildefonso, 15 September 1780, APRM, Reinados, Carlos III, 198-3.

15. APRM, Reinados, Carlos III, 198-3.

16. APRM, Reinados, Carlos III, 198-3. *Tercianas* is the Spanish term for intermittent fevers in which patients lapse into fever every forty-eight hours.

17. My account of the assessment of this disbursement of quina at the General Hospital and elsewhere comes from the description given by the Duque de Híjar. See Duque de Híjar to the Duque de Losada, Madrid, 4 September 1780, APRM, Reinados, Carlos III, 198-3.

18. This inspection was conducted under the auspices of the Tribunal of the Royal Protomedicato—the main regulatory and disciplinary board of the medical community in early modern Spain. Nine physicians signed the report, including Man[ue]l Prieto, Josef Solomon, Bartolome de Siles, Juan [Dayde?], Ant[oni]o Lorente, Vicente Velinchon, Nicolas Lopez y Valverde, Ygnacion Josef Serrano, and Eugenio Escolano. Manuel Prieto, et al. to el Conde de Mora, Madrid, 6 September 1779, APRM, Reinados, Carlos III, 198-3.

19. Duque de Híjar to the Duque de Losada, Madrid, 4 September 1780, APRM, Reinados, Carlos III, 198-3.

20. Duque de Híjar to the Duque de Losada, Madrid, 4 September 1780, APRM, Reinados, Carlos III, 198-3.

21. These reasons are summarized in the Duque de Híjar's letter of 1780. Híjar recounted the history of this dispute as part of his own request to the Duque de Losada, in which Híjar explicitly stated that he wanted fresh bark from a shipment recently arrived in Lisbon and not bark from the Royal Pharmacy's existing stores.

22. Duque de Híjar to the Duque de Losada, Madrid, 4 September 1780, APRM, Reinados, Carlos III, 198-3.

23. The fact that the boxes had the word "useless" written on them suggests that the pharmacists were well aware of the bark's quality. Given the increasing scarcity of the bark, however, the pharmacists may have had little choice but to send it to the General Hospital. It is likely that the Royal Pharmacy experienced not only a decline in the quantity of bark in the shipments but also in the quality of the bark especially because bark collectors had little choice but to take whatever cinchona was available in order to meet the terms of their contracts with merchants and landowners.

24. Mauricio de Echandi to Luis Blet, San Roque, 16 April 1780, APRM, Reinados, Carlos III, 198-3. Included with this document are the case histories of the use of the quina on patients by various army physicians from January to March 1780.

25. Luis Blet to the Duque de Losada, Algeciras, 20 April 1780, APRM, Reinados, Carlos III, 198-3. For Luis Blet's biography, see Eduardo Valverde Ruiz, "La Real Botica en el Siglo XIX," PhD diss., Universidad Complutense de Madrid, 1999, 6–7.

26. During times of military conflict, Spanish ships occasionally landed at Lisbon as in the case of the *Buen Consejo,* which in 1780 was carrying the Crown's annual supply of quina. See Miguel de Muzquiz to Duque de Losada, Aranjuez, 19 June 1780, APRM, Reinados, Carlos III, 197-2.

27. Duque de Losada to Duque de Híjar, San Ildefonso, 15 September 1780, APRM, Reinados, Carlos III, 198-3.

28. In a surprising reversal, the Duque de Híjar reported to the Duque de Losada that the junta of hospitals had agreed to reexamine the disputed quina based on the latest results from Alfonso Lope y Torralva's examinations in September and October that suggested the disputed quina was good quality. See Duque de Híjar to Duque de Losada, Madrid, 9 November 1780, APRM, Reinados, Carlos III, 198-3.

29. The Royal Pharmacy was asked to review "eighteen boxes and two sacks" of quina sent from Santa Fé de Bogotá by Sebastian López Ruiz. These samples

were those requested by Gómez Ortega in his 1777 report on the first samples sent to Madrid by López Ruiz. See Sebastian José López Ruiz, "Relacion de las muestras de Quina contenidas en los diez y ocho cajones y dos churlas numerados," Santa Fe, 19 October 1782, Archivo General de Simancas (AGS), Hacienda, 961-2, carpeta 101.

30. AGS, Hacienda, 961-2, carpeta 99, fol. 3r. The text of this summary of the head pharmacist's report is identical to the original; Juan Díaz to Marqués de Valdecarzana, El Pardo, 24 January 1785, AGS, Hacienda, 961-2, carpeta 112.

31. AGS, Hacienda, 961-2, carpeta 99, fol. 2v. Unfortunately, this document, which merely summarizes the tests conducted by the pharmacists and their results, does not explain what the "method prescribed in chemistry" was.

32. AGS, Hacienda, 961-2, carpeta 99, fol. 2v. Unfortunately, this document does not state specifically which works of "travelers" and "naturalists" were consulted.

33. AGS, Hacienda, 961–2, carpeta 99, fol. 3r.

34. AGS, Hacienda, 961–2, carpeta 99, fol. 3r.

35. AGS, Hacienda, 961–2, carpeta 99, fols. 3v–7v. I used the phrase "Quito or Peru" to indicate that some physicians compared their samples explicitly to quina from Quito while others compared their sample to quina from Peru. At this point, many in Europe would have considered the two categories synonymous.

36. AGS, Hacienda, 961-2, carpeta 99, fol. 8r.

37. The anonymous author of the summary wrote, "este es su parecer, en el que, no obstante, lo expuesto, deja al Publico el derecho de juzgar, y decidir este asunto más adelante con el tiempo." See AGS, Hacienda, 961-2, carpeta 99, fol. 9r.

38. Marqués de Valdecarzana to Pedro de Lerena, San Ildefonso, 3 August 1785, AGS, Hacienda, 961-2, carpeta 120.

39. Valdecarzana to Lerena, 3 August 1785, AGS, Hacienda, 961-2, carpeta 120.

40. Valdecarzana to Lerena, 3 August 1785, AGS, Hacienda, 961-2, carpeta 120.

41. On the rise of chemical medicine and use of chemical techniques in pharmacy, see Paula De Vos, "From Herbs to Alchemy: The Introduction of Chemical Medicine to Mexican Pharmacies in the Seventeenth and Eighteenth Centuries," *Journal of Spanish Cultural Studies* 8 (2007): 135–168; Ursula Klein, "Experimental History and Herman Boerhaave's Chemistry of Plants," *Studies in History and Philosophy of Biological and Biomedical Sciences* 34 (2003): 533–567; Jonathan Simon, "Analysis and the Hierarchy of Nature in Eighteenth-Century Chemistry," *British Journal of the History of Science* 35 (2002): 1–16; Paula De Vos, "The Art of Pharmacy in Seventeenth- and Eighteenth-Century México," PhD diss., University of California, Berkeley, 2001; Frederic Lawrence Holmes,

Eighteenth-Century Chemistry as an Investigative Enterprise (Berkeley: Office for History of Science and Technology, University of California, Berkeley, 1989).

42. See chapter 3 in this volume.

43. Pascual Iborra, *Historia del protomedicato en España (1477–1822)* (Valladolid: Universidad, Secretario de Publicaciones, 1987); John Tate Lanning, *The Royal Protomedicato: the Regulation of the Medical Professions in the Spanish Empire,* ed. John Jay Te-Paske (Durham: Duke University Press, 1985).

44. José de Gálvez to Duque de Losada, El Pardo, 9 January 1780, APRM, Reinados, Carlos III, 197-3.

45. For similar accounts of botanists and botanical institutions in other European states, see Emma C. Spary, *Utopia's Garden: French Natural History from Old Regime to Revolution* (Chicago: University of Chicago Press, 2000); John Gascoigne, *Science in the Service of Empire: Joseph Banks, the British State and the Uses of Science in the Age of Revolution* (Cambridge: Cambridge University Press, 1998); Alice Stroup, *A Company of Scientists: Botany, Patronage, and Community at the Seventeenth-Century Parisian Royal Academy of Science* (Berkeley: University of California Press, 1990).

46. Casimiro Gómez Ortega to Duque de Losada, 15 January 1777, APRM, Reindos, Carlos III, 197-3.

47. Sebastian José López Ruiz, "Cronología del descubrimiento de la Quina de Santa Fe de Bogotá," Santa Fe, 20 May 1784, AGI, Santa Fe 757, fols. 1216–1235. This submission by López Ruiz was just beginning of decades long dispute between López Ruiz and José Celestino Mutis over who first discovered cinchona trees near Santa Fé de Bogotá. See Gonzalo Hernández de Alba, *Quinas Amargas: El sabio Mutis y la discusión naturalista del siglo XVIII* (Bogotá: Tercer Mundo Editores, 1991); Marcelo Frías Núñez, *Tras El Dorado Vegetal: José Celestino Mutis y la Real Expedición Botánica del Nuevo Reino de Granada (1783–1808)* (Seville: Diputación de Sevilla, 1994).

48. Casimiro Gómez Ortega to José de Gálvez, Madrid, 24 April 1777, AGI, Santa Fe 757, fol. 58v.

49. Gómez Ortega to Gálvez, 24 April 1777, fol. 58v.

50. Gómez Ortega to Gálvez, 24 April 1777, fol. 59r.

51. Casimiro Gómez Ortega *Instrucción sobre el modo más seguro y económico de transportar plantas vivas* [1779], introduction by Francisco Javier Puerto Sarmiento (Madrid: Fundación de Ciencias de la Salud, 1992).

52. Javier Puerto Sarmiento, *Ciencia de Cámara: Casimiro Gómez Ortega (1741–1818) El Científico Cortesano* (Madrid: CSIC, 1992), 166–167. See also Henri-Louis Duhamel, *Avis pour transport par mer des arbres, des plantes vivaces, des semences, des animaux, et de differents autres morceaux d'Historia Naturelle* (1752); John Ellis, *Directions for bringing over seeds and plants from the East-Indies and other distant countries in a state of vegetation . . . to which is*

added the figure and botanical description of dionea muscipula (London, 1770); Casimiro Gómez Ortega to José de Gálvez, Madrid, 24 April 1777, AGI, Santa Fe 757, fols. 59r–v.

53. Puerto Sarmiento suggests that the instructions were intended for "botanical correspondents, those on expeditions and high functionaries overseas"; see Francisco Javier Puerto Sarmiento, *Ciencia de Cámara: Casimiro Gómez Ortega (1741–1818): El Científico Cortesano* (Madrid: CSIC, 1992), 166. Elsewhere, Puerto Sarmiento has noted that the instructions were also sent to colonial officials including the viceroys of New Spain, New Granada, and Peru; the governors of Puerto Rico, Santo Domingo, Yucatán, and Louisiana; and the intendants of Caracas and Habana; see Francisco Javier Puerto Sarmiento, "Introduction," to Casimiro Gómez Ortega, *Instrucción sobre el mode más seguro y económico*, xvi. Archival records show that Gómez Ortega's instructions were also sent to colonial officials, who had been asked in 1786 to send natural historical objects to the Crown; see Archivo Nacional del Ecuador, Quito (ANE/Q), Fondo Especial, box 99, vol. 240, no. 6090.

54. Instrucciónes were already in use in the Crown's royal reserve of quina as evidenced by the instrucción on the collection of cinchona bark written by the royal pharmacists for bark collectors in Loja in 1773 (see chapter 4 in this volume).

55. The *Instruction* included three main "articles" and an appendix followed the introduction. The first and second "articles" of the instructions explained how to dig up, package, and protect plants. The third article gave a list of those plants most desired by the Crown and the Royal Botanical Garden. According to Francisco Javier Puerto Sarmiento, Gómez Ortega solicited 172 plants in this section. His proposed uses of these plants, Sarmiento suggests, reflect the Crown's and Gómez Ortega's priorities in terms of the type of botanical commodities to develop. Uses included dyeing (4% of plants requested), miscellaneous (mining, cosmetics, industrial, etc.) (6.5% of plants requested), construction materials (13% of plants requested), foodstuffs (23% of plants requested), and medicine (53.5% of plants requested).

56. Gómez Ortega, *Instrucción*, 10–11.

57. Gómez Ortega, *Instrucción*, 1.

58. Gómez Ortega, *Instrucción*, 1–3.

59. Notably, Gómez Ortega makes no reference in his introduction to Francisco Hernández (1514–1587), royal physician to Philip II, who was sent to New Spain to conduct a survey of its flora. See Simon Varey, Rafael Chabrán, and Dora Weiner, eds., *Searching for the Secrets of Nature: The Life and Works of Dr. Francisco Hernández* (Stanford: Stanford University Press, 2000); José María López Piñero and José Pardo Tomás, *La influencia de Francisco Hernández (1515–1587) en la constitución de la botánica y la materia médica moderna* (Va-

lencia: Universidad de Valencia-CSIC, 1996); Simon Varey and Rafael Chabrán, "Medical Natural History in the Renaissance: The Strange Case of Francisco Hernández," *Huntington Library Quarterly* 57 (1994): 124–151.

60. Casimiro Gómez Ortega to José de Gálvez, Madrid, 23 February 1777, AGI, Indiferente General 1544, nn. 3r.

61. This move would effectively shift the center of gravity of the Crown's royal reserve of quina in Quito from Loja to Cuenca. Before he stepped down from his position as corregidor of Loja and commissioner of the forests, Pedro Valdivieso was asked to write instructions for officials in Cuenca on how to collect, prepare, and package quina for the Royal Pharmacy. See Pedro de Valdivieso to Royal Officials of Cuenca, Loja, 7 February 1783, ANE/Q, Fondo Especial, box 95, vol. 232, no. 54b; Pedro de Valdivieso to the Royal Officials of Cuenca, Loja, 7 March 1783, ANE/Q, Fondo Especial, box 95, vol. 232, no. 54c.

62. Casimiro Gómez Ortega to José de Gálvez, Madrid, 25 April 1785, ANE/Q, Fondo Especial, box 93, vol. 227, no. 5879, fols. 125r–127v.

63. José de Gálvez to Juan Josef de Villalengua, Aranjuez, 10 May 1785, ANE/Q, Fondo Especial, box 93, vol. 227, no. 5879, fols. 128r–129r.

64. Gálvez to Villalengua, 10 May 1785, fol. 128v.

65. Mutis has been the subject of a number of biographies and other studies over the years. Some of the most prominent works on Mutis and especially his involvement with quina include José Antonio Amaya, *Mutis, Apóstol de Linneo: Historia de la Botánica en el Virreinato de la Nueva Granada (1760–1783)* (Bogotá: Instituto Colombiano de Antropología e Historia, 2005); Mauricio Nieto Olarte, *Remedios para el imperio: Historia natural y la apropiación del nuevo mundo* (Bogotá: La Imprenta Nacional de Colombia, 2000); Marcelo Frías Núñez, *Tras El Dorado Vegetal: José Celestino Mutis y la Real Expedición Botánica del Nuevo Reino de Granada (1783–1808)* (Seville: Diputación de Sevilla, 1994); Hernández de Alba, *Quinas Amargas*; A. Federico Gredilla, *Biografía de Jose Celestino Mutis y Sus Observaciones sobre al Vigilias y Sueños de Algunas Plantas* (Bogotá: Plaza & Janes, [1911] 1982).

66. Frias Nuñez, *Tras El Dorado Vegetal,* 173.

67. Frías Núñez, *Tras El Dorado Vegetal,* 177.

68. Frías Núñez, *Tras El Dorado Vegetal,* 181. Other scholars have endorsed Frías Nuñez's interpretation of Mutis actions. See Santiago Castro-Gómez, *La hybris del punto cero: Ciencia, raza e ilustración en la Nueva Granada (1750–1816)* (Bogotá: Editorial Pontificia Universidad Javeriana, 2005), 223–227.

69. Bruno Latour, *Science in Action: How to Follow Scientists and Engineers through Society* (Milton Keynes: Open University Press, 1987), 215-257.

70. This debate between Mutis and López Ruiz has been covered in detail elsewhere. See Hernández de Alba, *Quinas Amargas,* 161–184. Indeed, most of

the archival documents at the Archivo General de Indias relating to quina from Santa Fé are the product of this dispute, see especially AGI, Santa Fe 757A.

71. Mutis made this recommendation in a 1784 letter to Eloy Valenzuela. See Frías Núñez, *Tras El Dorado Vegetal*, 180.

72. On the relationship between Mutis and Santisteban, see Bleichmar, *Visible Empire*, 143–146 and David J. Robinson's introduction to Miguel de Santisteban, *Mil leguas por América: De Lima a Caracas 1740–1741, Diario de don Miguel de Santisteban,* ed. David J. Robinson (Bogotá: Banco de la República, 1992).

73. Antonio Caballero y Góngora to José García de Leon y Pizarro, Santa Fé de Bógota, 2 October 1783, ANE/Q, Fondo Especial, box 80, vol. 203, n. 5294, fols. 36r–v. Caballero and Góngora described the collection of sample as "very important for the formation of regulations regarding the cutting and sending of quinas."

74. Valdivieso wrote, "no dudo de la Ynteligencia de los Botánicos de ese Reyno, pero sí, del actual cotejo, por no haverlo practicado." See Pedro Valdiviso to José García de Leon y Pizarro, Loja, 22 November 1783, ANE/Q, Fondo Especial, box 80, vol. 203, no. 5294, fols. 47r–v.

75. In August of this year, *visitador general* García de Leon implemented a royal order of March 21 that ordered the enclosure of the cinchona forests of Cuenca and Loja which effectively prohibited all collection and commerce in quina except for that bark destined for the Royal Pharmacy. See Luis de Cifuentes, "Testimonio en relacion del expediente formade acerca de los Montes de Cascarilla de esta Provincia de Quito," Quito, 1 December 1784, AGI, Quito 242, N. 136a.

76. San Andres was further motivated to win the contract because the José Gacía de Leon, the new president of Quito, had recently issued a moratorium on all collection and commerce in quina bark in the entire Audiencia of Quito.

77. Manuel Perfecto de San Andres to José García de Leon y Perfecto, Cuenca, 11 November 1783, ANE/Q, Fondo Especial, box 82, vol. 207, no. 142, fol. 167r.

78. Manuel Perfecto de San Andres to Juan José Villalengua, Cuenca, 26 May 1784, ANE/Q, Fondo Especial, box 87, vol. 217, no. 136, fol. 152r–v.

79. Perfecto de San Andres to Villalengua, 26 June 1784, fols. 183r–v.

80. José Celestino to Antonio Caballero y Góngora, Mariquita, 10 July 1784, in *Archivo Epistolar del Sabio Naturalist José Celestino Mutis,* ed. Gonzalo Hernández de Alba, vol. 1, 85.

81. Two samples of Perfecto de San Andres' quina had previously been examined by a group of three "physicians" in Quito in 1783. This group found one of the samples to be "superior" and the other to be "inferior"; see Juan Josef Villalengua, "Auto," Quito, 16 October 1784, AGI, Quito 242, N. 124a, nn. 1r–2r; Jo-

seph del Rosario, "Reconocimiento," Quito, 30 October 1784, AGI, Quito 242, N. 124a, nn. 2r–2v; Bernardo Delgado, "Reconocimiento," Quito, 30 October 1784, AGI, Quito 242, N. 124a, nn. 2v–3r; Francisco Eugenio Santa Cruz y Espejo, "Reconocimiento," Quito, 30 October 1784, AGI, Quito 242, N. 124a, nn. 3r–4r. In a follow up letter to Minister of the Indies José de Gálvez, President of Quito Juan Villalengua said that he did not trust the assessments of the local experts in Quito and would arrange to submit a sample of quina from Cuenca to Spain for further testing by "intelligent physicians and botanists"; see Juan José Villalengua to José de Gálvez, Quito, 18 November 1784, AGI, Quito 242, N. 124.

82. José Celestino Mutis to Francisco Martínez de Sobral, Mariquita, 19 December 1789, *Archivo Epistolar del Sabio Naturalist José Celestino Mutis,* ed. Gonzalo Hernández de Alba, vol. 1 (Bogotá: Instituto Colombiano de Cultural Hispanica, 1983), 505. Mutis also mentioned a "confidential correspondence" in a letter to José de Ezpeleta, viceroy of New Granada; see José Celestino Mutis to José de Ezpeleta, Mariquita, 24 February 1790, *Archivo Epistolar,* vol. 2, 19.

83. José Celestino Mutis to Francisco Martínez de Sobral, Mariquita, 19 December 1789, *Archivo Epistolar,* vol. 1, 505. The full passage from Mutis's letter is: "La muerte de Sonora sepultó las ideas confide[n]ciales en que habíamos convenido hasta publicar el progreso de la Real Administración, por evitar los clamores, aunque injustos, de los interesados en este comercio, y del público, tal vez sobresaltado a la voz de Estanco. Y ya que se volvió a proporcionar enderazar el asunto por la Real Orden que me remitió el Excelentísimo señor Porlier, quiso la desgracia que se cambiasen tres Virreyes en un año y duerma el importantísimo asunto de la *Quina* en el más profundo letargo."

84. Maehle, *Drugs on Trial,* 275–284.

Chapter 6. Imperial Reform, Local Knowledge, and the Limits of Botany in the Andean World

1. "Ynstruccion que han de observer el Corregidor de Loxa y el Botánico Chimico," Madrid, 26 August 1790, Archivo Nacional Histórico de Ecuador-Quito (ANE/Q), Fondo Especial, vol. 278, no. 6843, fols. 247r–250r. In the *Diccionario de Autoridades,* an eighteenth-century Spanish dictionary, *chimico* is listed as a noun meaning "someone who professes the art of chemistry and the same as Alchemist"; see Real Academia Española, *Diccionario de la Lengua Castellana,* vol. 2 (Madrid: La Imprenta de Francisco de Hierro, 1729), 319. This same dictionary only lists "*botanico*" as an adjective and not a noun, but other sources indicate that the term did exist as a noun for "botanist"; see: Real Academia Española, *Diccionario de la lengua castellana,* vol. 1 (Madrid: La Imprenta de Francisco de Hierro, 1729), 659.

2. Royal pharmacists in Madrid also noticed the declining quantities of quina in their annual shipments from Loja. Shipments reached a high of 18,000

pounds in 1785. Just four years later, in 1789, the pharmacy received slightly less than 5,000 pounds of bark. In 1785, officials in Spain thought that they had solved the problem of scarcity when the Royal Pharmacy approved of the quality of samples of cinchona bark from the forests near Santa Fé de Bogotá. However, in February 1789, experts in Madrid reclassified this bark as useless for royal purposes and the Crown permanently suspended shipments of quina from Santa Fé in 1790. See María Luisa de Andrés Turrión, "Quina del Peru para la Real Hacienda Española (1768–1807): Notas sobre su 'Estanco,'" in *La Expedición Botánica al Virreinto del Perú (1777–1788),* ed. Antonio González Bueno (Barcelona: Lunwerg Editores, 1988), 71–84.

3. In October 1786, the Crown sent an order to the president of Quito requiring him to send all available quina to Spain, via Lima, immediately; see Juan José Villalengua to Marqués de Sonora, Quito, 17 February 1787, AGI, Quito 245, no. 23. On disease and population in eighteenth-century Spain, see Jordi Nadal, *La Población Española (Siglos XVI a XX)* (Barcelona: Editorial Ariel, 1976), 84–142.

4. Matthew James Crawford, "An Empire's Extract: Chemical Manipulations of Cinchona Bark in the Eighteenth-Century Spanish Atlantic World," *Osiris* 29 (2014): 215–229.

5. Mark Honigsbaum, *The Fever Trail: In Search of the Cure for Malaria* (New York: Farrar, Straus and Giroux, 2001); Kavita Philip, "Imperial Science Rescues a Tree: Global Botanic Networks, Local Knowledge and the Transcontinental Transplantation of Cinchona." *Environment and History* 1 (1995): 173–200.

6. Antonio Porlier to Marqués de Valdecarzana, Madrid, 6 September 1789, Archivo General de Indias (AGI), Indiferente General 1555, fols. 361r–365v.

7. Hipólito Ruiz, *Relacion del viaje hecho a los Reinos del Perú y Chile.* Madrid: Real Academia de Ciencias Exactas, Físicas y Naturales, 1931, Apendices, 370, qtd. in Gonzalo Hernández de Alba, *Quinas Amargas: El sabio Mutis y la discusión naturalista del siglo XVIII* (Bogotá: Tercer Mundo Editores, 1991), 130 (my translation).

8. Daniela Bleichmar, "Atlantic Competitions: Botany in the Eighteenth-Century Spanish Empire," in *Science and Empire in the Atlantic World,* ed. James Delbourgo and Nicholas Dew (New York: Routledge, 2008), 225–252.

9. *Icho* was a Spanish transliteration of the Quechua word *ichu* a name for a type of grass found in the Andean regions of Peru. On mercury in the mining of silver in America, see Shawn William Miller, *An Environmental History of Latin America* (Cambridge: Cambridge University Press, 2007), 87–91.

10. Ruiz, *Relacion del viaje,* Apendices, 375, qtd. in Hernández de Alba, *Quinas Amargas,* 131 (my translation).

11. See chapter 4 in this volume.

12. James E. McClellan, *The Colonial Machine: French Science and Overseas Expansion in the Old Regime* (Turnout: Brepolis, 2011); Richard Drayton, *Nature's Government: Science, Imperial Britain, and the "Improvement" of the World* (New Haven: Yale University Press, 2000); Richard H. Grove, *Green Imperialism: Colonial Expansion, Tropical Island Edens and the Origins of Environmentalism, 1600–1860* (Cambridge: Cambridge University Press, 1995).

13. According to Arthur Steele, Ruiz and Pavón had originally planned to travel to the Audiencia of Quito and, presumably, to Loja, but changed their plans once they realized the riches of the flora of Peru. See Arthur Robert Steele, *Flowers for the King: The Expedition of Ruiz and Pavon and the Flora of Peru* (Durham: Duke University Press, 1964), 58 and 87.

14. Hipólito Ruiz, *The Journals of Hipólito Ruiz, Spanish Botanist in Peru and Chile 1777–1788*, trans. Richard Evans Schultes and María José Nemry von Thenen de Jaramillo-Arango (Portland: Timber Press, 1998), 144. Ruiz's account of the development of cinchona extraction in Huánuco is short on details such as biographical information about the individuals mentioned. In his journal, Ruiz did note that Renquifo had experience with cinchona trees in the "forests of Loja."

15. Ruiz, *Journals*, 144.

16. Cristina Ana Mazzeo, *El Comercio Libre en el Perú: Las estrategias de un comerciante criollo: José Antonio de Lavalle y Cortés, conde de Premio Real, 1775–1815* (Lima: Pontifica Universidad Católica del Perú, 1995); Cristina Ana Mazzeo, *Los Comerciantes Limeños a Fines del Siglo XVIII: Capacidad y Cohesión de una Elite, 1750–1825* (Lima: Pontificia Universidad Católica del Perú, 2000); Patricia H. Marks, "Confronting a Mercantile Elite: Bourbon Reformers and the Merchants of Lima, 1765–1796," *Americas* 60 (2004): 519–558.

17. Ruiz, *Journals*, 144. According to Cristina Mazzeo, Lavalle was a prominent dealer in cinchona as well as many other goods from Peru including cacao and copper.

18. Ruiz, *Journals*, 144.

19. Ruiz, *Journals*, 144.

20. Ruiz, *Journals*, 306.

21. On the repartimiento de mercancias, see Jeremy Baskes, *Indians, Merchants, and Markets: A Reinterpretation of the Repartimiento and the Spanish-Indian Economic Relations in Colonial Oaxaca, 1750–1821* (Stanford: Stanford University Press, 2000); Jeremy Baskes, "Coerced or Voluntary? The Repartimiento and Market Participation of Peasants in Late Colonial Oaxaca," *Journal of Latin American Studies* 28 (1996): 1–28.

22. On the repartimiento de mercancias system as a feature of economic production and social relations between colonizer and colonized in the colonial Andean World, see Susan Elizabeth Ramírez, *The World Upside Down:*

Cross-Cultural Contact and Conflict in Sixteenth-Century Peru (Stanford: Stanford University Press, 1996).

23. Ruiz noted, "when merchants are not of the priest's kin, they stand little chance of receiving their payments in full (see Ruiz, *Journals*, 306).

24. Ruiz, *Journals*, 306.

25. Luz del Alba Moya Torres, *El Arbol de la Vida: Auge y Crisis de la Cascarilla en la Audiencia de Quito, Siglo XVIII* (Quito: Facultad Latinoamericana de Ciencias Sociales Sede Ecuador, 1994).

26. Ruiz, *Journals*, 306, 307.

27. Hipólito Ruiz, *Quinología, o Tratado del Arbol de la Quina* (Madrid: En la Oficina de la Viuda é Hijo de Marin, 1792), 14. He says little about this method except to note that after "ten, twelve, or fifteen years, these trees do not sprout [new branches] or grow new trunks." This suggests that bark collectors were cutting down branches or whole trees simply to harvest the bark.

28. Ruiz, *Quinología*, 14.

29. Ruiz, *Quinología*, 15.

30. Ruiz, *Quinología*, 14–16.

31. Marqués de Valdecarzana to Antonio Porlier, Madrid, 15 August 1789, AGI, Indiferente General 1555, fols. 336r–358r; Marqués de Valdecarzana to Antonio Porlier, Madrid, 30 September 1789, AGI, Indiferente General 1555, 368r–391r.

32. [Antonio Porlier] to Marqués de Valdecarzana, Madrid, 6 September 1789, AGI, Indiferente General 1555, fols. 361r–365v.

33. Albert O. Hirschman, *The Passions and the Interests: Political Arguments for Capitalism Before Its Triumph* (Princeton: Princeton University Press, 1997); J. G. A. Pocock, *Virtue, Commerce, and History. Essays on Political Thought and History, Chiefly in the Eighteenth Century* (Cambridge: Cambridge University Press, 1986); John H. R. Polt, "Jovellanos and His English Sources: Economic, Philosophical, and Political Writings," *Transactions of the American Philosophical Society* 54 (1964): 1–74.

34. Valdecarzana's observation was, in part, accurate for places like Loja where the shortage of money meant that cinchona bark became a de facto currency.

35. Valdecarzana to Porlier, 15 August 1789, fol. 344v.

36. Valdecarzana to Porlier, 30 September 1789, fols. 370r–v.

37. Valdecarzana to Porlier, 15 August 1789, fol. 348r.

38. Valdecarzana to Porlier, 30 September 1789, fol. 377v.

39. Valdecarzana to Porlier, 30 September 1789, fol. 371r.

40. Valdecarzana to Porlier, 30 September 1789, fol. 378r.

41. Valdecarzana to Porlier, 30 September 1789, fol. 374v.

42. He explained that if the Crown paid bark collectors in cash, they could

use the money to purchase goods from merchants. Cash from the Crown would become a substitute for cinchona bark. Valdecarzana described his proposed plan as a way to break that "other kind of Estanco which currently exists among the powerful vendor [*tenedor*] of goods, the Commerce of Lima, and the foreign [merchant]"; see Valdecarzana to Porlier, 30 September 1789, fol. 382r. This passage may be an oblique reference to the notorious system of the repartimiento de mercáncias in which corregidores forced indigenous laborers to purchase goods at inflated prices; see John Fisher, *Bourbon Peru, 1750–1824* (Liverpool: Liverpool University Press, 2003), 33–34, 43–45, 58–59; Alba Moya, *El Arbol de la Vida,* 94–104.

43. Valdecarzana to Porlier, 15 August 1789, fol. 351r.

44. Valdecarzana to Porlier, 15 August 1789, fol. 351v.

45. On the Bourbon reforms, see Gabriel B. Paquette, *Enlightenment, Governance, and Reform in Spain and Its Empire, 1759–1808* (New York: Palgrave Macmillan, 2008); John Lynch, *Bourbon Spain, 1700–1808* (Oxford: Basil Blackwell, 1989); D. A. Brading, "Bourbon Spain and Its American Empire," in *The Cambridge History of Latin America*, vol. 1, ed. Leslie Bethell (Cambridge: Cambridge University Press, 1984).

46. Antonio Porlier to the president of Quito et. al., San Ildefonso, 7 September 1790, ANE/Q, FE, c. 118, v. 278, n. 6843, fol. 251r. Porlier wrote: "no se trate de el estanco de Quina, sino q[u]e solo se tome el arbitrio de remitir toda la q[u]e [b]rindan los Montes de Loja, Calisaya, y otros q[u]e la producen igual y aprobada p[o]r de superior calidad de cuenta de S[u] M[ajestad]."

47. Valdecarzana may have written these instructions himself. The authorship of these instructions is unclear in the documentary record.

48. This shift reflects a significant hardening of Spanish officials' view that the quality of the bark was not determined by geographical location. This was a further attack on the local view that the bark's origin was key to determining its quality.

49. "Ynstrucción que han de observar el Corregidor de Loxa y el Botanico Chimico . . . ," Madrid, 26 August 1790, ANE/Q, FE, c. 118, v. 278, n. 6843, fol. 247r, 251v. The original text for the second quote reads, "dejando al Comercio la libertad de hacer de su Cuenta y riesgo y p[o]r via de negociacion toda la [quina] de el Peru."

50. Marks, "Confronting a Mercantile Elite"; Mazzeo, *El Comercio Libre en el Perú;* John T. S. Melzer, *Bastion of Commerce in the City of Kings: The Consulado de Comercio de Lima, 1593–1887* (Lima: Consejo Nacional de Ciencia y Technología, 1991); John P. Moore, *The Cabildo in Peru under the Bourbons* (Durham: Duke University Press, 1966).

51. Casimiro Gómez Ortega to Marqués de Baxamar, Madrid, 26 May 1791, AGI, Indiferente General 1555, fols. 623v–624r.

52. "Ynstrucción que han de observar el Corregidor de Loxa y el Botanico Chimico . . . ," 26 August 1790, fol. 249r.

53. David Mackay, "Agents of Empire: the Banksian Collectors and Evaluation of New Lands," in *Visions of Empire: Voyages, Botany, and Representations of Nature*, ed. D. P. Miller and P. H. Reill (Cambridge: Cambridge University Press, 1996), 38–57.

54. "Ynstrucción que han de observar el Corregidor de Loxa y el Botanico Chimico . . . ," 26 August 1790, fol. 248v.

55. "Ynstrucción que han de observar el Corregidor de Loxa y el Botanico Chimico . . . ," 26 August 1790, fol. 247v.

56. Crawford, "An Empire's Extract."

57. "Ynstrucción que han de observar el Corregidor de Loxa y el Botanico Chimico . . . ," 26 August 1790, fol. 247v–248r.

58. "Ynstrucción que han de observar el Corregidor de Loxa y el Botanico Chimico . . . ," 26 August 1790, fol. 247r.

59. Tómas Ruiz de Quevedo and Vicente Olmedo to Marqués de Valdecarzana, Loja, 18 January 1793, ANE/Q, Fondo Especial, Box 136, vol. 313, no. 7637-10, fol. 15v.

60. Ruiz de Quevedo and Olmedo to Valdecarzana, 18 January 1793, fol. 15v.

61. David J. Weber, *Bárbaros: Spaniards and Their Savages in the Age of Enlightenment* (New Haven: Yale University Press, 2005), 19–51.

62. Vicente Olmedo to Tomás Ruiz de Quevedo, Loja, 10 December 1792, AGI, Indiferente General 1556, fols. 260r-261v.

63. Vicente Olmedo to Tomás Ruiz de Quevedo, Loja, 17 December 1793, ANE/Q, Fondo Especial, box 136, vol. 313, no. 7637–190, fol. 239r.

64. Olmedo to Ruiz de Quevedo, 17 December 1793, fol. 239r–v.

65. Olmedo did note in his report that bark collectors had some complaints about the harvesting of the bark. In particular, they reported that Olmedo still rejected too much of their bark as poor quality, that good cinchona trees were scarce, and the price offered by the Crown was too low. See Olmedo to Ruiz de Quevedo, 17 December 1793, fol. 240r.

66. Vicente Olmedo to Tomás Ruiz de Quevedo, Loja, 15 May 1794, ANE/Q, Fondo Especial, box 137, vol. 316, no. 7644-126, fol. 151r. I have not yet found independent confirmation that Olmedo's techniques, communicated by instructions, were as effective as he claimed.

67. Olmedo and other colonial officials framed their discussions of increasing the numbers of cinchona trees and boosting production of the bark primarily in the language of conservation. In addition, Olmedo's claims to success in his reports to Quevedo in 1793 and 1794 is notably at odds with the observations on the ignorance of bark collectors and their seeming inability to learn new techniques of extraction that Olmedo and Quevedo made to the chamberlain of

the Royal Household in the 18 January 1793 report. See Ruiz de Quevedo and Olmedo to Valdecarzana, 18 January 1793, fol. 15v.

68. Tomás Ruiz de Quevedo and Vicente Olmedo to Diego de Gardoqui, Loja, 11 June 1794, AGI, Indiferente General 1556, fol. 247v.

69. Kenneth Andrien, *The Kingdom of Quito, 1690–1830: The State and Regional Development* (Cambridge: Cambridge University Press, 1995), 37–44, 51–54.

70. Tomás Ruiz de Quevedo and Vicente Olmedo to Diego de Gardoqui, Loja, 25 November 1796, ANE/Q, Fondo Especial, Box 147, vol. 336, no. 8127-160, fol. 197r.

71. Ruiz de Quevedo and Olmedo to Gardoqui, 25 November 1796, fol. 197r.

72. See articles 5 and 13 of "Ynstrucción que han de observar el Corregidor de Loxa y el Botanico Chimico . . . ," 26 August 1790, fols. 247r–250r.

73. "Ynstrucción que han de observar el Corregidor de Loxa y el Botanico Chimico . . . ," 26 August 1790, fol. 247r–249r.

74. An early twentieth-century study of Dutch plantations of cinchona on the island of Java noted that the Dutch had similar difficulty balancing the planting of cinchona trees for export agriculture and other plants for the subsistence of the laboring population. See the introduction by Peter Hönig to Norman Taylor, *Cinchona in Java: The Story of Quinine* (New York: Greenberg, 1945).

75. Ruiz de Quevedo and Olmedo to Gardoqui, 11 June 1794, fol. 248v.

76. Ruiz de Quevedo and Olmedo to Gardoqui, 25 November 1796, fol. 197v.

77. Diego Gardoqui to the president of Quito, Aranjuez, 16 March 1796, ANE/Q, Fondo Especial, box 147, vol. 336, no. 8127-54, fol. 68r.

78. Diego Gardoqui to the president of Quito, Aranjuez, 16 March 1796, ANE/Q, Fondo Especial, box 147, vol. 336, no. 8127-54, fols. 67r–73r.

79. Francisco de Hector, Baron de Carondelet, to Miguel Cayetano Soler, Quito, 21 August 1800, ANE/Q, Fondo Especial, box 159, vol. 367, no. 8652, fol. 5r.

80. Franciso José de Caldas, "Memoria sobre el estado de las quinas en general y en particular sobre la de Loja," Quito, 16 March 1805, in *Obras Completas de Francisco Jose de Caldas* (Bogotá: Imprenta Nacional, 1966), 241–260.

81. Andreas-Holger Maehle, *Drugs on Trial: Experimental Pharmacology and Therapeutic Innovation in the Eighteenth Century* (Amsterdam: Rodopi, 1999).

82. Casimiro Gómez Ortega to the Marqués de Baxamar, Madrid, 26 May 1791, AGI, Indiferente General 1555, fols. 623r–625v; Antoine Fourcroy, "Analyse du Quinquina de Saint-Domingue," *Anales de Chimie* 8 (1791): 113–183.

83. Ruiz, *Quinologia*, 46–48.

84. Ruiz, *Quinologia*, 46–48.

85. Ruiz, *Quinologia*, 46–52.

86. Vicente Olmedo to Tomás Ruiz de Quevedo, Loja, 7 September 1794, ANE/Q, Fondo Especial, box 137, vol. 316, no. 7644–46, fol. 293v.

87. Olmedo to Ruiz de Quevedo, 7 September 1794, fol. 293v.

88. On the technology of pharmaceutical and chemical extracts in Spain and Spanish America, see Paula De Vos, "The Art of Pharmacy in Seventeenth- and Eighteenth-Century México," PhD diss., University of California, Berkeley, 2001, 192–258.

89. Jorge Cañizares-Esguerra, "Iberian Science in the Renaissance: Ignored How Much Longer?" *Perspectives on Science* 12 (2004): 86–124

90. In the early nineteenth century, accounts of Olmedo's activities appeared in Spanish documents translated into English and printed in England as part of the efforts to disseminate knowledge about the cinchona tree and its bark, which had become objects of interest to the British Empire especially after the collapse of the Spanish Empire in the Americas. See Aylmer Bourke Lambert, *An Illustration of the Genus Cinchona; Comprising Descriptions of the all the Official Peruvian Barks, including several new species. Baron de Humboldt's Account of the Cinchona Forests of South America: and Labert's Memoir on the different species of Quina. To which are added several dissertation of Don Hippolito Ruiz* (London: John Searle, 1821).

91. Matthew James Crawford, "Science as Statecraft: Imperial Ideology, Botany, and Monopoly in the Spanish Atlantic World (1742–1790)," in *Global Economies, Cultural Currencies in the Eighteenth Century*, ed. Michael Rotenberg-Schwartz and Tara Czechowski (New York: AMS Press, 2012), 121–144; Paula De Vos, "The Science of Spices: Empiricism and Economic Botany in the Early Spanish Empire," *Journal of World History* 17 (2006): 399–427.

Chapter 7. Regalist and Mercantilist Visions of Empire in the "War of the Quinas"

1. Francisco Antonio Zea, "Memoria sobre la quina según los principios del Sr. Mutis por D. Francisco Antonio Zea, Botánico de la expedicion de Santa Fe, y discípulo del mismo Sr. Mutis, Director de ella," *Anales de Historia Natural* 2, no. 5 (September 1800): 196–235.

2. Jeremy Adelman, *Sovereignty and Revolution in the Iberian Atlantic* (Princeton: Princeton University Press, 2006); Anthony McFarlane, "Identity, Enlightenment and Political Dissent in Late Colonial Spanish America," *Transactions of the Royal Historical Society, Sixth Series* 8 (1998): 309–335.

3. Hipólito Ruiz, *Quinología* (Madrid: En La Oficina de la Viuda é Hijo de Marin, 1792).

4. Hipólito Ruiz and José Pavón, *Flora Peruviana et Chilensis*, vol. 2 (Madrid: Typis Gabrielis de Sanchis, 1799). The first volume of the *Flora* appeared in 1798 and was preceded by the publication of an introductory volume, *Prodro-*

mus, in 1794; see Hipólito Ruiz and José Pavón, *Flora Peruviana et Chilensis. Prodromus, descripciones y láminos de los nuevos géneros de plantas de la Flora del Perú y Chile* (Madrid, 1794). A third volume of the *Flora Peruviana et Chilensis* appeared in 1802.

5. [José Celestino Mutis], *Instrucción formada por un facultative existente por muchos años en el Perú, relativa de las especies y virtudes de la Quina* (Cádiz: Don Manuel Ximenez Carrero, 1792). It is not clear whether this work was published with Mutis's permission or knowledge. The title notes that the author has "resided for many years in Peru." Mutis lived and worked in the Viceroyalty of New Granada and never visited Peru. It is likely that the publisher assumed that Mutis lived in Peru because most Europeans knew quina as the "Peruvian bark" since the late seventeenth century. See A. W. Haggis, "Fundamental Errors in the Early History of Cinchona," *Bulletin of the History of Medicine* 10 (1941): 417–459, 568–592; Saul Jarcho, *Quinine's Predecessor: Francesco Torti and the Early History of Cinchona* (Baltimore: Johns Hopkins University Press, 1993), 192–210.

6. José Celestino Mutis, "El Arcano de la Quina, revalado a beneficio de la humanidad o discurso de la parte médica de la Quinologia de Bogotá," *El Papel Periódico de la ciudad de Santafé de Bogotá* no. 89 (10 May 1793)–no. 129 (14 February 1794).

7. José Celestino Mutis, "Observaciones y conocimiento de la quina, debidas al doctor Celestino Mutis, comisionado por su Majestad para éste y otros importantes asuntos,"*El Mercurio Peruano* nos. 608–611 (1795): 211–286; José Celestino Mutis, "Extracto de una memoria del Dr. D. Joseph Celestino Mutis, célebre medico y botánico de Santa Fe de Bogotá," *Seminario de agricultura y artes dirigido a los párrocos* no. 85 (16 August 1798): 101–110 and no. 86 (23 August 1798): 119–123.

8. Jarcho, *Quinine's Predecessor.* The topic of fevers as well as their cause and treatment was addressed in a steady stream of published texts in eighteenth-century Spain. These include Miguel Marcelino Boix y Moliiner, *Hippócrates defendido de las imposturas y calumnias que algunos médicos poco cautos le imputan* (Madrid: Matheo Blanco, 1711); Antonio Díaz del Castillo, *Hypocrates desagraviado, de las ofensas por Hypocrates defendido: en particular en la curacion de calenturas agudas, de dolor de costado y tercianas* (Alcala: Julian Garcia Briones, 1713); Martin Martínez, *Medicina sceptica, y Cirugia moderna: con un tratado de operaciones chirurgicas . . .* Compuesto por el Doctor Don Martín Martínez, 2 vols. (Madrid, 1722–1725); Francisco Suárez de Ribera, *Medicina invencible, legal o theatro de fiebres intermitentes complicadas* (Madrid, 1726); Francisco Sanz de Dios Guadalupe, *Medicina Práctica* (Madrid, 1730); Félix Pachecho Ortiiz, *Rayos de luz prácitica con que desvanece las sombras con que el Dr. D. Francisco Sanz intentó obscurecer la hypóthesis de Fiebres del Dr. D. Martin Martínez, y hace resplandecer la particular hipóthesis y debida cura-*

ción de las fiebres intermitentes del Dor. D. Luis Enriquez ([Madrid?], 1731); Antonio José Rodríguez, *Palestra crítico-médica* (Pamplona: Oficina de Joseph Joaquín Martínez y Zaragoza, 1734–1749); Francisco García Hernández, *Tratado de fiebres malignas, con su curacion acomodado á la mas racional prácitca* (Madrid: M. F. Rodriguez, 1747); Pasqual Francisco Virrey y Mange, *Palma febril, medico-prácitca, hypocratica-chymica, methodico-galenica, seguro methodo de curar las fiebres,* 2 vols. (Madrid, 1756); André Piquer y Arrufat, *Tratado de las calenaturas* (Madrid: Ibarra, 1760); Luis José Pereyra, *Tratado completo de calenturas, fundado sobre las leyes de la infamación y putrefacción. Compuesto con méthodo geométrico y caracteres botánicos* (Madrid: B.R.A. de M. de Madrid, 1768); Juan Sastre y Puig, *Reflexiones instructivas apologéticas sobre el eficaz y seguro remedio de curar las calenturas pútridas y malignas, inventado por el ilustre Sr. Dr. D. Josef Masdeval* [sic] (Cervera, 1787–1788); Joseph Masdevall, *Reflexiones instructivo-apologenéticas sobre el eficas y seguro método de curar las calenturas pútridas malignas* (Cervera, 1788).

9. For Ruiz's description of the many uses of quina in medicine, see *Quinología,* 39–40.

10. The mythic status and lack of factual basis for the story of the Countess of Chinchón has been shown by a number of authors, including Jaime Jaramillo Arango, *Estudio Critico acera de los hechos basicos de la historia de la Quina* (Quito: Imp. de la Universidad, 1950); Haggis, "Fundamental Errors," 417–459, 568–592

11. Mauricio Nieto Olarte, *Remedios para el imperio: Historia natural y la apropiación del nuevo mundo* (Bogotá: La Imprenta Nacional de Colombia, 2000), 201.

12. Arthur Robert Steele, *Flowers for the King: The Expedition of Ruiz and Pavon and the Flora of Peru* (Durham: Duke University Press, 1964).

13. Nieto Olarte, *Remedios para el imperio,* 201.

14. José Celestino Mutis, *El Arcano de la Quina. Discursoque contiene la parte médica de las cuatro especies de Quinas oficinales, sus virtudes eminentes y su legítima preparacion,* ed. Manuel Hernandez de Gregorio (Madrid: Ibarra, 1828), 5–6. Unless otherwise noted, citations from Mutis's *El Arcano de la Quina* are from this 1828 posthumous edition. I have compared this version with the edition that appeared in the *Papel Periódico de Santa Fé de Bogotá* and found no significant difference between the texts other than the 1828 editor's notes and appendix. Orange quina was a "direct" febrifuge, meaning that it treated the root cause of fevers; Mutis designed red, yellow, and white quina as "indirect" febrifuges, meaning that they treated the occasional or accidental causes of fevers.

15. Mutis, *El Arcano,* 16.

16. Mutis, *El Arcano,* 16.

17. Tore Frängsmyr, ed., *Linnaeus: The Man and His Work* (Canton: Science History Publications, 1994); Londa Schiebinger, *Plants and Empire: Colonial Bioprospecting in the Atlantic World* (Cambridge: Harvard University Press, 2004), chap. 5.

18. Gonzalo Hernández de Alba, *Quinas Amargas: El sabio Mutis y la discusión naturalista del siglo XVIII* (Bogotá: Tercer Mundo Editores, 1991), 243.

19. "Prologue," in Ruiz, *Quinología,* nn. 6r.

20. Zea, "Memoria sobre la quina."

21. Hipólito Ruiz and José Pavón, *Suplemento a la Quinología* (Madrid: En La Imprenta de la Viuda e Hijo de Marin, 1801).

22. Zea, "Memoria sobre la quina," 51.

23. Mutis found such errors in the works of several prominent botanists and scientific travelers of his time, including Nikolaus von Jacquin (1727–1817), Georg Forster (1754–1794), Olaf Swartz (1760–1818), and Johann König (1728–1785). Jacquin claimed to have discovered cinchona in the Caribbean while Forster claimed to have found it in the Pacific and König and Swartz in the East Indies. All of the botanists' descriptions of cinchona species were made in print; this is probably how Mutis learned about them. The one exception is Georg Forster, who, according to Mutis, described his observations of cinchona species found on islands in the Pacific in a letter to Linnaeus in 1775. This letter was mentioned in the publication of the Academy of Uppsala; see Mutis, *El Arcano,* 14–16. See also Nikolaus von Jacquin, *Selectarum Stirpium Americanarum Historia* (Vindobonae: Ex Officina Krausiana, 1763); Nikolaus von Jacquin, *Observationum botanicarum iconibus ab auctore delineatis illustratam,* 4 vols. (Vindobonae: Ex Officina Krausiana, 1764–1771); Olaf Swartz, *Nova Genera & Species Plantarum seu Prodromus descriptionem vegetabilium, maximem partem incognitorum quae sub itinere in Indiam Occidentalem* (Holmiae: In Bibliopoliis acad. M. Swederi, 1788); Andreae Johannis Retzii, *Observationes botanicae: sex fasciculis comprehensae quibus accedunt Joannis Gerhardi Koenig Descriptiones monandrarum et epidendrorum in India Orientali factae* (Liepzig: Siegfried Lebrecht Crusium, 1779–1791).

24. Mutis, *El Arcano,* 15.

25. Ruiz equated Linnaeus's *C. offinalis* with his own *C. officinalis* while equating La Condamine's *quinquina* to his *C. glabra.*

26. Zea, "Memoria," 227, 299, 231.

27. Hipólito Ruiz and José Pavón, "Defensa que hacen de la Quinas finas Peruvianas y de las de Loxa los Botánicos de la Expedición del Perú Don Hipolito Ruiz y Don Josef Pavon, respondiendo a la Memorio que Don Francisco Antonio Zea insertó en los Anales de Historia Natural Quaderno numero 5 sobre las Quinas de Santa Fe y demostracion de que estas son muy inferiores a aquellas," in *Suplemento a la Quinología,* 21–105. Ruiz and Pavón also included a "Sum-

mary" of their main arguments for readers unwilling to slog through the eighty pages of tedious refutation; see *Suplemento a la Quinología,* 110–114.

28. Ruiz and Pavón, *Suplemento,* 21.

29. Ruiz and Pavón, *Suplemento,* 25.

30. Ruiz and Pavón, *Suplemento,* 77.

31. Mary Terrall, "Heroic Narratives of Quest and Discovery," *Configurations* 6 (1998): 223–242; Mary Louis Pratt, *Imperial Eyes: Travel Writing and Transculturation* (New York: Routledge, 1992).

32. Ruiz and Pavón, *Suplemento,* 24.

33. Ruiz and Pavón would have recognized that they could not make such a comparison either, which is probably another reason why they were more reserved in their critique of Mutis's system.

34. Ruiz and Pavón, *Suplemento,* 30.

35. Ruiz and Pavón, *Suplemento,* 54.

36. Ruiz and Pavón were careful to limit their own claims about the classification of cinchona within the limits of their botanical knowledge of the species. Consider their discussion of "orange quina" and "yellow quina." In the case of "orange quina," Ruiz and Pavón were certain that this species was not equivalent to any of theirs as suggested by Zea (*C. officinals, glabra,* and *fusca*) because they possessed the "skeleton" of "orange quina" from Sebastian López Ruiz. By contrast, Ruiz and Pavón were less certain that Mutis's "yellow quina" was equivalent to their *C. pallescens* because they only had "incomplete Skeletons in bad condition." "We cannot be sure [of the equivalence]," they write, "without seeing complete and well conditioned Skeletons." As to their other species (*C. tenuis, micrantha* and *purpurea*) that Zea suggested were not only the same as Mutis's "yellow quina" but also equivalent to each other, Ruiz and Pavón again pointed to the observable differences between the dried specimens of the species that they had in their possession at the Botanical Office in Madrid. Direct comparison and observation was essential to achieving certainty. To this effect, they wrote, "in order to have assurance in a matter as delicate as this, it is indispensable to observe living plants or at least dried plants and compare them reciprocally"; see Ruiz and Pavón, 56 and 67.

37. Ruiz and Pavón, *Suplemento,* 93.

38. Ruiz and Pavón, *Suplemento,* 24.

39. For other case studies in the debates and tensions that arose around classification, see D. Graham Burnett, *Trying Leviathan: The Nineteenth-Century New York Case That Put the Whale on Trial and Challenged the Order of Nature* (Chicago: University of Chicago Press, 2010); Harriet Ritvo, *The Platypus and the Mermaid and Other Figments of the Classifying Imagination* (Cambridge, MA: Harvard University Press, 1997); Michel Foucault, *The Order of Things: An Archaeology of the Human Sciences* (New York: Vintage Books, 1973).

40. As the language here suggests, my analysis parallels that of Simon Schaffer and Steven Shapin regarding the seventeenth-century debate between Thomas Hobbes and Robert Boyle over the use of experiments in the study of the natural world. See Steven Shapin and Simon Schaffer, *Leviathan and the Air-Pump: Hobbes, Boyle and the Experimental Life* (Princeton: Princeton University Press, 1985).

41. Mutis wrote, "this Science [i.e., natural history] would have never obtained the perfection, with which it is admired in our century, if sovereigns and other distinguished persons had not conceived the idea of liberally promoting, supporting, and rewarding various Learned Naturalists." See José Celestino Mutis, "Representación hecha al Rey soliticitando la formación de la Historia Natural de América, remitida desde Cartagena en el mes de Mayo de 1763, esforzada yrepetidad en Junio en 1764, con el adjunto informe que hizo de oficio á S.M. el Virrey de este Reino el Excmo. Sr. D. Fray Pedro Mesía de la Cerda," Santa Fe de Bogotá, 20 June 1764 in A. Federico Gredilla, *Biografía de Jose Celestino Mutis y sus observaciones sobre las vigilias y sueños de algunas plantas* (Bogotá: Plaza & Janes, [1911] 1982), 41.

42. Mutis, "Representación hecha al Rey," 44.

43. Mutis, "Representación hecha al Rey," 44.

44. Mutis projected that such practices would "make this tree just as unknown in Peru as in Norway"; see Mutis, "Representación hecha al Rey," 44–45.

45. Mutis, "Representación hecha al Rey," 44.

46. Marcelo Frías Núñez, *Tras El Dorado Vegetal: José Celestino Mutis y la Real Expedición Botánica del Nuevo Reino de Granada (1783–1808)* (Seville: Diputación de Sevilla, 1994), 196–206. For eighteenth-century views of the viceroys of New Granda, see the summaries of documents in "Desde el año de 1773 a 1784. El expediente original formado en el Consejo de Indias sobre el estanco," Archivo General de Indias (AGI), Indiferente General 1554, fols. 777r–1024r.

47. This quote is from a larger excerpt of Caballero y Góngora's report cited in Hermann A. Schumacher, *Mutis, un forjador de la cultura,* translated by Ernesto Guhl (Bogotá: Emprese Colombiana de Petróleos, 1984), 81–82. The original document is listed as "Oficio de Caballero y Góngora al Marqué de Sonora" Cartagena, 16 April 1787 in Frías Núñez, *Tras El Dorado Vegetal,* p. 205, n. 151.

48. Frías Núñez, *Tras El Dorado Vegetal,* 205–207.

49. Gabriel Paquette, *Enlightenment, Governance, and Reform in Spain and Its Empire, 1759–1808* (New York: Palgrave Macmillan, 2008); Stanley J. Stein, and Barbara H. Stein, *Apogee of Empire: Spain and New Spain in the Age of Charles III, 1759–1789* (Baltimore: Johns Hopkins University Press, 2003); John Fisher, *Commercial Relations Between Spain and Spanish America in the Era of Free Trade, 1778–1796* (Liverpool: Centre for Latin-American Studies, University of Liverpool, 1985).

50. Miguel de Piedra, Loja, 14 Abril 1752, ANH/Q, Cascarilla, Caja 1, Expediente 2, fols. 1r-3v; Miguel de Santisteban, "Relación informativa práctica de la quina de la ciudad de Loxa," Santa Fe, 4 June 1753, Biblioteca del Palacio Real de Madrid (BPRM), II/2823, fols. 82r–88v.

51. José Antonio Manso de Velaso, "[Dictamen a Marqués de Esenada]," Lima, 4 November 1753, AGI, Indiferente General 1552, fol. 337r–v.

52. Zea, "Memoria," 230, n. 1.

53. Ruiz and Pavón, *Suplemento*, 37.

54. Ruiz and Pavón, *Suplemento*, 37.

55. Ruiz and Pavón, *Suplemento*, 21.

56. Ruiz and Pavón, "Aviso al Lector," in *Suplemento*, nn. 4r.

57. Ruiz and Pavón, *Suplemento*, p. 21

58. For example, Ruiz and Pavón made the following contrast: "We recognize a difference in the climate and land [*temperamento y suelo*] of Loja and that of Santa Fe, while Loja is situated almost 4 degrees South [of the Equator], Santa Fe is 4.5 degrees North [of the Equator]. The Mountains [in Loja] are much closer to the South Sea than those of Santa Fe"; see *Suplemento*, 101.

59. Ruiz and Pavón noted that Huánuco and Loja were separated by "five degrees" of latitude while Santa Fé and Loja were separated by "nine degrees" of latitude. Huánuco was also "almost as close to the South Sea" as Loja and had a comparable altitude relative to the "Kingdom of Santa Fe"; see *Suplemento*, 101.

60. Ruiz and Pavón, *Suplemento*, 37.

61. Ruiz and Pavón, *Suplemento*, 51–52.

62. On patronage and science in the early modern period, see Mary Terrall, *The Man Who Flattened the Earth: Maupertuis and the Sciences in the Enlightenment* (Chicago: University of Chicago Press, 2002); Mario Biagioli, *Galileo Courtier: The Practice of Science in the Culture of Absolutism* (Chicago: University of Chicago Press, 1994), 11–102; Richard S. Westfall, "Science and Patronage: Galileo and the Telescope," *Isis* 76 (1985): 11–30.

63. Qtd. in Steele, *Flowers for the King*, 225.

64. Royal Order, San Lorenzo, 17 September 1791, qtd. in Steele, *Flowers for the King*, 212, n. 2.

65. Chapter 12 in Steele's *Flowers for the King* is titled "America Rescues the *Flora*," 212.

66. Eduardo Estrella, "Expediciones botánicas," in *Carlos III y la ciencia de la Ilustración*, ed. Manuel Sellés, José Luis Peset, and Antonio Lafuente (Madrid: Alianza Editorial, 1987), 331–351.

67. As a concession to the commercial interests of Peru and the Audiencia of Quito, Viceroy Góngora proposed in the late 1780s that the quina trade be divided between the two viceroyalties with New Granada supplying Europe via its port of Cartagena and Quito and Peru supplying New Spain, the Philippines,

and Asia. Unfortunately, Hernandez de Alba does not give the exact date the viceroy's proposal for the division of the quina trade. See Hernández de Alba, *Quinas Amargas*, 191.

68. A variety of historians of science and science studies scholars have offered many different iterations of this solution. See Steven Shapin and Simon Schaffer, *Leviathan and the Air-Pump: Hobbes, Boyle, and the Experimental Life* (Princeton: Princeton University Press, 1985); Bruno Latour, *The Pasteurization of France* (Cambridge: Harvard University Press, 1988); Peter Galison, *Image and Logic: A Material Culture of Microphysics* (Chicago: University of Chicago Press, 1997).

69. Londa Schiebinger and Claudia Swan, "Introduction," In *Colonial Botany: Science, Commerce, and Politics in the Early Modern World* (Philadelphia: University of Pennsylvania Press, 2005), 2.

70. Sheila Jasanoff, "The Idiom of Co-production," in *States of Knowledge: The Co-production of Science and Social Order,* ed. Sheila Jasanoff (New York: Routledge, 2004), 2–3.

71. Shapin and Schaffer, *Leviathan and the Air-Pump,* 332.

72. Mutis, *El Arcano,* 5.

73. Matthew James Crawford, "'Para Desterrar las Dudas y Adulteraciones': Scientific Expertise and the Attempts to Make a Better Bark for the Royal Monopoly of *Quina* (1751–1790)," *Journal of Spanish Cultural Studies* 8, no. 2 (2007): 193–212.

Conclusion: The Natures of Empire before the "Drapery" of Modern Science

1. Carlos Suarez to Torivio Montes, Loja, 26 September 1814, Archivo Nacional de Ecuador, Quito (ANE/Q), Cascarilla, Box 5, Expediente 13.

2. William Stevenson, *Historical and Descriptive Narrative of Twenty Years' Residence in South America,* vol. 2 (London: Hurst, Robinson and Co., 1825), 66–67. Stevenson's quote has erroneously been attributed to Alexander von Humboldt; see Mark Honigsbaum, *The Fever Trail: In Search of the Cure for Malaria* (New York: Farrar, Straus and Giroux, 2001), 62

3. Clements Markham, *Travels in Peru and India* (London: John Murray, 1862), 44.

4. Lucile Brockway, *Science and Colonial Expansion: The Role of the British Royal Botanic Garden* (New York: Academic Press, 1979), 111.

5. Brockway, *Science and Colonial Expansion,* 112.

6. M. L. Duran-Reynals, *The Fever Bark Tree: The Pageant of Quinine* (Garden City, NY: Doubleday, 1946), 206.

7. Kavita Philip, "Imperial Science Rescues a Tree: Global Botanic Networks, Local Knowledge and the Transcontinental Transplantation of Cinchona," *En-*

vironment and History 1 (1995): 173–200; Kavita Philip, "Global Botanical Networks, Environmentalist Discourses and the Political Economy of Cinchona Transplantation to British India," *Revue Francaise d'Historie d'Outre-Mer* 86 (1999): 119–142.

8. Stuart McCook has convincingly argued that the global transplantations of the nineteenth century are best understood as a "neo-Colombian" exchange, see Stuart McCook, "The Neo-Colombian Exchange: The Second Conquest of the Greater Caribbean, 1720–1930," *Latin American Research Review* 46 (2011): 11–31.

9. Honigsbaum, *The Fever Trail*; Philip, "Imperial Science Rescues a Tree"; Brockway, *Science and Colonial Expansion.*

10. Daniel Headrick, *Power over Peoples: Technology, Environments, and Western Imperialism, 1400 to the Present* (Princeton: Princeton University Press, 2010); Philip Curtin, *Death by Migration: Europe's Encounter with the Tropical World in the Nineteenth Century* (Cambridge: Cambridge University Press, 1989); Daniel R. Headrick, *The Tools of Empire: Technology and European Imperialism in the Nineteenth Century* (New York: Oxford University Press, 1981).

11. Teodoro S. Kaufman and Edmundo A. Rúveda, "The Quest for Quinine: Those Who Won the Battles and Those Who Won the War," *Angewandte Chemie International Edition* 44 (2005), 854–885; Saul Jarcho, *Quinine's Predecessor: Francesco Torti and the Early History of Cinchona* (Baltimore: Johns Hopkins University Press, 1993); Duran-Reynals, *The Fever Bark Tree;* Norman Taylor, *Cinchona in Java: The Story of Quinine* (New York: Greenberg, 1945).

12. These findings about the mobility of science are similar to the observations in the recent volume of traveling facts edited by Peter Howlett and Mary S. Morgan, *How Well Do Facts Travel?: The Dissemination of Reliable Knowledge* (Cambridge: Cambridge University Press, 2010).

13. Christine Daniels and Amy Turner Bushnell, eds., *Negotiated Empires: Centers and Peripheries in the Americas, 1500–1820* (New York: Routledge, 2002); Richard Drayton, *Nature's Government: Science, Imperial Britain, and the "Improvement" of the World* (New Haven: Yale University Press, 2000); Michael Adas, *Machines as the Measure of Men: Science, Technology, and Ideologies of Western Dominance* (Ithaca: Cornell University Press, 1989). Daniel R. Headrick, *The Tools of Empire: Technology and European Imperialism in the Nineteenth Century* (New York: Oxford University Press, 1981).

14. Roy MacLeod, "'On Visiting the 'Moving Metropolis': Reflections on the Architecture of Imperial Science," in *Scientific Colonialism: A Cross-Cultural Comparison,* ed. Nathan Reingold and Marc Rothenberg (Washington, DC: Smithsonian Institution Press, 1987), 217–249.

15. Jim Endersby, *Imperial Nature: Joseph Hooker and the Practices of Victorian Science* (Chicago: University of Chicago Press, 2008).

16. Endersby, *Imperial Nature,* 34.

17. Endersby, *Imperial Nature,* chap. 1. Here, Endersby provides evidence from both the papers presented at the meetings of the British Association for the Advancement of Science (BAAS) and the funding provided by BAAS as well as reviews of botanical texts in popular publications of the time.

18. William Whewell, *BAAS Report* (1841), xxxiii, qtd. in Endersby, *Imperial Nature,* 41.

19. Endersby, *Imperial Nature,* 41–42.

20. Classification was one of the central interests of eighteenth-century naturalists, but it was not the only one.

21. Loïc Charles and Paul Cheney, "The Colonial Machine Dismantled: Knowledge and Empire in the French Atlantic," *Past & Present* 219 (2013): 127–163

22. Commercial networks were also essential for the production of natural knowledge in the Atlantic World. See James Delbourgo and Nicholas Dew, eds., *Science and Empire in the Atlantic World* (New York: Routledge, 2008).

23. Arndt Brendecke, *Imperio e información: funciones del saber en el dominio colonial español* (Madrid: Iberoamericana, 2012); Gabriel B. Paquette, *Enlightenment, Governance, and Reform in Spain and Its Empire, 1759–1808* (New York: Palgrave Macmillan, 2008); Gabriel B. Paquette, "Empire, Enlightenment and Regalism: New Directions in Eighteenth-Century Spanish History," *European History Quarterly* 35 (2005): 107–117; Colin M. MacLachlan, *Spain's Empire in the New World: The Role of Ideas in Institutional and Social Change* (Berkeley: University of California Press, 1988); D. A. Brading, "Bourbon Spain and Its American Empire," in *The Cambridge History of Latin America,* ed. Leslie Bethell (Cambridge: Cambridge University Press, 1984).

24. MacLachlan, *Spain's Empire in the New World.*

25. Michael Gibbons et al., *The New Production of Knowledge: The Dynamics of Science and Research in Contemporary Societies* (Thousand Oaks, CA: Sage, 1994), 1–9.

26. The phrasing here intentionally echoes that of Steven Shapin and Simon Schaffer, in *Leviathan and the Air-Pump: Hobbes, Boyle and the Experimental Life* (Princeton: Princeton University Press, 1985), 332.

27. In this way, the epistemic culture of the Spanish colonial government appears similar to what some science policymakers called "Mode 2" knowledge production, which is knowledge produced in a direct context of application (as opposed to dividing the production of knowledge and its application into two steps known as Mode 1). There has been much debate about this scheme of knowledge production and its history. For example, some historians have disputed the notion that Mode 1 knowledge production was ever put into practice, and claim, instead, that this conception of knowledge production represents a modernist

ideology of knowledge production rather than an actual form of knowledge production. See Lissa Roberts, Simon Schaffer, and Peter Dear, eds., *The Mindful Hand: Inquiry and Invention from the Late Renaissance to Early Industrialisation* (Amsterdam: Koninklijke Nederlandse Akademie van Wetenschappen, 2007); Michael Gibbons et al., *The New Production of Knowledge: The Dynamics of Science and Research in Contemporary Societies* (Thousand Oaks, CA: Sage, 1994).

28. Roberts et al., *The Mindful Hand.*

29. Matthew James Crawford, "Science as Statecraft: Imperial Ideology, Botany, and Monopoly in the Spanish Atlantic World (1742–1790)," in *Global Economies, Cultural Currencies in the Eighteenth Century,* ed. Michael Rotenberg-Schwartz and Tara Czechowski (New York: AMS Press, 2012), 121–144.

30. Daniela Bleichmar, "Atlantic Competitions: Botany in the Eighteenth-Century Spanish Empire," in *Science and Empire in the Atlantic World,* ed. James Delbourgo and Nicholas Dew (New York: Routledge, 2008), 225–252; Paula De Vos, "Natural History and the Pursuit of Empire in Eighteenth-Century Spain," *Eighteenth-Century Studies* 40 (2007): 209–239; Paula De Vos, "The Science of Spices: Empiricism and Economic Botany in the Early Spanish Empire," *Journal of World History* 17 (2006): 399–427.

31. Harold J. Cook, *Matters of Exchange: Commerce, Medicine, and Science in the Dutch Golden Age* (New Haven: Yale University Press, 2007); Londa Schiebinger and Claudia Swan, eds., *Colonial Botany: Science, Commerce, and Politics in the Early Modern World* (Philadelphia: University of Pennsylvania Press, 2005); Pamela Smith and Paula Findlen, eds., *Merchants & Marvels: Commerce, Science, and Art in Early Modern Europe* (New York: Routledge, 2002).

32. This claim is true at least for the generation of botanists that came of age in the late eighteenth century. In the nineteenth century, Antonio Cavanilles, director of the Royal Botanical Garden, worked tirelessly to eliminate the utilitarian emphasis of Spanish botany that flourished under Casimiro Gómez Ortega. See Antonio González Bueno, *Tres botánicos de la Ilustración: Gómez Ortega, Zea, Cavanilles: la ciencia al servicio del poder* (Nivola: Tres Cantos, 2002).

33. The original reports from Ruiz, Pavón, and Ruiz del Cerro remain to be found. This account is taken from an 1808 report to the head pharmacist of the Royal Pharmacy by Gregorio Bañares, a chamber pharmacist to the king. See Gregorio Bañares, "Informe sobre la Memorio hech apor el Doctor D[o]n Fran[cis]co Josef Caldas del estado de las Quinas en general y en particular de la de Loxa," Madrid, 28 January 1808, AGI, Indiferente General 1557, fols. 874r–905v.

34. Here—as explained by Bañares, Ruiz, and Pavón—we are following the opinion of their "disciple," Juan Tafalla, a botanist in Peru who continued the

operations of the botanical expedition to Peru after Ruiz and Pavón returned to Spain in 1788. On 15 August 1806, Ruiz and Pavón published a summary of a letter from Tafalla in the *Gaceta de Madrid,* in which they noted that Tafalla noted that the quina sent from Loja to the Royal Pharmacy was a mixture of "cascarilla colorada y amarilla." See *Gaceta de Madrid,* no. 67 (15 August 1806): 699–701.

35. Bañares listed his affiliations on the title page of a published pamphlet on quina. See Gregorio Bañares, *Memoria sobre las ventajas y utilidades de la quina Buena y perjuicios de la mala* (Madrid: Imprenta Real, 1807).

36. Gregorio Bañares, "Informe sobre la Memoria hecha por el Doctor D[o]n Fran[cis]co Josef Caldas del estado de las Quinas en general y en particular de la de Loxa," Madrid, 28 January 1808, AGI, Indiferente 1557, fols. 874r–905v.

37. Bañares writes, "al parecer, que no es decir nada," fol. 882.

38. Ruiz and Pavón were required by the Crown to follow the Linnaean method.

39. Ruiz and Pavón had emphasized precisely this methodological point in their debate with José Celestino Mutis.

40. Jim Endersby makes the same claim about natural historical specimens in Victorian Britain; see Endersby, *Imperial Nature,* 18.

Bibliography

Archives

Archivo del Palacio Real de Madrid (APRM)
Biblioteca del Palacio Real de Madrid (BPRM)
Archivo General de Indias (AGI)
Archivo General de Simancas (AGS)
Archivo Nacional del Ecuador, Quito (ANE/Q)

Primary Sources

Alsinet, José. *Nuevas utilidades de quina*. Madrid: A. Muñoz del Valle, 1763.

Arriaga, Pablo José de. *La Extirpación de la Idolatria del Piru*. Lima: Geronymo de Contreras, 1621.

Badus, Sebastian. *Anastasis corticis Peruviae, seu chinae chinae defensio, Sebastian Badi Genviensis, patrii utriusque nosochomij olim medici, et publicae sanitatis in ciutate consultoris. Contra ventilationes Iacobi Chifletii, gemitusque Vopisci Fortunatii Plempii, illustrium medicorum*. Genoa: Calenzani, 1663.

Bañares, Gregorio. *Memoria sobre las ventajas y utilidades de la quina buena y perjuicios de la mala*. Madrid: Imprenta Real, 1807.

Boix y Moliiner, Miguel Marcelino. *Hippócrates defendido de las imposturas y calumnias que algunos médicos poco cautos le imputan*. Madrid: Matheo Blanco, 1711.

Calancha, Antonio de la. *Corónica Moralizada del Orden de San Augustin en el Peru con sucesos [y] egemplares vistos en esta monarquia.* Barcelona: Pedro Lacavalleria, 1639.

Caldas, Francisco José de. *Obras Completas de Francisco Jose de Caldas.* Bogotá: Imprenta Nacional, 1966.

Caldera de Heredia, Gaspar. *De pulvere febrifugo Occidentalis Indiae (1663) y la introducción de la quina en Europe.* Ed. José María López Piñero and Francisco Calero. Valencia: Instituto de Estudios Documentales e Históricos sobre la Ciencia, 1992.

Campillo y Cosío, José del. *Nuevo Sistema de Gobierno Económico para América.* Madrid: En la imprenta de Benito Cano, 1789.

Condamine, Charles Marie de la. *Journal du voyage fait par ordre du Roi, a l'équateur servant d'introduction historique a la mesure des trois premiers degres du méridien.* Paris: De L'Imprimerie Royale, 1751.

Condamine, Charles Marie de la. "Sur l'arbre de quinquina." *Historie de l'Académie Royale des Sciences* (1738): 226–243.

Díaz del Castillo, Antonio (O.F.M.). *Hypocrates desagraviado, de las ofensas por Hypocrates defendido: en particvlar en la cvracion de calenturas agudas, de dolor de costado y tercianas.* Alcala: Julian Garcia Briones, 1713.

Dictionnaire universel françois et latin, contenant la signification et la définition tant des mots de l'une & de l'autre langue, avec leurs différens usages. Nouvelle édition corrigée et considerablement augmentée. Paris: Chez la Vueve Delaune, rue S. Jacques, 1743.

Duhamel du Monceau, Henri-Louis. *Avis pour transport par mer des arbres, des plantes vivaces, des semences, des animaux, et de differens autres morceaux d'Histoire Naturelle.* 1752.

Ellis, John. *Directions for bringing over seeds and plants from the East-Indies and other distant countries in a state of vegetation.* London, 1770.

Feijoo, Jerónimo Benito. *Cartas erudítas, y curiosas, en que por la mayor parte se continúa designio de el Theatro critico universal, impugnando, o reduciendo a dudosas varias opiniones comunes.* 5 vols. Madrid: F. del Hierro, 1742–1760.

Fourcroy, Antoine. "Analyse du Quinquina de Saint-Domingue." *Anales de Chimie* 8 (1791): 113–183.

Gaceta de Madrid. No. 67 (15 August 1806): 693–704.

García Hernández, Francisco. *Tratado de fiebres malignas, con su curacion acomodado á la mas racional prácitca.* Madrid: M. F. Rodriguez, 1747.

Gómez Ortega, Casimiro. *Instrucción sobre el modo más seguro y económico de transportar plantas vivas* [1779]. Introduction by Francisco Javier Puerto Sarmiento. Madrid: Fundación de Ciencias de la Salud, 1992.

Gómez Ortega, Casimiro. "Instrucción a que deberán arreglarse los sujetos destinados por S.M. para pasar a la América Meridional en compañía del médico Josef

Dombey a fin de reconocer las plants y yerbas y de hacer observaciones botánicas en aquellos países" (1776), in *Relación histórica del viage, que hizo a los reinos del Perú y Chile el botánico D. Hipólito ruiz en el año 1777 hasta el de 1788, en cuya época regresó a Madrid*. Ed. Jaime Jaramillo Arango. 2 vols. Madrid: Real Academica de Ciencias Exactas Físicas y Naturales, 1952.

Gray, John, William Arrot, and Phil. Miller. "An Account of the Peruvian or Jesuits Bark, by Mr. John Gray, F.R.S. Now at Cartagena in the Spanish West-Indies; Extracted from Some Papers Given Him by Mr. William Arrot, a Scotch Surgeon, Who Had Gather'd it at the Place Where it Grows in Peru. Communicated by Phil. Miller, F.R.S., & c." *Philosophical Transactions (1683–1775)* 40 (1737–1738): 81–86.

Guaman Poma de Ayala, Felipe. *El primer nueva corónica y buen gobierno*. København, Det Kongelige Bibliotek, GKS 2232 4. 1615/1616.

Humboldt, Alexander von. *Ensayo político sobre el Reino de la Nueva España*. México: Porum, 1966.

Jacquin, Nikolaus von. *Observationum botanicarum iconibus ab auctore delineatis illustratam*. 4 vols. Vindobonae: Ex Officina Krausiana, 1764–1771.

Jacquin, Nikolaus von. *Selectarum Stirpium Americanarum Historia*. Vindobonae: Ex Officina Krausiana, 1763.

Juan, Jorge, and Antonio Ulloa. *Noticias Secretas de America*. London: La Imprenta de R. Taylor, 1826.

Juan, Jorge, and Antonio Ulloa. *Relacion Historica del Viage a la America Meridional*. 4 vols. Madrid: Antonio Marin, 1748.

Laguna, Andrés de. *Pedacion Dioscórides Anazerbeo (1555)*. Madrid: Instituto de España, 1968.

Lambert, Aylmer Bourke. *An Illustration of the Genus Cinchona*. London: John Searle, 1821.

Leeuwenhoek, Anthony Van. "Microscopical Observations on the Cortex Peruvianus." *Philosophical Transactions* 312 (1707): 2446–2455.

Linnaeus, Carolus. *Species Plantarum*. 2 vols. Holmiae: Imprensis Laurentii Salvii 1753.

Markham, Clements Robert. *Travels in Peru and India*. London: John Murray, 1862.

Martínez, Martin. *Medicina sceptica, y Cirugia moderna: con vn tratado de operaciones chirurgicas*. 2 vols. Madrid, 1722–1725.

Mendoza, Diego de. *Chronica de la Provincia de S. Antonio de los Charcas del Orden de n[uest]ro seraphico P.S. Francisco en la Indias Occidentales reyno del Peru*. Madrid: Villafranca Sculptor Regius, 1665.

Masdevall, Joseph. *Reflexiones instructivo-apologenéticas sobre el eficaz y seguro método de curar las calenturas pútridas malignas*. Cervera, 1788.

Monardes, Nicolas. *Segunda parte del libro de las cosas que se traen nuestras Indias*

Occidentales, que sirven al uso de medicina. Seville: En Casa Alonso Escriuano Impressor, 1571.

Mutis, José Celestino. *Archivo Epistolar del Sabio Naturalista Don José Celestino Mutis.* Ed. Gonzalo Hernández de Alba. 4 vols. Bogotá: Instituto Colombiano de Cultural Hispanica, 1983.

Mutis, José Celestino. *El Arcano de la Quina. Discurso que contiene la parte médica de las cuatro especies de Quinas oficinales, sus virtudes eminentes y su legítima preparacion.* Ed. Manuel Hernandez de Gregorio. Madrid: Ibarra, 1828.

Mutis, José Celestino. "Extracto de una memoria del Dr. D. Joseph Celestino Mutis, célebre médico y botánico de Santa Fé de Bogotá." *Seminario de agricultura y artes dirigido a los párrocos* no. 85 (16 August 1798): 101–110 and no. 86 (23 August 1798): 119–123.

Mutis, José Celestino. "Observaciones y conocimiento de la quina, debidas al doctor Celestino Mutis, comisionado por su Majestad para éste y otros importantes asuntos." *El Mercurio Peruano* no. 608–611 (1795): 211–286.

Mutis, José Celestino. "El Arcano de la Quina, revalado a beneficio de la humanidad o discurso de la parte médica de la Quinologia de Bogotá." *El Papel Periódico de la ciudad de Santafé de Bogotá* no. 89 (10 May 1793)–no. 129 (14 February 1794).

[Mutis, José Celestino]. *Instruccion formada por un facultativo existente por muchos años en el Perú, relativa de las especies y virtudes de la Quina.* Cádiz: Don Manuel Ximenez Carrero, 1792.

Pachecho Ortiiz, Félix. *Rayos de luz prácitica con que desvanece las sombras con que el Dr. D. Francisco Sanz intentó obscurecer la hypóthesis de Fiebres del Dr. D. Martin Martínez, y hace resplandecer la particular hipóthesis y debida curaci'on de las fiebres intermitentes del Dor. D. Luis Enriquez.* [Madrid?], 1731.

Pereyra, Luis José. *Tratado completo de calenturas, fundado sobre las leyes de la inflamación y putrefacción. Compuesto con méthodo geométrico y caracteres botánicos.* Madrid: B.R.A. de M. de Madrid, 1768.

Piquer y Arrufat, Andrés. *Tratado de las calenaturas.* Madrid: Ibarra, 1760.

Real Academia Española. *Diccionario de la Lengua Castellana.* Madrid: La Imprenta de Francisco de Hierro, 1729.

Retzii, Andreae Johannis. *Observationes botanicae: sex fasciculis comprehensae quibus accedunt Joannis Gerhardi Koenig Descriptiones monandrarum et epidendrorum in India Orientali factae.* Liepzig: Siegfried Lebrecht Crusium, 1779–1791.

Rodríguez, Antonio José. *Palestra crítico-médica.* Pamplona: Oficina de Joseph Joaquín Martínez y Zaragoza, 1734–1749.

Ruiz, Hipólito. *The Journals of Hipólito Ruiz, Spanish Botanist in Peru and Chile, 1777–1788.* Trans. Richard Evans Schultes and María José Nemry von Thenen de Jaramillo-Arango. Portland: Timber Press, 1998.

Ruiz, Hipólito. *Quinología, o Tratado del Arbol de la Quina*. Madrid: En la Oficina de la Viuda é Hijo de Marin, 1792.

Ruiz, Hipólito, and José Pavón. *Suplemento a la Quinología*. Madrid: En La Imprenta de la Viuda e Hijo de Marin, 1801.

Ruiz, Hipólito, and José Pavón. *Flora Peruviana et Chilensis*. 3 vols. Madrid, 1798–1802.

Ruiz, Hipólito, and José Pavón. *Flora Peruviana et Chilensis. Prodromus, descripciones y láminos de los nuevos géneros de plantas de la Flora del Perú y Chile*. Madrid, 1794.

Salomon, Frank, and George Urioste, eds. *The Huarochirí Manuscript: A Testament of Ancient and Colonial Andean Religion*. Austin: University of Texas Press, 1991.

Santisteban, Miguel de. *Mil leguas por América: De Lima a Caracas 1740–1741, Diario de don Miguel de S antisteban*. Ed. David J. Robinson. Bogotá: Banco de la República, 1992.

Sanz de Dios Guadalupe, Francisco. *Medicina Práctica*. Madrid, 1730.

Sastre y Puig, Juan. *Reflexiones instructivas apologéticas sobre el eficaz y seguro remedio de curar las calenturas pútridas y malignas, inventado por el ilustre Sr. Dr. D. Josef masdeval*. Cervera, 1787–1788.

Stevenson, William. *Historical and Descriptive Narrative of Twenty Years' Residence in South America*. 2 vols. London: Hurst, Robinson and Co., 1825.

Suarez de Ribera, Francisco. *Clave Botanica, o Medicina Botanica Nueva y Novissima*. Madrid: Manuel de Moya, 1738.

Suarez de Ribera, Francisco. *Medicine invencible, legal o theatro de fiebres intermitentes complicadas*. Madrid, 1726.

Swartz, Olaf. *Nova Genera & Species Plantarum seu Prodromus descriptionem vegetabilium, maximem partem incognitorum quae sub itinere in Indiam Occidentalem*. Holmiae: In Bibliopoliis acad. M. Swederi, 1788.

Torti, Francesco. *Therapeutice specialis*. Mutinae: Typis B. Soliani, 1712.

[Unanue, Hipólito]. "Introduccion á la Descripcion Científica de las Plantas del Peru." *Mercurio Peruano* 2, no. 43 (29 May 1791): 68–76.

Vega, Garcilaso de la. *Primera Parte de los Comentarios Reales, que tartan, de el origen de los Incas, reies, que fueron del Perú, de su idolatria, leies, y govierno, en pas, y en guerra,de sus vidas, y conquistas, y de todo lo que fue aquel imperio*. 2nd ed. Madrid: Oficina Real y á Costa de Nicolas Rodriguez Franco, Impresor de Libros, 1723.

Vega, Garcilaso de la. *Primera Parte de los Comentarios reales*. Lisbon: En la oficina de Pedro Crasbeeck, 1609.

Velasco, Juan de. *Historia del reino de Quito en la América meridional*. Quito: Casa de la Cultura Ecuatoriana "Benjamin Carrión," 1994.

Virrey y Mange, Pasqual Francisco. *Palma febril, medico-prácitca, hypocratica-*

chymica, methodico-galenica, seguro methodo de curar las fiebres. 2 vols. Madrid, 1756.

Ward, Bernardo. *Proyecto económico: en que se proponen varias providencias dirigidas a promover los intereses de España*. Madrid: Joachin Ibarra, 1779.

Zea, Francisco Antonio. "Memoria sobre la quina según los principios del Sr. Mutis, por D. Francisco Antonio Zea, Botánico de la expedicion de Santa Fe, y discípulo del mismo Sr. Mutis, Director de ella." *Anales de Historia Natural* 2, no. 5 (September 1800): 196–235.

Secondary Sources

Achim, Miruna, ed. "Science in Translation: The Commerce of Facts and Artifacts in the Transatlantic Spanish World." Special Issue of *Journal of Spanish Cultural Studies* 8, no. 2 (2007).

Adas, Michael. *Machines as the Measure of Men: Science, Technology, and Ideologies of Western Dominance*. Ithaca: Cornell University Press, 1989.

Adelman, Jeremy. *Sovereignty and Revolution in the Iberian Atlantic*. Princeton: Princeton University Press, 2006.

Adorno, Rolena. *Guaman Poma: Writing and Resistance in Colonial Peru*. Austin: University of Texas Press, 1988.

Alchon, Suzanne Austin. *Native Society and Disease in Colonial Ecuador*. Cambridge: Cambridge University Press, 1991.

Alegre Pérez, M. E. "La asistencia social en la Real Botica durante el último cuarto del siglo XVIII." *Boletín de la Sociedad Española de Historia de la Farmacia* 139 (1984): 199–211.

Alegre Pérez, M. E. "La Real Botica y las especies americanas (siglo XVIII)." *Boletín de la Sociedad Española de Historica de la Farmacia* 140 (1984): 225–244.

Alvarez Peláez, Raquel. *La conquista de la naturaleza americana*. Madrid: Consejo Superior de Investigaciones Científicas, 1993.

Amaya, José Antonio. *Mutis, apóstol de Linneo. Historia de la botánica en el virreinato de la Nueva Granada (1760–1783)*. 2 vols. Bogotá: Instituto Columbiano de Antropologiá e Historia, 2005.

Anagnostou, Sabine. "Jesuits in Spanish America: Contributions to the Exploration of the American Materia Medica." *Pharmacy in History* 47 (2005): 3–17.

Anagnostou, Sabine. "Jesuit Missionaries in Spanish America and the Transfer of Medical-Pharmaceutical Knowledge." *Archives Internationales d'Histoire des Sciences* 52 (2002): 176–197.

Anda Aguirre, Alfonso. *Corregidores y Servidores Públicos de Loja*. Quito: Banco Central del Ecuador, 1987.

Anderson, E. N. "Why Is Humoral Medicine So Popular?" *Social Science and Medicine* 25 (1987): 331–337.

Andersson, Lennart. "Tribes and Genera of the Cinchoneae Complex (Rubiaceae)." *Annals of the Missouri Botanical Garden* 82 (1995): 409–427.

Andrés del Turrión, María. "Quina del Perú para la Real Hacienda Española (1768–1807): notas sobre su estanco." In *La expedición botánica al Virreinato del Peréu (1777–1788)*, ed. Antonio González Bueno, vol. 1, 71–84. Barcelona: Lunwerg, 1988.

Andrés del Turrión, María Luisa de, and Maria Rosario Terreros Gómez. "Organización administrativa del Ramo de la Quina para la Real Hacienda española en el virreinato de Nueva Granada." In *Medicina y Quina en la España del Siglo XVIII*, ed. Juan Riera Palmero, 37–43. Salamanca: EUROPA Artes Gráficas, 1997.

Andrien, Kenneth. *Andean Worlds: Indigenous History, Culture, and Consciousness under Spanish Colonial Rule, 1532–1825*. Albuquerque: University of New Mexico Press, 2001.

Andrien, Kenneth. "The *Noticias secretas de America* and the Construction of a Governing Ideology for the Spanish American Empire." *Colonial Latin American Review* 7 (1998): 175–192.

Andrien, Kenneth. *The Kingdom of Quito, 1690–1830: The State and Regional Development*. Cambridge: Cambridge University Press, 1995.

Añon Feliu, Carmen. *Real Jardína Botánico de Madrid: Sus Origines, 1755–1781*. Madrid: Real Jardín Botánico, CSIC, 1987.

Appadurai, Arjun, ed. *The Social Life of Things: Commodities in Cultural Perspective*. Cambridge: Cambridge University Press, 1986.

Aranda, Marcelo, Katherine Arner, Lina del Castillo, Helen Cowie, Matthew Crawford, Joseph Cullon, Marcelo Figueroa, Claire Gherini, Melissa Grafe, Sarah Irving, Ryan Kashanipour, Carla Lois, AdriánLópez-Denis, Bertie Mandelblatt, Iris Montero Sobrevilla, Kathleen Murphy, Eric Otremba, Christopher Parsons, Heather Peterson, Emily Senior, Teresa Vergara, Kelly Wisecup, and Anya Zilberstein. "The History of Atlantic Science: Collective Reflections from the 2009 Harvard Seminar on Atlantic History." *Atlantic Studies* 7 (2010): 493–509.

Ash, Eric, ed. "Expertise: Practical Knowledge and the Early Modern State." *Osiris* 25 (2010).

Ash, Eric. *Power, Knowledge and Expertise in Elizabethan England*. Baltimore: Johns Hopkins University Press, 2004.

Axtel, Brian Keith, ed. *From the Margins: Historical Anthropology and Its Futures*. Durham: Duke University Press, 2002.

Baratas Díaz, Luis Alfredo. *Conocimiento botánico de las especies de Cinchona entre 1750 y 1850: relevencia de la obra botánica española en América*. Salamanca: Consejería de Educación y Cultura de la Junta de Castilla y León, 1998.

Barrera-Osorio, Antonio. *Experiencing Nature: The Spanish American Empire and the Early Scientific Revolution*. Austin: University of Texas Press, 2006.

Barrera-Osorio, Antonio. "Empire and Knowledge: Reporting from the New World." *Colonial Latin American Review* 15 (2006): 39–54.

Basalla, George. "The Spread of Western Science." *Science* 156 (May 1967): 611–622.

Baskes, Jeremy. *Indians, Merchants, and Markets: A Reinterpretation of the Repartimiento and the Spanish-Indian Economic Relations in Colonial Oaxaca, 1750–1821*. Stanford: Stanford University Press, 2000.

Baskes, Jeremy. "Coerced or Voluntary? The Repartimiento and Market Participation of Peasants in Late Colonial Oaxaca." *Journal of Latin American Studies* 28 (1996): 1–28.

Bastien, Joseph W. "Differences between Kallawaya-Andean and Greek-European Humoral Theory." *Social Science and Medicine* 28 (1989): 45–51.

Bastien, Joseph. *Healers of the Andes: Kallawaya Herbalists and the Plants*. Salt Lake City: University of Utah Press, 1987.

Bastien, Joseph "Qollahuaya-Andean Body Concepts: A Topographical-Hydraulic Model of Physiology." *American Anthropologist* 87 (1985): 595–611.

Bastien, Joseph W. "Exchange between Andean and Western Medicine." *Social Science and Medicine* 16 (1982): 795–803.

Berquist Soule, Emily. *The Bishop's Utopia: Envisioning Improvement in Colonial Peru*. Philadelphia: University of Pennsylvania Press, 2014.

Biagioli, Mario. *Galileo Courtier: The Practice of Science in the Culture of Absolutism*. Chicago: University of Chicago Press, 1994.

Bleichmar, Daniela. *Visible Empire: Botanical Expeditions and Visual Culture in the Hispanic Enlightenment*. Chicago: University of Chicago Press, 2012.

Bleichmar, Daniela. "Painting as Exploration: Visualizing Nature in Eighteenth-Century Colonial Science." *Colonial Latin American Review* 15 (2006): 81–104.

Bleichmar, Daniela, and Peter Mancall, eds. *Collecting across Cultures: Material Exchanges in the Early Modern Atlantic World*. Philadelphia: University of Pennsylvania Press, 2013.

Bloch, Marc. *The Royal Touch*. Trans. J. E. Anderson. New York: Dorset Press, [1924] 1989.

Brading, D. A. "Bourbon Spain and Its American Empire." In *The Cambridge History of Latin America*, edited by Leslie Bethell, vol. 1. Cambridge: Cambridge University Press, 1984.

Brendecke, Arndt. *Imperio e información. Funciones del saber en el dominio colonial español*. Madrid: Iberoamericana, 2012.

Brockway, Lucile. *Science and Colonial Expansion: The Role of the British Royal Botanic Gardens*. New York: Academic Press, 1979.

Brosseder, Claudia. *The Power of Huancas: Change and Resistance in the Andean World of Colonial Peru*. Austin: University of Texas Press, 2014.

Brotóns, Víctor Navarro, and William Eamon, eds. *Mas allá de la leyenda Negra:*

España y la revolución científica/Beyond the Black Legend: Spain and the Scientific Revolution. Valencia: Instituto de Historia de la Ciencia y Documentación López Piñero, 2007.

Burke, Peter. *A Social History of Knowledge: From Gutenberg to Diderot*. Oxford: Polity Press, 2000.

Burnett, D. Graham. *Trying Leviathan: The Nineteenth-Century New York Case That Put the Whale on Trial and Challenged the Order of Nature*. Chicago: University of Chicago Press, 2010.

Burns, Kathryn. *Into the Archives: Writing and Power in Colonial Peru*. Durham: Duke University Press, 2010.

Bussman, R. W., and A. Glenn. "Cooling the Heat: Traditional Remedies for Malaria and Fever in Northern Peru." *Ethnobotany Research and Application* 8 (2010): 125–134.

Bussman, Rainer, and Douglas Sharon. "Shadows of the Colonial Past—Diverging Plant Use in Northern Peru and Southern Ecuador." *Journal of Ethnobiology and Ethnomedicine* 5 (2009): 1–17.

Bussman, Rainer, and Douglas Sharon. "Traditional Medicinal Plant Use in Northern Peru: Tracking Two Thousand Years of Healing Culture." *Journal of Ethnobiology and Ethnomedicine* 2 (2006): 47–65.

Bussman, Rainer, and Douglas Sharon. "Traditional Medicinal Plant Use in Loja Province, Southern Ecuador." *Journal of Ethnobiology and Ethnomedicine* 2 (2006): 44–55.

Bynum, William F., and Vivian Nutton, eds. *Theories of Fever from Antiquity to the Enlightenment*. London: Wellcome Institute for the History of Medicine, 1981.

Camino, Lupe. *Cerros, plantas, y lagunas poderosas: La medicina al norte del Perú*. Piura: Cipca, 1992.

Cañizares-Esguerra, Jorge. *Nature, Empire, and Nation: Explorations of the History of Science in the Iberian World*. Stanford: Stanford University Press, 2006.

Cañizares-Esguerra, Jorge. "Iberian Science in the Renaissance: Ignored How Much Longer?" *Perspectives on Science* 12 (2004): 86–124.

Cañizares-Esguerra, Jorge. *How to Write the History of the New World: Histories, Epistemologies and Identities in the Eighteenth-Century Atlantic World*. Stanford: Stanford University Press, 2001.

Carey, Mark. "Commodities, Colonial Science and Environmental Change in Latin America." *Radical History Review* 107 (2010): 185–194.

Carey, Mark. "Latin American Environmental History: Current Trends, Interdisciplinary Insights and Future Directions." *Environmental History* 14 (2009): 221–252.

Carey, Mark. "The Nature of Place: Recent Research on Environment and Society in Latin America." *Latin American Research Review* 42 (2007): 251–264.

Castro-Gómez, Santiago. *La hybris del punto cero: Ciencia, raza e ilustración en la Nueva Granada (1750–1816)*. Bogotá: Editorial Pontificia Universidad Javeriana, 2005.

Charles, Loïc, and Paul Cheney. "The Colonial Machine Dismantled: Knowledge and Empire in the French Atlantic." *Past & Present* 219 (2013): 127–163.

Classen, Constance. *Inca Cosmology and the Human Body*. Salt Lake City: University of Utah Press, 1993.

Cohn, Bernard S. *Colonialism and Its Forms of Knowledge*. Princeton: Princeton University Press, 1996.

Contreras, Carlos. *El Sector Exportador de una Economía Colonial: La Costa del Ecuador entre 1760 y 1820*. Quito: Facultad Latinoamericana de Ciencias Sociales/ABYA-YALA, 1990.

Cook, Harold J. "Markets and Cultures: Medical Specifics and the Reconfiguration of the Body in Early Modern Europe." *Transactions of the Royal Historical Society* 21 (2011): 123–145.

Cook, Harold J. *Matters of Exchange: Commerce, Medicine, and Science in the Dutch Golden Age*. New Haven: Yale University Press, 2007.

Cook, Harold J., and Timothy Walker. "Circulation of Medicine in the Early Modern Atlantic World." *Social History of Medicine* 26 (2013): 337–351.

Cook, Noble David. *Born to Die: Disease and New World Conquest, 1492–1650*. Cambridge: Cambridge University Press, 1998.

Cooper, Frederick. *Colonialism in Question: Theory, Knowledge, History*. Berkeley: University of California Press, 2006.

Cowie, Helen. *Conquering Nature in Spain and Its Empire, 1750–1850*. New York: Manchester University Press, 2011.

Crawford, Matthew James. "A Cure for Empire? An Andean Wonder Drug and the Politics of Knowledge in the Eighteenth-Century Spanish Empire." In *Eighteenth Century: Theory and Interpretation* (forthcoming).

Crawford, Matthew James. "An Empire's Extract: Chemical Manipulations of Cinchona Bark in the Eighteenth-Century Spanish Atlantic World." *Osiris* 29 (2014): 215–229.

Crawford, Matthew James. "Science as Statecraft: Imperial Ideology, Botany, and Monopoly in the Spanish Atlantic World (1742–1790)." In *Global Economies, Cultural Currencies in the Eighteenth Century*, ed. Michael Rotenberg-Schwartz and Tara Czechowski, 121–144. New York: AMS Press, 2012.

Crawford, Matthew James. "A 'Reasoned Proposal' against 'Vain Science': Creole Negotiations of an Atlantic Medicament in the *Audiencia* of Quito (1776–1792)." *Atlantic Studies* 7, no. 4 (2010): 397–419.

Crawford, Matthew James. "'Para Desterrar las Dudas y Adulteraciones': Scientific Expertise and the Attempts to Make a Better Bark for the Royal Monopoly of Quina (1751–1790)." *Journal of Spanish Cultural Studies* 8 (2007): 193–212.

Crespo Ortiz, Fernando. "Fragoso, Monardes, and Pre-Cinchona Knowledge of Cinchona." *Archives of Natural History* 22 (1995): 169–181.

Crosby, Alfred. *The Colombian Exchange: Biological and Cultural Consequences of 1492*. Westport, CT: Greenwood Press, 1973.

Curtin, Philip. *The Rise and Fall of the Plantation Complex: Essays in Atlantic History*. Cambridge: Cambridge University Press, 1998.

Curtin, Philip. *Death by Migration: Europe's Encounter with the Tropical World in the Nineteenth Century*. Cambridge: Cambridge University Press, 1989.

Daniels, Christine, and Amy Turner Bushnell, eds. *Negotiated Empires: Centers and Peripheries in the Americas, 1500–1820*. New York: Routledge, 2002.

Daston, Lorraine, ed. *Things That Talk: Object Lessons from Art and Science*. New York: Zone Books, 2004.

Daston, Lorraine, ed. *Biographies of Scientific Objects*. Chicago: University of Chicago Press, 2000.

De Vos, Paula. "From Herbs to Alchemy: The Introduction of Chemical Medicine to Mexican Pharmacies in the Seventeenth and Eighteenth Centuries." *Journal of Spanish Cultural Studies* 8 (2007): 135–168.

De Vos, Paula. "Research, Development, and Empire: State Support of Science in the Later Spanish Empire." *Colonial Latin American Review* 15 (2006): 55–79.

De Vos, Paula. "The Science of Spices: Empiricism and Economic Botany in the Early Spanish Empire." *Journal of World History* 17 (2006): 399–427.

De Vos, Paula. "The Art of Pharmacy in Seventeenth- and Eighteenth-Century México." PhD diss., University of California, Berkeley, 2001.

Delbourgo, James, and Nicholas Dew, eds. *Science and Empire in the Atlantic World*. New York: Routledge, 2008.

Desowitz, Robert S. *The Malaria Capers: Tales of Parasites and People*. New York: W. W. Norton, 1993.

Díez Torre, Alejandro R., et al., eds. *La ciencia española en ultramar. Actas de las I Jornadas sobre "España y las expediciones científicas en América y Filipinas."* Madrid: Doce Calles, 1991.

Domínguez Ortiz, Antonio. *Carlos III y la España de la Ilustración*. Madrid: Alianza Editorial, 1988.

Drayton, Richard. *Nature's Government: Science, Imperial Britain, and the "Improvement" of the World*. New Haven: Yale University Press, 2000.

Dugatkin, Lee Alan. *Mr. Jefferson and the Giant Moose: Natural History in Early America*. Chicago: University of Chicago Press, 2009.

Duran-Reynals, M. L. *The Fever Bark Tree: The Pageant of Quinine*. Garden City, NY: Doubleday, 1946.

Elliott, J. H. *Empires of the Atlantic: Britain and Spain in America, 1492–1830*. New Haven: Yale University Press, 2007.

Elshakry, Marwa. "When Science Became Western: Historiographical Reflections." *Isis* 101 (2010): 98–109.

Endersby, Jim. *Imperial Nature: Joseph Hooker and the Practices of Victorian Science*. Chicago: University of Chicago Press, 2008.

Engstrand, Iris. *Spanish Scientists in the New World: The Eighteenth-Century Expeditions*. Seattle: University of Washington Press, 1981.

Estrella, Eduardo. "Ciencia ilustrada y saber popular en el conocimiento de la quina en el siglo XVIII." In *Saberes Andinos: Ciencia y tecnología en Bolivia, Ecuador, y Perú*, ed. Marcos Cueto. Lima: Instituto de Estudios Peruanos, 1995.

Estrella, Eduardo. "Expedición Geodesica: Mito y Realidad de la Quina." *Anales de las II Jornadas de Historia de la Medicina Hispanoamericana* (1989): 25–32.

Estrella, Eduardo. *Medicina Aborigen: La Practica Medica Aborigen De La Sierra Ecuatoriana*. Quito: Editorial Epoca, 1977.

Ewalt, Margaret. *Peripheral Wonders: Nature, Knowledge and Enlightenment in the Eighteenth-Century Orinoco*. Lewisburg: Bucknell University Press, 2008.

Fernández-Santamaria, J. H. "Reason of State and Statecraft in Spain (1595–1640)." *Journal of the History of Ideas* 41 (1980): 353–379.

Ferreiro, Larrie D. *Measure of the Earth: The Enlightenment Expedition That Reshaped Our World*. New York: Basic Books, 2011.

Findlen, Paula. *Possessing Nature: Museums, Collecting, and Scientific Culture in Early Modern Italy*. Berkeley: University of California Press, 1994.

Fisher, John. *Bourbon Peru, 1750–1824*. Liverpool: Liverpool University Press, 2003.

Fisher, John. *Commercial Relations between Spain and Spanish America in the Era of Free Trade, 1778–1796*. Liverpool: Centre for Latin-American Studies, University of Liverpool, 1985.

Fisher, John R., Allan J. Kuethe, and Anthony McFarlane, eds. *Reform and Insurrection in Bourbon New Granada and Peru*. Baton Rouge: Louisiana State University Press, 1990.

Foster, George M. "The Validating Role of Humoral Theory in Traditional Spanish-American Therapeutics." *American Ethnologist* 15 (1988): 120–135.

Foster, George M. "On the Origin of Humoral Medicine in Latin America." *Medical Anthropology Quarterly* 1 (1987): 355–393.

Foucault, Michel. *The Order of Things: An Archaeology of the Human Sciences*. New York: Vintage, 1994

Frängsmyr, Tore, ed. *Linnaeus: The Man and His Work*. Canton: Science History Publications, 1994.

Frías Núñez, Marcelo. *Tras el Dorado Vegetal: José Celestino Mutis y la Real Expedición Botánica del Nuevo Reino de Granada (1783–1808)*. Seville: Imprenta A. Pinelo, 1994.

Galison, Peter. *Image and Logic: A Material Culture of Microphysics*. Chicago: University of Chicago Press, 1997.

Gänger, Stefanie. "World Trade in Medicinal Plants from Spanish America, 1717–1815." *Medical History* 59 (2015): 44–62.

García-Baquero González, Antonio. *Cádiz y el Atlántico (1717–1778): El comercio colonial español bajo el monopolio gaditano.* 2 vols. Cadíz: Diputación Provincial de Cádiz, [1976] 1988.

García de Yébenes Torres, María del Pilar. "La Real Botica durante el reinado de Felipe V (1700–1746)." PhD diss., Universidad Complutense de Madrid, 1994.

Garofalo, Leo. "Conjuring with Coca and the Inca: The Andeanization of Lima's Afro-Peruvian Ritual Specialists, 1580–1690." *Americas* 63 (2006): 53–80.

Gascoigne, John. *Science in the Service of Empire: Joseph Banks, the British State and the Uses of Science in the Age of Revolution.* Cambridge: Cambridge University Press, 1998.

Gerbi, Antonello Gerbi. *The Dispute over the New World: The History of a Polemic, 1750–1900.* Trans. Jeremy Moyle. Pittsburgh: University of Pittsburgh Press, 2010.

Gibbons, Michael, et al. *The New Production of Knowledge: The Dynamics of Science and Research in Contemporary Societies.* Thousand Oaks, CA: Sage, 1994.

Gilles, Herbert M., and David A. Warrell. *Bruce-Chwatt's Essential Malariology.* London: Edward Arnold, 1993.

González Bueno, Antonio. *Tres botánicos de la Ilustración: Gómez Ortega, Zea, Cavanilles: la ciencia al servicio del poder.* Nivola: Tres Cantos, 2002.

González Bueno, Antonio, ed. *La expedición botánica al Virreinato del Perú (1777–1788).* Volume 1. Barcelona: Lunwerg, 1988.

Goodman, David. *Power and Penury: Government, Technology, and Science in Philip II's Spain.* Cambridge: Cambridge University Press, 1988.

Goodman, Jordan, and Vivian Walsh. *The Story of Taxol: Nature and Politics in the Pursuit of an Anti-Cancer Drug.* Cambridge: Cambridge University Press, 2001.

Goodman, Jordan, Paul E. Lovejoy, and Andrew Sherratt, eds. *Consuming Habits: Drugs in History and Anthropology.* London: Routledge, 1995.

Grafton, Anthony, April Shelford, and Nancy Siraisi. *New Worlds, Ancient Texts: The Power of Tradition and the Shock of Discovery.* Cambridge: Belknap Press, 1992.

Gredilla, A. Federico. *Biografia de José Celestino Mutis y Sus Observaciones sobre al Vigilias y Sueños de Algunas Plantas.* Bogotá: Plaza & Janes, [1911] 1982.

Griffiths, Nicholas, and Fernando Cervantes, eds. *Spiritual Encounters: Interactions between Christianity and Native Religions in Colonial America.* Birmingham: University of Birmingham Press, 1999.

Grove, Richard H. *Green Imperialism: Colonial Expansion, Tropical Island Edens and the Origins of Environmentalism, 1600–1860.* Cambridge: Cambridge University Press, 1995.

Haggis, A. W. "Fundamental Errors in the Early History of Cinchona." *Bulletin of the History of Medicine* 10 (1941): 417–459, 568–592.

Hamlin, Christopher. *More Than Hot: A Short History of Fever.* Baltimore: Johns Hopkins University Press, 2014.

Harding, Sandra. *Is Science Multicultural?* Bloomington: Indiana University Press, 1998.

Harris, Steven J. "Long-Distance Corporations, Big Sciences, and the Geography of Knowledge." *Configurations* 6 (1998): 269–304.

Headrick, Daniel R. *Power over Peoples: Technology, Environments, and Western Imperialism 1400 to the Present.* Princeton: Princeton University Press, 2012.

Headrick, Daniel R. *The Tools of Empire: Technology and European Imperialism in the Nineteenth Century.* New York: Oxford University Press, 1981.

Hernández de Alba, Gonzalo. *Quinas Amargas: El sabio Mutis y la discusión naturalista del siglo XVIII.* Bogotá: Tercer Mundo Editores, 1991.

Hirschman, Albert O. *The Passions and the Interests: Political Arguments for Capitalism before Its Triumph.* Princeton: Princeton University Press, 1997.

Holmes, Frederic Lawrence. *Eighteenth-Century Chemistry as an Investigative Enterprise.* Berkeley: Office for History of Science and Technology, University of California, Berkeley, 1989.

Honigsbaum, Mark. *The Fever Trail: In Search of the Cure for Malaria.* New York: Farrar, Straus and Giroux, 2001.

Howlett, Peter, and Mary S. Morgan, eds. *How Well Do Facts Travel?: The Dissemination of Reliable Knowledge.* Cambridge: Cambridge University Press, 2010.

Iborra, Pascual. *Historia del protomedicato en España (1477–1822).* Valladolid: Universidad, Secretario de Publicaciones, 1987.

Ishikawa, Chiyo, ed. *Spain in the Age of Exploration.* Lincoln: University of Nebraska Press, 2004.

Jaramillo Arango, Jaime. *Estudio Critico acera de los hechos basicos de la historia de la Quina.* Quito: Imp. de la Universidad, 1950.

Jarcho, Saul. *Quinine's Predecessor: Francesco Torti and the Early History of Cinchona.* Baltimore: Johns Hopkins University Press, 1993.

Jasanoff, Sheila, ed. *States of Knowledge: The Co-production of Science and Social Order.* New York: Routledge, 2004.

Kaufman Teodoro S., and Edmundo A. Rúveda. "The Quest for Quinine: Those Who Won the Battles and Those Who Won the War." *Angewandte Chemie International Edition* 44 (2005): 854–885.

Klein, Ursula. "Technoscience avant la lettre." *Perspectives on Science* 13 (2005): 226–266.

Klein, Ursula. "Experimental History and Herman Boerhaave's Chemistry of Plants." *Studies in History and Philosophy of Biological and Biomedical Sciences* 34 (2003): 533–567.

Knorr-Cetina, Karin. *Epistemic Cultures: How the Sciences Make Knowledge.* Chicago: University of Chicago Press, 1999.

Koerner, Lisbet. *Linnaeus: Nature and Nation.* Cambridge: Harvard University Press, 2001.

Kuethe, Allan J., and Kenneth J. Andrien. *The Spanish Atlantic World in the Eighteenth Century: War and the Bourbon Reforms, 1713–1796.* Cambridge: Cambridge University Press, 2014.

Lafuente, Antonio. "Enlightenment in an Imperial Context: Local Science in the Late-Eighteenth-Century Hispanic World." *Osiris* 15 (2000): 155–173.

Lafuente, Antonio, and Antonio Mazuecos. *Los caballeros del punto fijo: ciencia, política y aventura en la expedición geodésica hispanofrancesa al virreinato del Perú en el siglo XVIII.* Barcelona: Serbal-CSIC, 1987.

Lanning, John Tate. *The Royal Protomedicato: the Regulation of the Medical Professions in the Spanish Empire.* Ed. John Jay Te-Paske. Durham: Duke University Press, 1985.

Latour, Bruno. *The Pasteurization of France.* Cambridge: Harvard University Press, 1988.

Latour, Bruno. *Science in Action: How to Follow Scientists and Engineers through Society.* Cambridge: Harvard University Press, 1987.

Law, John. "On the Social Explanation of Technical Change: The Case of the Portuguese Maritime Expansion." *Technology and Culture* 28 (1987): 227–252.

Law, John. "On the Methods of Long-Distance Control: Vessels, Navigation and the Portuguese Route to India." In *Power, Action and Belief: A New Sociology of Knowledge?* ed. John Law, 234–263. London: Routledge & Kegan Paul, 1986.

Lima, Manuel. *The Book of Trees: Visualizing Branches of Knowledge.* New York: Princeton Architectural Press, 2014.

Livingstone, David N. *Putting Science in Its Place: Geographies of Scientific Knowledge.* Chicago: University of Chicago Press, 2003.

López Piñero, José María, and José Pardo Tomás. *La influencia de Francisco Hernández (1515–1587) en la constitución de la botánica y la materia médica moderna.* Valencia: Universidad de Valencia-CSIC, 1996.

Lynch, John. *Bourbon Spain, 1700–1808.* Oxford: Basil Blackwell, 1989.

MacLeod, Roy, ed. "Nature and Empire: Science and the Colonial Enterprise" *Osiris* 15 (2000).

MacLeod, Roy. "On Visiting the 'Moving Metropolis:' Reflections on the Architecture of Imperial Science." In *Scientific Colonialism: A Cross-Cultural Comparison,* ed. Nathan Reingold and Marc Rothenberg, 217–249. Washington, DC: Smithsonian Institution Press, 1987.

MacLachlan, Colin M. *Spain's Empire in the New World: The Role of Ideas in Institutional and Social Change.* Berkeley: University of California Press, 1988.

Maehle, Andreas-Holger. *Drugs on Trial: Experimental Pharmacology and Therapeutic Innovation in the Eighteenth Century.* Amsterdam: Rodopi, 1999.

Mann, Charles C. *1491: New Revelations of the Americas before Columbus.* New York: Vintage, 2005.

Mapp, Paul. *The Elusive West and the Contest for Empire, 1713–1763.* Chapel Hill: University of North Carolina Press, 2013.

Marciada, José Ramón, and Juan Pimentel. "Green Treasures and Paper Floras: The Business of Mutis in New Granada (1783–1808)." *History of Science* 52 (2012): 277–296.

Margoscy, Daniel. *Commercial Visions: Science, Trade and Visual Culture in the Dutch Golden Age.* Chicago: University of Chicago Press, 2014.

Marks, Patricia H. "Confronting a Mercantile Elite: Bourbon Reformers and the Merchants of Lima, 1765–1796." *Americas* 60 (2004): 519–558.

Márquez, Graciela. "Commercial Monopolies and External Trade." In *The Cambridge Economic History of Latin America*, 928–1004. Cambridge: Cambridge University Press, 2006.

Maxwell, Kenneth. "The Atlantic in the Eighteenth Century: A Southern Perspective on the Need to Return to the 'Big Picture.'" *Transactions of the Royal Historical Society*, Sixth Series 3 (1993): 209–236.

Mazzeo, Cristina Ana. *Los Comerciantes Limeños a Fines del Siglo XVIII: Capacidad y Cohesión de una Elite, 1750–1825.* Lima: Pontificia Universidad Católica del Perú, 2000.

Mazzeo, Cristina Ana. *El Comercio Libre en el Perú: Las estrategias de un comerciante criollo: José Antonio de Lavalle y Cortés, conde de Premio Real, 1775–1815.* Lima: Pontifica Universidad Católica del Perú, 1995.

McClellan, James E. *The Colonial Machine: French Science and Overseas Expansion in the Old Regime.* Turnout: Brepolis, 2011.

McCook, Stuart. "The Neo-Colombian Exchange: The Second Conquest of the Greater Caribbean, 1720–1930." *Latin American Research Review* 46 (2011): 11–31.

McFarlane, Anthony. "Identity, Enlightenment and Political Dissent in Late Colonial Spanish America." *Transactions of the Royal Historical Society, Sixth Series* 8 (1998): 309–335.

McNeill, J. R. *Mosquito Empires: Ecology and War in the Greater Caribbean, 1620–1914.* Cambridge: Cambridge University Press, 2010.

McNeill, J. R. "Observations on the Nature and Culture of Environmental History." *History and Theory* 42 (2003): 5–43.

Melzer, John T. S. *Bastion of Commerce in the City of Kings: the Consulado de Comercio de Lima, 1593–1887.* Lima: Consejo Nacional de Ciencia y Technología, 1991.

Miller, D. P., and P. H. Reill, eds. *Visions of Empire: Voyages, Botany, and Representations of Nature.* Cambridge: Cambridge University Press, 1996.

Miller, Shawn William. *An Environmental History of Latin America.* Cambridge: Cambridge University Press, 2007.

Mills, Kenneth. *Idolatry and Its Enemies: Colonial Andean Religion and Extirpation, 1640–1750*. Princeton: Princeton University Press, 1997.

Minchom, Martin. "The Making of a White Province: Demographic Movement and Ethnic Transformation in the South of the *Audiencia* de Quito (1670–1830)." *Bulletin de l'Institut Française d'Estudes Andines* 12 (1983): 23–39.

Mintz, Sidney. *Sweetness and Power: The Place of Sugar in Modern History*. New York: Viking, 1985.

Missouri Botanical Garden. *Proceedings of the Celebration of the Three Hundredth Anniversary of the First Recognized Use of Cinchona*. St. Louis: Missouri Botanical Garden, 1931.

Moore, John P. *The Cabildo in Peru under the Bourbons*. Durham: Duke University Press, 1966.

Moya Torres, Luz del Alba. *El Arbol de la Vida: Auge y Crisis de la Cascarilla en la Audiencia de Quito, Siglo XVIII*. Quito: Facultad Latinoamericana de Ciencias Sociales Sede Ecuador, 1994.

Müller-Wille, Staffan. "Nature as Marketplace: The Political Economy of Linnaean Botany." *History of Political Economy* 35 (2003): 154–172.

Mundy, Barbara. *The Mapping of New Spain: Indigenous Cartography and the Maps of the Relaciones Geográficas*. Chicago: University of Chicago Press, 1996.

Nadal, Jordi. *La Población Española (Siglos XVI a XX)*. Barcelona: Editorial Ariel, 1976.

Newson, Linda. *Life and Death in Early Colonial Ecuador*. Norman: University of Oklahoma Press, 1995.

Nieto Olarte, Mauricio. *Remedios para el imperio: Historia natural y la apropiación del nuevo mundo*. Bogotá: La Imprenta Nacional de Colombia, 2000.

Norton, Marcy. *Sacred Gifts, Profane Pleasures: A History of Tobacco and Chocolate in the Atlantic World*. Ithaca: Cornell University Press, 2010.

Packard, Randall. *The Making of a Tropical Disease: A Short History of Malaria*. Baltimore: Johns Hopkins University Press, 2011.

Pagden, Anthony. *Spanish Imperialism and the Political Imagination: Studies in European and Spanish-American Social and Political Theory, 1513–1830*. New Haven: Yale University Press, 1990.

Paquette, Gabriel, ed. *Enlightened Reform in Southern Europe and Its Atlantic Colonies, 1750–1830*. Burlington: Ashgate, 2009.

Paquette, Gabriel B. *Enlightenment, Governance, and Reform in Spain and Its Empire, 1759–1808*. New York: Palgrave Macmillan, 2008.

Paquette, Gabriel B. "Empire, Enlightenment and Regalism: New Directions in Eighteenth-Century Spanish History." *European History Quarterly* 35 (2005): 107–117.

Parrish, Susan Scott. *American Curiosity: Cultures of Natural History in the Colonial British Atlantic World*. Chapel Hill: University of North Carolina Press, 2006.

Parsons, Christopher, and Kathleen S. Murphy. "Ecosystems under Sail: Specimen Transport in the Eighteenth-Century French and British Atlantics." *Early American Studies* 10 (2012): 503–539.

Petitjean, M., and Y. Saint-Geours. "La economía de la cascarilla en el corregimiento de Loja." *Cultura: Revista del Banco Central del Ecuador* V (1983): 171–207.

Petitjean, Patrick, Catherine Jami, and Anne Marie Moulin, eds. *Science and Empires: Historical Studies about Scientific Development and European Expansion.* Dordrecht: Kluwer Academic Publishers, 1992.

Philip, Kavita. "Global Botanical Networks, Environmentalist Discourses and the Political Economy of Cinchona Transplantation to British India." *Revue Francaise d'Historie d'Outre-Mer* 86 (1999): 119–142.

Philip, Kavita. "Imperial Science Rescues a Tree: Global Botanic Networks, Local Knowledge and the Transcontinental Transplantation of Cinchona." *Environment and History* 1 (1995): 173–200.

Pickstone, John V. "Thinking over Wine and Blood: Craft-Products, Foucault, and the Reconstruction of Enlightenment Knowledges." *Social Analysis* 41 (1997): 99–108.

Pimentel, Juan. *El rinoceronte y el megaterio: un ensayo de morfologia historica.* Madrid: Abada, 2010.

Pimentel, Juan. "The Iberian Vision: Science and Empire in the Framework of a Universal Monarchy, 1500–1800." *Osiris* 15 (2000): 17–30.

Pocock, J. G. A. *Virtue, Commerce, and History. Essays on Political Thought and History, Chiefly in the Eighteenth Century.* Cambridge: Cambridge University Press, 1986.

Polt, John H. R. "Jovellanos and His English Sources: Economic, Philosophical, and Political Writings." *Transactions of the American Philosophical Society* 54 (1964): 1–74.

Portuondo, María M. *Secret Science: Spanish Cosmography and the New World.* Chicago: University of Chicago Press, 2009.

Pratt, Mary Louise. *Imperial Eyes: Travel Writing and Transculturation.* London: Routledge, 1992.

Prieto, Andrés I. *Missionary Scientists: Jesuit Science in Spanish South America, 1570–1810.* Nashville: Vanderbilt University Press, 2011.

Puerto Sarmiento, Francisco Javier. *Ciencia de Cámara. Casimiro Gómez Ortega (1741–1818). El Científico Cortesano.* Madrid: CSIC, 1992.

Puerto Sarmiento, Francisco Javier. *La Ilusión Quebrada: Botánica, Sanidad y Política Científica en la España Ilustrada.* Madrid: CSIC, 1988.

Pumfrey, Stephen, Paolo Rossi, and Maurice Slawinski, eds. *Science, Culture and Popular Belief in Renaissance Europe.* Manchester: Manchester University Press, 1991.

Raj, Kapil. *Relocating Modern Science: Circulation and the Construction of Knowledge in South Asia and Europe, 1650–1900*. New York: Palgrave Macmillan, 2007.

Ramírez, Susan Elizabeth. *The World Upside Down: Cross-Cultural Contact and Conflict in Sixteenth-Century Peru*. Stanford: Stanford University Press, 1996.

Revene, Z., R. W. Bussman, and D. Sharon. "From Sierra to Coast: Tracing the Supply of Medicinal Plants in Northern Peru—a Plant Collector's Tale." *Ethnobotany Research and Application* 6 (2008): 15–22.

Riddle, John. *Dioscorides on Pharmacy and Medicine*. Austin: University of Texas Press, 1985.

Ritvo, Harriet. *The Platypus and the Mermaid and Other Figments of the Classifying Imagination*. Cambridge: Harvard University Press, 1997.

Roberts, Lissa, Simon Schaffer, and Peter Dear, eds. *The Mindful Hand: Inquiry and Invention From the Late Renaissance to Early Industralisation*. Amsterdam: Koninklijke Nederlandse Akademie van Wetenschappen, 2007.

Rocco, Fiammetta. *The Miraculous Fever-Tree: Malaria and the Quest for a Cure That Changed the World*. New York: HarperCollins, 2003.

Rostworowski, Maria. *A History of the Inca Realm*. Cambridge: Cambridge University Press, 1998.

Safier, Neil. *Measuring the New World: Enlightenment Science and South America*. Chicago: University of Chicago Press, 2008.

Saldaña, Juan José, ed. *Cross Cultural Diffusion of Science: Latin America*. Mexico City: Cuadernos de Quipu, 1987.

Salomon, Frank. "Shamanism and Politics in Late-Colonial Ecuador." *American Ethnologist* 10 (1983): 413–428.

Sánchez Tellez, Carmen. "Estudio histórico de la botica del Palacio como institución real." PhD diss., Universidad de Granada, 1979.

Schabas, Margaret, and Neil De Marchi, eds. *Oeconomies in the Age of Newton*. Durham: Duke University Press, 2003.

Schaffer, Simon, Lissa Roberts, Kapil Raj, and James Delbourgo, eds. *The Brokered World: Go-betweens and Global Intelligence, 1770–1820*. Sagamore Beach: Science History Publications, 2009.

Schiebinger, Londa, ed. "Focus: Colonial Science." *Isis* 96 (2005): 52–63.

Schiebinger, Londa. *Plants and Empire: Colonial Bioprospecting in the Atlantic World*. Cambridge: Harvard University Press, 2004.

Schiebinger, Londa. *Nature's Body: Gender in the Making of Modern Science*. Boston: Beacon Press, 1993.

Schiebinger, Londa, and Claudia Swan, eds. *Colonial Botany: Science, Commerce, and Politics in the Early Modern World*. Philadelphia: University of Pennsylvania Press, 2005.

Schumacher, Hermann A. *Mutis, un forjador de la cultura.* Trans. Ernesto Guhl. Bogotá: Emprese Colombiana de Petróleos, 1984.

Scott, Heidi. *Contested Territory: Mapping Peru in the Sixteenth and Seventeenth Centuries.* Notre Dame: University of Notre Dame Press, 2009.

Sellés, Manuel, José Luis Peset, and Antonio Lafuente, eds. *Carlos III y la ciencia de la Ilustración.* Madrid: Alianza Editorial, 1988.

Seth, Suman. "Putting Knowledge in Its Place: Science, Colonialism and the Postcolonial." *Postcolonial Studies* 12 (2009): 373–388.

Shapin, Steven. "Placing the View from Nowhere: Historical and Sociological Problems in the Location of Science." *Transactions of the Institute of British Geographers* 23 (1998): 5–12.

Shapin, Steven. "The House of Experiment in Seventeenth-Century England." *Isis* 79 (1988): 373–404.

Shapin, Steven, and Simon Schaffer. *Leviathan and the Air-Pump: Hobbes, Boyle, and the Experimental Life.* Princeton: Princeton University Press, 1985.

Simon, Jonathan. "Analysis and the Hierarchy of Nature in Eighteenth-Century Chemistry." *British Journal of the History of Science* 35 (2002): 1–16.

Silverblatt, Irene. *Modern Inquisitions: Peru and the Colonial Origins of the Civilized World.* Durham: Duke University Press, 2004.

Silverblatt, Irene. "The Evolution of Witchcraft and the Meaning of Healing in Colonial Andean Society." *Culture, Medicine, and Psychiatry* 7 (1983): 417–418.

Siraisi, Nancy. *Medieval and Early Renaissance Medicine.* Chicago: University of Chicago Press, 1990.

Sivasundaram, Sujit, ed. "Focus: Global Histories of Science." *Isis* 101 (2010): 95–158.

Slater, John, and Andrés Prieto. "Introduction: Was Spanish Science Imperial?" *Colorado Review of Hispanic Studies* 7 (2009): 3–10.

Smith, Pamela, and Paula Findlen, eds. *Merchants & Marvels: Commerce, Science, and Art in Early Modern Europe.* New York: Routledge, 2002.

Soto-Lavega, Gabriela. *Jungle Laboratories: Mexican Peasants, National Projects and the Making of the Pill.* Durham: Duke University Press, 2009.

Sowell, David. *The Tale of Healer Miguel Pedromo Neira: Medicine, Ideologies, and Power in Nineteenth-Century Andes.* Lanham, MD: Rowman & Littlefield, 2001.

Spary, Emma C. *Utopia's Garden: French Natural History from Old Regime to Revolution.* Chicago: University of Chicago Press, 2000.

Stearns, Raymond Phineas. "Colonial Fellows of the Royal Society of London, 1661–1788." *Notes and Records of the Royal Society of London* 8, no. 2 (1951): 178–246.

Steele, Arthur Robert. *Flowers for the King: The Expedition of Ruiz and Pavon and the Flora of Peru.* Durham: Duke University Press, 1964.

Stein, Stanley J., and Barbara H. Stein. *Apogee of Empire: Spain and New Spain in the Age of Charles III, 1759–1789.* Baltimore: Johns Hopkins University Press, 2003.

Stein, Stanley J., and Barbara H. Stein. *Silver, Trade, and War: Spain and America in the Making of Early Modern Europe*. Baltimore: Johns Hopkins University Press, 2000.

Stoler, Ann Laura. *Along the Archival Grain: Epistemic Anxieties and Colonial Common Sense*. Princeton: Princeton University Press, 2010.

Stoll, Jacob. *The Information Master: Jean-Baptiste Colbert's Secret State Intelligence System*. Ann Arbor: University of Michigan Press, 2011.

Stroup, Alice. *A Company of Scientists: Botany, Patronage, and Community at the Seventeenth-Century Parisian Royal Academy of Science*. Berkeley: University of California Press, 1990.

Sweet, James. *Domingos Álvares, African Healing, and the Intellectual History of the Atlantic World*. Chapel Hill: University of North Carolina Press, 2011.

Talbot, Charles. "America and the European Drug Trade." In *First Images of America: The Impact of the New World on the Old*. ed. Fredi Chiappelli et al., vol. 2, 833–844. Berkeley: University of California Press, 1976.

Taussig, Michael. *Shamanism, Colonialism and the Wild Man: A Study in Terror and Healing*. Chicago: University of Chicago Press, 1987.

Taussig, Michael. "Folk Healing and the Structure of Conquest in Southwest Colombia." *Journal of Latin American Lore* 6 (1980): 217–241.

Taylor, Norman. *Cinchona in Java: The Story of Quinine*. New York: Greenberg, 1945.

Teigen, P. M. "Taste and Quality in Fifteenth- and Sixteenth-Century Galenic Pharmacology," *Pharmacy in History* 29 (1987): 60–68.

Terrall, Mary. *The Man Who Flattened the Earth: Maupertuis and the Sciences in the Enlightenment*. Chicago: University of Chicago Press, 2002.

Terrall, Mary. "Heroic Narratives of Quest and Discovery." *Configurations* 6 (1998): 223–242.

Tilley, Helen. "Global Histories, Vernacular Science and African Genealogies; Or, Is the History of Science Ready for the World?" *Isis* 101 (2010): 110–119.

Topik, Steven, Carlos Marichal, and Zephyr Frank, eds. *From Silver to Cocaine: Latin American Commodity Chains and the Building of the World Economy, 1500–2000*. Durham: Duke University Press, 2006.

Tyrer, Robson. "The Demographic and Economic History of the Audiencia of Quito: Indian Population and the Textile Industry, 1600–1800," PhD diss., University of California, Berkeley, 1978.

Valdivieso, Marcia Stacey de. *La polemica sangre de los Riofrío: la Casa de Riofrío en Segovia, Ecuador, Perú, Chile*. Quito: M. Stacey Ch., 1997.

Valverde Ruiz, Eduardo. "La Real Botica en el Siglo XIX." PhD diss., Universidad Complutense de Madrid, 1999.

Van Kessel, J. "Ayllu y ritual terapéutico en la medicine andina." *Chungara: Revista de Antropología Chilena* 10 (1983): 165–176.

Varey, Simon, Rafael Chabrán, and Dora Weiner, eds. *Searching for the Secrets of*

Nature: The Life and Works of Dr. Francisco Hernández. Stanford: Stanford University Press, 2000.

Varey, Simon, and Rafael Chabrán. "Medical Natural History in the Renaissance: The Strange Case of Francisco Hernández." *Huntington Library Quarterly* 57 (1994): 124–151.

Wallis, Patrick. "Exotic Drugs and English Medicine: England's Drug Trade, c. 1550–c. 1800." *Social History of Medicine* 25 (2011): 25–46.

Walker, Geoffrey J. *Spanish Politics and Imperial Trade, 1700–1789*. London: Macmillan, 1979.

Walker, Timothy D. "The Medicines Trade in the Portuguese Atlantic World: Acquisition and Dissemination of Healing Knowledge from Brazil (c. 1580–1800)." *Social History of Medicine* 26 (2013): 403–431.

Webb, Jr., James L. *Humanity's Burden: A Global History of Malaria*. Cambridge: Cambridge University Press, 2008.

Weber, David J. *Bárbaros: Spaniards and Their Savages in the Age of Enlightenment*. New Haven: Yale University Press, 2005.

Westfall, Richard S. "Science and Patronage: Galileo and the Telescope." *Isis* 76 (1985): 11–30.

Wisecup, Kelly. *Medical Encounters: Knowledge and Identity in Early American Literature*. Amherst: University of Massachusetts Press, 2013.

Withers, Charles W. J. *Placing the Enlightenment: Thinking Geographically about the Age of Reason*. Chicago: University of Chicago Press, 2007.

World Health Organization. *World Malaria Report 2014*. Geneva: World Health Organization, 2014.

Index

142, 146, 182, 183; quality of, 74, 77,
82, 92, 101, 107, 111, 114, 128, 139,
166, 181, 227n23, 237n48, 238n65;
samples of, 9, 83–87, 94, 107, 113,
114, 117, 143; shipping, 19, 78–79,
100–101, 106, 111, 113, 177, 234n2;
study of, 30, 154; substitute for,
237n42; trade in, 54, 96, 147; using,
33, 34, 49, 57, 95, 101, 204n3;
volume of, 56
Cinchona microantha, 160, 244n36
Cinchona officinalis, 155, 158, 160,
244n36
Cinchona pallescens, 160, 244n36
Cinchona purpurea, 160, 244n36
Cinchona tenuis, 160, 244
cinchona trees, 10, 42, 61, 63, 66, 68,
74, 76, 79, 85, 87, 97, *104*, 109,
125, 129, 139; conservation of,
16, 18; cultivation of, 131, 136,
144, 145, 149; disappearance of,
58, 145, 177; domestication of,
207n41; exploitation of, 131, 135;
finding, 14, 75, 108, 110; as gift
from God, 214n43; illustrations of,
61; knowledge about, 17, 26, 68, 78,
86, 203n80, 240n90; monopoly on,
67, 181; as natural resource, 149;
number of, 238n67; reproduction
of, 57; scarcity of, 97, 110, 111, 112,
127, 149, 238n65; species of, 7, 185,
197n13, 243n23; stands of, 142;
study of, 23, 24, 100, 126, 154, 168;
thinking about, 136; transplanting,
11, 74, 177–78; understanding, 152;
varieties of, 15
cinnamon, 120, 133, 164
classification, 4, 130, 151, 154, 155, 165,
204n2, 244n36; natural phenomena
and, 180; naturalists and, 249n20;
pharmacological, 204n3; principles/
laws of, 179; quina, 18, 101–2, 152,
153, 157, 173; systems of, 156, 157,
158, 162
*Clave Botanica, o Medicina Botanica
Nueva y Novissima* (Suarez de
Ribera), 64
colonial government, 9, 12, 46, 94, 110,
212n29; Andean healing under,
34–36, 38–42; empiricism in, 79;
epistemic culture of, 10, 15, 17, 71,
72, 77–78, 80, 81, 83, 87, 107,
108, 117, 128, 152, 153, 166, 173,
178, 180, 181, 249n27; knowledge/
information collected by, 210n13
colonial society, 15, 42, 78, 199n25;
challenges of, 18; reorganization of,
149
colonization, 8, 15, 19, 32, 40;

epidemiological frontiers and,
26; European, 27, 41; indigenous
communities and, 27; malaria and,
52; Spanish, 19, 26, 38, 39, 46
commerce, 73, 183, 184; knowledge and,
61, 63–68
commodities, 67–68; botanical, 3, 16, 17,
44, 70, 71, 130, 133, 183; medical,
47, 54, 57, 61; production/circulation
of, 213n39
corregidor, 3, 74, 97, 139, 140, 141
cosmography, 8, 12
cosmology, 35; medical, 26–30, 34,
201n54
costrón, case of, 98–100
Council of the Indies, 8, 9, 71
Cowie, Helen, 70
Creoles, 16, 138; elites, 4, 70, 78, 200n44
Crespo Ortiz, Fernando, 30
Cuenca, 60, 63, 79, 84, 105, 135, 137,
139, 231n61, 232n75; cinchona trees
in, 126; forests of, 141, 142; quina
from, 96, 123, 127, 140, 233n81
culture, 12, 180; Andean, 42, 199n25;
popular, 203n82. *See also* epistemic
culture
curanderos, 7, 25, 28, 32, 39, 40, 42, 108;
Andean, 34, 38; knowledge of, 38;
medical cosmology of, 29; personal,
36; types of, 27; work of, 26, 29

Dayde, Juan, 226n18
Daza y Fominaya, Manuel, 97, 21n19
De Vos, Paula, 71
Díaz, Juan, 118
Dictionnaire universel françois et latin,
80
Diguja, José, 91, 106, 221n19, 221n46,
223n66, 223n68; cinchona bark and,
105; *costrón* and, 98, 99; Valdivieso
and, 97, 102
Dioscorides, Pedanius: work of, 204n3
disease, 31, 34, 59; epidemiology
of, 47; spread of, 27, 47, 49–53;
supernatural causes of, 29, 30
drug trade, 44, 54, 55
Duhamel du Monceau, Henri-Louis, 121
Duran-Reynals, Margaret, 177

East India Company, 164
East India Medical Board, 177
economic development, 15, 16, 86, 137,
144, 147, 174, 184; quina industry
and, 57–61
El Arcano de la Quina (Mutis), 153,
154, 155, 156, 157, 164, 242n14;
"Prospectus" from, *159*
Elliott, J. H., 13
Ellis, John, 121

empire: botany and, 112, 123–29, 131, 132, 140, 150, 175; epistemic culture of, 85, 129, 175, 178; *esqueletos* of, 3–5; knowledge and, 73; mercantile vision of, 137, 165–69, 171–75; regionalist vision of, 162–65; science and, 5–12, 18, 20, 87, 95, 112, 122, 147, 149, 173, 179, 181–84, 186, 193n38
Empires of the Atlantic (Elliott), 13
empiricism, 9, 78–81, 86
Endersby, Jim, 249n17, 251n40
Enlightenment, 6, 13, 14, 29, 38, 39, 41, 67, 77, 111, 112, 113, 119, 127, 131, 137, 140, 161
environmental conditions, 15, 16, 50, 51
epidemiology, 16, 26, 30–34, 47, 50–51
epistemic culture, 8, 10, 13, 15, 17, 71, 86, 87, 107, 112, 128, 129, 152, 153, 166, 173, 175, 178, 180, 181, 184; characteristics of, 77–78, 85, 108; empiricism in, 78–81; hierarchy in, 77–78; political nature of, 9, 72
epistemology, 73, 92, 93, 180; imperial, 93, 106, 107, 184–86
Escolano, Eugenio, 226n18
Esqueletos, 3–5
Estrella, Eduardo, 203n80
Eten, 31; axis of health and, 34
ethnobotanists, 31, 42, 108
ethnomedicine, 27
Extirpation of Idolatry in Peru, The (Arriaga), 34

Farquhar, Judith, 225n8
Feijóo, Benito Jerónimo, 98, 219n26
Ferdinand VI, Migas Calientes and, 215n59
fevers, 34, 59, 131, 204n3; causes of, 242n14; early modern, 50 (table); malaria and, 33, 49, 53; quartan, 49, 115; quotidian, 49; tertian, 49, 114, 115; treating, 4, 44, 53
Flora Peruviana et Chilensis (Ruiz and Pavón), 153, 157, 160, 161, 167, 168, 169, 171; contributions from Spanish America/publication of, 169 (table); contributions from Viceroyalty of Peru/publication of, 170 (table)
flota, 55, 207n36
Forster, Georg, 243n23
Fourcroy, Antoine, 146
free blacks, population of, 4
Frías Núñez, Marcelo, 124, 231n68

Gaceta de Madrid, Tafalla and, 251n34
Galen, 27

Gálvez, José de, 119, 121, 122, 123; death of, 127–28
García de Cáceres, Miguel, 110
García de Leon y Pizarro, José, 187n1, 232nn75–76
Garofalo, Leo, 201n63
General Hospital, 113; cinchona bark and, 115, 120; complaints from, 226n14; junta of, 114, 115, 116; quina at, 226n17, 227n23; testing at, 226n12
genuses, classification of, 157
Gherini, Claire, 208n58
globalization, 20, 50
Gómez Ortega, Casimiro, 112–13, 120–22, 123, 126, 140, 167, 182, 215n59, 230n53, 230n55, 230n59, 250n32; botany and, 122; chemical analysis and, 146; cinchona bark and, 119–20; colonial government, 128; experiences of, 127; quina and, 133; science/empire and, 122
Gongora, Viceroy, 246n67
Gray, John, 63, 64, 65, 208n58
Grove, Richard, 195n54
Guayaquil, 30, 60, 79, 144
Gutiérrez de Piñeres, Juan Francisco, 123–24

Hasskaral, Justus Charles, 177–78
healers, 108, 198n19, 202n68; Afro-Peruvian, 201n63; Andean, 25, 26–30, 31, 32–33, 38, 40, 42, 53, 201n54, 201n56; axis of health and, 33–34; bark and, 7, 33; European, 27, 32; folk, 198n21; general/specialized, 27; idolatry and, 34–35; indigenous, 15, 25, 35, 38, 39, 68, 98; master/novice, 27
healing practices, 26, 27, 198n15; Andean, 30–34, 34–36, 38–42; spiritual/magical, 201n63
health, 27; Andean conceptions of, 28; axis of, 31, 33–34; maintaining, 28; public, 6
herbalists, 27, 35, 36
Hernández de Alba, quina trade and, 247n67
Hernández, Francisco, 8, 230n59
Híjar, Pedro de Alcántara Fernández de Híjar y Abarca de Bolea, Duque de, 116, 227n21, 227n28
Hippocrates, 27
Historical and Descriptive Narrative of Twenty Years' Residence in South America (Stevenson), 177
Hobbes, Thomas, 173, 245n40
hombres péritos, 15, 79, 195n51

Hooker, Joseph, 179
hospitals, 115; royal, 105, 113, 117, 168, 220n39, 225n9. *See also* General Hospital
hot and cold, geography of, 33 (table)
House of Trade, 8, 55
huánuco, 133, 134, 235n14, 246n59
Huarochirí Manuscript, 36, 202n71
Humboldt, Alexander von, 14, 147
humoral theory, 27, 34, 199n25

icho, 133, 234n9
identification, 7, 18, 38, 68, 130, 180, 183, 185
idolatry, 34–35, 38, 201n56
imperial ideology, 136–42; botany/empire and, 150
imperialism, 8, 18, 108, 122, 128, 175, 176, 186; European, 11, 19, 177, 178; knowledge and, 81–83, 179
imperial reform, 15, 148
Incas, 36, 39, 40, 47
indigenous peoples, 4, 8, 19, 41, 202n71; colonization and, 27; malaria and, 51; migration of, 60; nature and, 42
Instrucción formada por un facultativo existente por muchos años en el Peru, relativa de las especies y virtudes de la Quina (Mutis), 153, 154
Instruction (Gómez Ortega), 122, 230n55
"Introduction to the Scientific Description of the Plants of Peru" (Unanue), 38
Isthmus of Panama, 47, 63, 65, 215n61

Jacquin, Nikolaus von, 243n23
Jaen, 105, 137, 139, 223n59; cascarilla from, 222n55; cinchona bark from, 103; forests of, 106, 141, 142; quina from, 140
Jarcho, Saul, 10, 204n4
Jasanoff, Sheila: coproduction and, 173
Jefferson, Thomas, 93
Jesuits, 69–70, 80
Jesuits' Bark, 46, 54, 55, 63
Juan, Jorge, 73, 74, 211n20
Jungas region, 84, 85
Jussieu, Joseph de, 39, 87, 215n62

Kallawaya, 29, 30, 36, 38
knowledge, 15, 17, 26, 84, 86, 111, 132, 137, 138, 182; Andean, 30, 31, 34, 42; botanical, 41, 100; cartographic, 8; circulation of, 66, 68, 181; collecting, 12, 181, 210n13; commerce and, 9, 46, 61, 63–68; European, 93; geography of, 190n20;

216n5; historical, 18–20; imperial governance and, 73, 81–83, 179; indigenous, 9, 36, 41, 42, 46, 81–83; medical, 16, 30, 31, 34, 36, 39, 40, 41, 92, 100, 108, 203n80; plant, 94; politics and, 10, 14–16, 18–20, 83–87, 179–81, 182–83, 186; power and, 184; production of, 8, 19, 41, 42, 63, 66, 68, 71, 85, 96–98, 101, 107, 172, 173, 179, 180, 181, 183, 184; scientific, 9, 41, 92, 108, 112; secular, 35; social history of, 93
König, Johann, 243n23

La Calancha, Antonio de, 31, 35–36, 38
La Condamine, Charles Marie de, 23, 38, 42, 63, 64, 65, 73, 74, 80, 87, 154, 155, 160, 165; cinchona and, 40, 61, 67; illustration from, *62*; Loja and, 61; quinquina tree and, 61, 66, 158; royal order and, 67; Santisteban and, 79; study by, 24; Vega and, 39
labor, 93, 207n49; indigenous, 142–44; skilled, 149
Lafarga, Joseph, 114
Laguna, Andrés de, 204n3
Latour, Bruno, 108, 216n7
Lavalle y Córtes, Jose Antonio de, Conde de Premio Real, 134, 235n17
Leeuwenhoek, Antonie van, 61; cinchona bark and, 43–44; illustrations by, *45*
Lima, 58, 79, 134, 139, 165, 167, 172; earthquakes in, 59; merchant guild in, 183
Linnaeus, 70, 155, 185, 243n23; cinchona trees and, 24; principles of, 158, 160, 161; taxonomy of, 67, 156, 160, 251n38
Loja, 3, 39, 42, 60, 61, 63, 74, 78, 79, 84, 91, 97, 102, 128, 131, 135, 136, 137; bark collectors in, 65, 182, 183, 230n54; cascarilla trees from, 73, 222n55; cinchona bark from, 16, 31, 34, 92, 95, 98, 105, 106, 107, 116, 138, 143, 214n46; cinchona trees from, 23, 25, 31, 57, 58, 64, 125, 126, 149, 214n46; climate/land of, 168, 246n58; consensus in, 105–9; expedition to, 64; forests of, 97, 106, 123, 139, 141, 142, 144; as healing center, 31, 32; Huánuco and, 134, 246n58; identity of, 166; knowledge from, 107, 108; map of, *104*; quina from, 4, 5, 6, 17, 41, 71, 75, 95, 96, 105, 109, 111, 129, 133, 140–41, 164, 165, 166, 167, 168, 171, 172, 176, 188n6, 207n43, 224n76, 233n2;

Loja (*cont.*); royal reserve in, 67–68,
75–77, 93, 96, 100, 111, 129, 130,
136, 143, 148, 152, 166, 168
López Ruiz, Sebástian, 120, 124, 125,
227n29, 229n47, 244n36
López y Valverde, Nicolas, 226n18
Lope y Torralva, Alfonso, 115, 227n28
Lorente, Antonio, 226n18
Losada, Duque de, 114, 115–17, 119,
227n21, 227n28; communiqué by,
226n10

Maca Uisa, 38, 202n71
Madrid, 66, 69, 73, 76, 86, 87, 119,
121, 127, 128, 129, 130, 132, 138;
knowledge from, 108, 117; medical
testing in, 113–17; royal reserve in,
184
Maehle, Andreas-Holger, 191n28
Malacatos, 23, 24, 39, 102, 105, 221n49
malaria, 31, 206n18; bark and, 26;
epidemiology of, 50–51; eradication
of, 11; etiology of, 50; fevers and, 33,
49, 53; indigenous populations and,
51; parasites of, 47, 49, 50 (table),
51, 52, 205nn11–12; plasmodium
and, 206n17; spread of, 27, 44, 47,
49–53; treating, 4, 49
Manso de Velasco, José, 78, 81, 83, 84,
166, 214n46; cinchona bark and, 82;
letter from, 69; quina and, 85
Markham, Clements, 177
Martínez Toledano, José, 113, 126,
220n39, 225n9; bark of, 215n57;
cinchona bark and, 101
materia medica, 54, 69, 70, 71, 72, 76,
93
Mazzeo, Cristina, 235n17
McCook, Stuart, 248n8
medical testing, 113–17
medical theory, 25, 26, 198n15
medicine, 4, 11, 16, 53, 75, 76, 87, 96;
Andean, 17, 26, 27, 28, 30, 34, 40,
42, 44, 47, 67, 73, 198n15; botanical,
44, 52, 54; European, 28, 34, 43,
111; history of, 10; hybridization of,
28; preparation of, 64; quina and,
153–56
Mémoires (Royal Academy of Sciences),
La Condamine and, 24
mercantilism, 75–77, 95
merchants, 8, 15, 72, 85, 111, 149, 169,
171, 178, 183; knowledge of, 68;
quina and, 163, 165; Spanish, 76
Mercurio Peruano, 38, 154
Mestizos, 4, 38, 145
"Microscopical Observations on the
Cortex Peruvianus" (Leeuwenhoek),
43; illustrations from, *45*

Migas Calientes, founding of, 215n59
Ministry of the Indies, 9, 70, 97, 106,
118, 119, 123, 127, 136, 137, 138,
174; hombre peritos and, 195n51;
quina trade and, 77
Miraculous Fever-Tree, The (Rocco),
191n28
missionaries, 15, 34, 38, 72, 178,
195n51, 208n59; Augustinian, 35;
European, 35, 39, 53; Jesuit, 53,
84
Missouri Botanical Garden, 198n23
Monardes, Nicolas, 30
Montes, Torivio, 176
Mora, Jose Antonio de Rojas Ibarra y
Vargas, Conde de, 114–15, 116
*Moral Chronicle of the Order of Saint
Augustine in Peru* (Calancha),
30–31
mosquitos, 32, 47, 50, 52
Mountain of Caxanuma, 23, 40, 81, 84,
96, 214n46
Murphy, Kathleen, 208n58
Mutis, José Celestino, 113, 123, 124,
126, 229n47, 233n83, 241n5,
242n14, 242n23; attack on, 161;
Caballero y Góngora and, 125,
164; cinchona trees and, 155, 156;
classification and, 151, 154, 158;
colonial government and, 128; on
Gálvez, 127–28; interpretation of,
231n68; monopoly and, 163; Pavón
and, 244n33; political power and,
171; political/social network of,
166–67; "Prospectus" by, *159*; quina
and, 123, 127, 151, 153, 154, 157,
160, 161, 162–65, 163 (table), 174,
231n65, 244n36; recommendation
by, 232n71; research methods of,
160–61; Ruiz and, 155, 165, 166,
167, 172, 173–74, 224n33; Spanish
Crown and, 171; work of, 154, 165;
Zea and, 160, 161

Napoleon, 176
natural history, 40, 70, 152;
cosmography and, 12; museums of,
69; shortcomings of, 14
naturalists, 7, 68, 152, 184, 228n32,
249n20
natural phenomena, 24, 71, 87, 95;
classification of, 179, 180
natural resources, 6, 9, 44, 63, 71, 72,
73–75, 76, 86, 111, 149, 181, 183;
controlling, 132, 164; development
of, 67; exploitation of, 16, 19, 139;
imperial, 14, 68; knowledge of, 78,
81; useful, 112
natural world, 12, 15, 40, 174, 181

Quevedo, 238n67

quina, 3, 9, 11–12, 23, 41, 42, 68, 71, 72, 75, 82, 83, 98, 99, 106, 113, 117; administering, 114; assessment of, 7, 14, 15, 43, 116, 118, 137, 183; botanical/medical traditions and, 153–56; challenges of, 112; classification of, 18, 101–2, 152, 153, 157, 173; collection of, 60, 141; contests of, 10; controlling, 44, 136–37, 182; distribution of, 66, 73, 77, 81, 133, 137; efficacy of, 44, 76; experience with, 79; exploiting, 164, 165; as finite resource, 131; harvesting, 77, 83, 105; identification of, 7, 18, 68, 183; imperial policy on, 128–29, 132; import of, 56, 76; interest in, 152; knowledge of, 17, 74, 84, 98, 99, 111, 125, 137, 138, 147, 182, 183, 194n51; medicinal, 25, 161, 242n9; monopoly on, 100–101, 139, 162, 165, 174, 214n55; as natural resource, 7, 73–75, 136; official, 164, 171, 174; orange, 160, 167, 242n14, 244n36; overlooking, 66–67; perception of, 74, 149; portrayal of, 44; preservation of, 177; problems associated with, 18; production of, 10, 59, 73, 76, 79, 133, 134, 137, 141, 143, 145, 146–50, 163, 174, 182; quality of, 5, 20, 83, 87, 92, 102, 141, 143, 166, 167, 168, 171, 224n76; quantity of, 82; quinine and, 11; red, 100, 160, 167, 219n23; royal reserve of, 6, 15, 71, 83, 110, 113, 129, 131, 147–49, 152, 164, 165, 176, 179, 181–83, 185, 186, 230n54; salts/extracts of, 126; samples of, 85, 112, 214n54; science and, 20; selecting, 102; shipments of, 92, 100–101, 109, 115, 125, 140; sources of, 85, 96, 110; supply of, 92, 132; transformation of, 16, 73; types of, 84, 96, 100, 111, 153, 156; understanding, 44, 67, 144; value of, 4, 113; white, 160, 167; yellow, 100, 160, 167, 219n23, 244n36

quina industry: economic development and, 57–61; expansion of, 61, 136

quina trade, 17, 40, 58, 63, 66, 73, 77, 80, 81, 83, 91, 99, 103, 107, 110, 123, 139, 142, 155, 162, 164, 166; balance in, 75; challenges for, 46, 111; expansion of, 53–57, 60; intervention in, 67, 82; knowledge of, 147; private, 148, 149; problems in, 75–76; regulation of, 148

quina trees, 3, 64, 74, 79, 80; conservation of, 142; cutting down, 163; knowledge about, 66; leaves/bark from, 5; remembering, 132–36; skeletons of, 4

quinine, 10, 11, 24, 178, 205n11

Quinine's Predecessor (Jarcho), 10, 204n4

Quinología o tratado del arbol de la Quina (Ruiz), 136, 146, 153, 154, 155, 156, 157, 158, 160, 161, 162, 167, 174

quinquina, 23, 39, 64, 67, 80, 158

Quito, 39, 77, 107, 117, 118, 123; axis of health and, 34; earthquakes in, 59; expedition to, 108; President of, 98, 140, 221n48; quina in, 231n61; textile industry in, 58, 59–60

Rate Book for London, 65

Renquifo, Don Francisco, 133–34, 235n14

repartimiento de mercáncias, 134, 183, 207n43

Resaque, 100

Robinson, David, 212n33

Rocco, Fiammetta, 191n28

Royal Academy of Medicine, 185

Royal Academy of Sciences, 23, 24, 39, 65, 79, 154, 155

Royal Botanical Expedition, 3, 113, 120, 123, 124, 126–27, 132, 151, 153, 171

Royal Botanical Garden, 70, 94, 113, 119, 123, 124, 126, 127, 132, 140, 149, 151, 152, 167, 178; botanical expeditions and, 218n17; establishment of, 14; Olmedo and, 147

Royal Cabinet of Natural History, 14, 94, 152

Royal Charity Hospital, 105

Royal Chemical Laboratory, 120

Royal College of Physicians, 185

Royal Pharmacy, 4, 9, 11, 18, 46, 69, 83, 84, 85, 86, 87, 91, 97, 112, 116, 136, 138, 140, 148, 149, 164, 176, 178; analysis at, 117–29; assessments by, 95, 101, 114; bark collectors and, 103, 105, 107; cinchona bark and, 92–93, 95, 96, 98, 99–100, 102, 106, 114, 117, 120, 125, 227n21, 227n23; credibility of, 115; establishment of, 217n8; as imperial institution, 94–96, 190n25; instructions/standards from, 141; medical knowledge and, 108; Olmedo and, 141; quina and, 96, 101–2, 105, 106, 109, 111, 113, 117, 123, 185; royal reserve and, 17,

www.ingramcontent.com/pod-product-compliance
Lightning Source LLC
Chambersburg PA
CBHW030455210326
41597CB00013B/674